METHODS IN MOLECULAR BIOLOGY™

Series Editor
John M. Walker
School of Life Sciences
University of Hertfordshire
Hartfield, Hertfordshire, AL10 9AB, UK

For other titles published in this series, go to
www.springer.com/series/7651

Plant Embryo Culture

Methods and Protocols

Edited by

Trevor A. Thorpe and Edward C. Yeung

Department of Biological Sciences, University of Calgary, Calgary, Alberta, Canada

Humana Press

Editors
Trevor A. Thorpe
Department of Biological Sciences
University of Calgary
Calgary, Alberta
Canada
tthorpe@ucalgary.ca

Edward C. Yeung
Department of Biological Sciences
University of Calgary
Calgary, Alberta
Canada
yeung@ucalgary.ca

ISSN 1064-3745
e-ISSN 1940-6029
ISBN 978-1-4939-5789-7
ISBN 978-1-61737-988-8 (eBook)
DOI 10.1007/978-1-61737-988-8
Springer New York London Dordrecht Heidelberg

Softcover re-print of the Hardcover 1st edition 2011

Printed on acid-free paper

Humana Press is part of Springer Science+Business Media (www.springer.com)

Preface

Embryo development is a great fascination for biologists, irrespective of their discipline. While technically challenging, the study of embryo development provides indispensable information concerning the origins of the various forms and structures that make up the organism. Moreover, many useful applications have been derived from the knowledge gained through the study of plant embryology. The introduction of in vitro culture of zygotic embryos has greatly facilitated the study of embryo development and has allowed for studies not possible in vivo. In vitro culture has in and of itself proven invaluable as a method in plant science for both applied and basic research. The main purpose of this book is to provide a ready source of information for culturing zygotic embryos for different types of studies, both theoretical and practical. Although some procedures described here are standard, we expect that the assembly under the same theme will provide a quick reference source for our readers.

A range of related topics have been selected. Our intention is that the protocols in this volume will serve as reference materials that can be used to help others develop their own customized methods for different species and for different purposes. It is not intended to be exhaustive. The book chapters are divided into five main sections: (1) protocols focusing on the culture of zygotic embryos for developmental studies, (2) application of embryo culture techniques focusing on embryo rescue methods, (3) cryopreservation of zygotic embryos, (4) the use of zygotic embryos as explants for somatic embryogenesis and organogenesis, and (5) transformation protocols using zygotic embryos as starting material.

The first chapter for each section is longer, but provides a general overview of the topic. Hence, it departs slightly from the recommended format. The main reason for organizing the protocols into sections is that the technique itself, zygotic embryo isolation, is a relatively simple one and is similar for different species. What determines how and when you excise the embryo is its final use, as will be evident in the various chapters. The value of the technique lies in its various applications, which we show to be very broad-based. Thus, we expect that the book will appeal to a wide array of researchers.

We would like to thank all the authors who contributed to this book project, Ms. Stephanie Yeung for her diligent editorial work, and Professor John M. Walker of Humana Press for his valuable advice during the assembly of the book.

Calgary, Alberta, Canada

Trevor A. Thorpe
Edward C. Yeung

Contents

Contributors

Jean-Pierre Baudoin • *Unité de Phytotechnie tropicale et d'Horticulture, Gembloux Agro-Bio Tech, Université de Liège, Gembloux, Belgium*
Kylie Bucalo • *Institute of Paper Science and Technology, Georgia Institute of Technology, Atlanta, GA, USA*
Elena Corredoira • *Instituto de Investigaciones Agrobiológicas de Galicia, CSIC, Santiago de Compostela, Spain*
Anna De Carlo • *IVALSA/Istituto per la Valorizzazione del Legno e delle Specie Arboree, CNR, Sesto Fiorentino, Firenze, Italy*
Natalia Dolce • *Facultad de Ciencias Agrarias (UNNE), Instituto de Botánica del Nordeste (IBONE), CONICET – UNNE, Corrientes, Argentina*
Nosisa Dube • *CSIR Biosciences, Pretoria, South Africa*
Mohamed Elhiti • *Department of Plant Science, University of Manitoba, Winnipeg, MB, Canada*
Florent Engelmann • *Institut de recherche pour le développement (IRD), UMR DIAPC, 911 Avenue Agropolis, BP 64501, 34394 Montpellier Cedex 5, France; Bioversity International, Rome, Italy*
Bronwyn Frame • *Plant Science Institute, Department of Agronomy, Center for Plant Transformation, Iowa State University, Ames, IA, USA*
Richard Fratini • *Departamento de Biología Molecular, Area de Genética, Universidad de León, León, Spain*
Malgorzata D. Gaj • *Department of Genetics, Silesian University, Katowice, Poland*
Pascal Geerts • *Life Science Department, Walloon Agricultural Research Center, Gembloux, Belgium*
Fred G. Gmitter Jr • *IFAS, Citrus Research and Education Center, University of Florida, Lake Alfred, FL, USA*
Ana Gonzalez • *Facultad de Ciencias Agrarias (UNNE), Instituto de Botánica del Nordeste (IBONE), CONICET – UNNE, Corrientes, Argentina*
Allen G. Good • *Department of Biological Sciences, University of Alberta, Edmonton, AB, Canada*
Jude W. Grosser • *IFAS, Citrus Research and Education Center, University of Florida, Lake Alfred, FL, USA*
Eun Joo Hahn • *Research Center for the Development of Advanced Horticultural Technology, Chungbuk National University, Cheong-ju, Republic of Korea*
Tegan M. Haslam • *Department of Biological Sciences, University of Calgary, Calgary, AB, Canada*
Pankaj Kaushal • *Central Rice Research Institute, Cuttack, Orissa, India*
Maurizio Lambardi • *IVALSA/Istituto per la Valorizzazione del Legno e delle Specie Arboree, CNR, Sesto Fiorentino, Firenze, Italy*
Yung-I Lee • *Botany Department, National Museum of Natural Science, Taiwan, ROC*

Claudia Luna • *Facultad de Ciencias Agrarias (UNNE), Instituto de Botánica del Nordeste (IBONE), CONICET – UNNE, Corrientes, Argentina*
Marcy Main • *Department of Agronomy, Center for Plant Transformation, Plant Science Institute, Iowa State University, Ames, IA, USA*
Bernard Malaurie • *Institut de recherche pour le développement (IRD), UMR DIAPC, Montpellier, France*
Devendra Ram Malaviya • *Indian Grassland and Fodder Research Institute, Jhansi, India*
Nalini Mallikarjuna • *International Crops Research Institute for the Semi-Arid Tropics (ICRISAT), Andhra Pradesh, India*
Guy Mergeai • *Unité de Phytotechnie tropicale et d'Horticulture, Gembloux Agro-Bio Tech, Université de Liège, Gembloux, Belgium*
Luis Mroginski • *Facultad de Ciencias Agrarias (UNNE), Instituto de Botánica del Nordeste (IBONE), CONICET – UNNE, Corrientes, Argentina*
Fred J. Muehlbauer • *Agricultural Research Service, United States Department of Agriculture, Washington State University, Pullman, WA, USA*
Oulo N'Nan • *Cocody University, Abidjan, Ivory Coast*
Hector Gordon Nuñez-Palenius • *Departamento de Ingeniería Genética de Plantas, Centro de Investigación y de Estudios Avanzados del Instituto Politécnico Nacional (Cinvestav)-Unidad Irapuato, Irapuato, Gto, Mexico*
Sergio J. Ochatt • *INRA, CR Dijon, UMR LEG, Dijon, France*
Neftalí Ochoa-Alejo • *Departamento de Ingeniería Genética de Plantas, Centro de Investigación y de Estudios Avanzados del Instituto Politécnico Nacional (Cinvestav)-Unidad Irapuato, Irapuato, Gto, Mexico*
Takashi Okamoto • *Department of Biological Sciences, Tokyo Metropolitan University, Tokyo, Japan*
Martha M. O'Kennedy • *CSIR Biosciences, Pretoria, South Africa*
Elif Aylin Ozudogru • *Biology Department, GYTE, Gebze Institute of Technology, Gebze, Kocaeli, Turkey*
Kee Yoeup Paek • *Research Center for the Development of Advanced Horticultural Technology, Chungbuk National University, Cheong-ju, Republic of Korea*
So Young Park • *Biotechnology Division, Korea Forest Research Institute, Suwon, Republic of Korea*
Gerald S. Pullman • *School of Biology and Institute of Paper Science and Technology, Georgia Institute of Technology, Atlanta, GA, USA*
Nicole S. Ramesar-Fortner • *Department of Biological Sciences, University of Calgary, Calgary, AB, Canada*
Rafael Ramírez-Malagón • *Departamento de Agronomía. División de Ciencias de la Vida, Campus Irapuato-Salamanca, Universidad de Guanajuato, Irapuato, Gto, México*
Hebe Rey • *Facultad de Ciencias Agrarias (UNNE), Instituto de Botánica del Nordeste (IBONE), CONICET – UNNE, Corrientes, Argentina*
Ajoy Kumar Roy • *Indian Grassland and Fodder Research Institute, Jhansi, India*

MARIA L. RUIZ • *Area de Genética, Departamento de Biología Molecular, Universidad de León, León, Spain*
M. CARMEN SAN-JOSÉ • *Instituto de Investigaciones Agrobiológicas de Galicia, CSIC, Santiago de Compostela, Spain*
PEDRO SANSBERRO • *Facultad de Ciencias Agrarias (UNNE), Instituto de Botánica del Nordeste (IBONE), CONICET – UNNE, Corrientes, Argentina*
ROSEMARIE SCHICK • *Department of Agronomy, Center for Plant Transformation, Plant Science Institute, Iowa State University, Ames, IA, USA*
XIULI SHEN • *IFAS, Citrus Research and Education Center, University of Florida, Lake Alfred, FL, USA*
ASHOK K. SHRAWAT • *Department of Biological Sciences, University of Alberta, Edmonton, AB, Canada*
HESTER C. STARK • *CSIR Biosciences, Pretoria, South Africa*
CLAUDIO STASOLLA • *Department of Plant Science, University of Manitoba, Winnipeg, MB, Canada*
MUHAMMAD TAHIR • *Department of Plant Science, University of Manitoba, Winnipeg, MB, Canada*
ANDRÉ TOUSSAINT • *Unité de Phytotechnie tropicale et d'Horticulture, Gembloux Agro-Bio Tech, Université de Liège, Gembloux, Belgium*
ANA M. VIEITEZ • *Instituto de Investigaciones Agrobiológicas de Galicia, CSIC, Santiago de Compostela, Spain*
KAN WANG • *Department of Agronomy, Center for Plant Transformation, Plant Science Institute, Iowa State University, Ames, IA, USA*
EJAZ A. WARAICH • *Department of Plant Science, University of Manitoba, Winnipeg, MB, Canada*
EDWARD C. YEUNG • *Department of Biological Sciences, University of Calgary, Calgary, AB, Canada*

Part I

Culture of Zygotic Embryos: Fundamental Studies

Chapter 1

Zygotic Embryo Culture: An Overview

Tegan M. Haslam and Edward C. Yeung

Abstract

Zygotic embryo culture has proven itself an invaluable method in plant science for both pure and applied research. The composition of medium used to sustain embryos is a key to successful culture. Optimal composition of the medium changes during embryonic development; generally, the younger the embryo, the more complex is its nutritional requirements. Feeder cell and "double medium" culture methods have been developed to improve the survival of zygotes and proembryos in vitro. In this chapter, we discuss the nutritional requirements of cultured embryos and the importance of the osmotic environment for nurturing young embryos. Specific methodological adaptations used in the culture of *Capsella* are outlined to demonstrate how standard protocols can be manipulated to suit one's needs.

Key words: *Capsella*, Embryo culture, Endosperm, Feeder cells, Osmoticum, *Phaseolus*, Proembryo, Suspensor, Zygote

1. Introduction

Biologists are fascinated by how embryos develop in vivo. The study of embryo development provides indispensable information on the origins of various forms and structures of plants (1). Useful applications of knowledge gained through the study of plant embryology were recognized early (2). In vitro culture of zygotic embryos has allowed for study of embryo development that would not be possible in in vivo studies and has greatly facilitated research methods. Furthermore, it has of itself proven invaluable as a method in plant science in both applied and pure research. We begin this chapter by outlining some important examples of these applications.

Work by Hannig (3) established zygotic embryo culture (ZEC) as an experimental tool to study embryo development in vitro. Subsequent studies using the technique of ZEC have

Trevor A. Thorpe and Edward C. Yeung (eds.), *Plant Embryo Culture: Methods and Protocols*, Methods in Molecular Biology, vol. 710, DOI 10.1007/978-1-61737-988-8_1,

been instrumental in optimizing embryo growth, overcoming embryo inviability, and bypassing seed dormancy (2, 4). In 1925, Laibach (5) found the first of many practical applications for ZEC when he systematically varied medium components and their concentrations to suit the precise needs of developing embryos. In doing so, Laibach discovered that very fastidious embryos could be grown if culture conditions were optimized. ZEC as a method for embryo rescue of fastidious progeny is especially relevant to cross-breeding in agriculture and horticulture. Breeders often cross inbred individuals having low genetic diversity to wild individuals as the latter are usually more hardy or disease resistant. However, these crosses are frequently unsuccessful because genomic incompatibility between parents disallows double fertilization. This incompatibility prevents development of the endosperm, thereby leading to starvation of the embryo. By placing the embryo in nutrient medium immediately following fertilization, essential nourishment is provided and the embryo may survive. This application is discussed by Asif et al. (6) and Bakry et al. (7) using banana as a model organism; Clarke et al. (8) also discuss this topic in relation to chickpea production. Applications for propagation of citrus fruits derived from interploid crosses are discussed by Viloria et al. (9). Pérez-Tornero and Porras (10) also discuss crosses between citrus fruit, though it limits itself to applications for embryo rescue by avoiding polyembryony in lemon. Methods for raising fastidious zygotic embryos produced from cross-pollination of cucumber and melon (*Cucumis*) are presented by Ondrej and Navratilova (11). Again, all of these studies were performed with the aim of raising hardy or disease-resistant crops.

Another common problem in fruit production is that many fruits ripen at a rate far exceeding that of the embryo maturation. In 1933, Tukey (12) used ZEC to solve this problem in peach. By transferring peach embryos to in vitro culture, he allowed them to finish developing independently of fruit development. Several chapters in this book provide standard protocols that may be applied practically to agriculture.

Seed dormancy and its triggers can also be investigated using ZEC, as cultured embryos skip this stage of the plant life cycle. Dormancy is preceded by desiccation, metabolic reduction, and finally quiescence (13). There is obvious value for pure research into the physiology of this process. Furthermore, the fact that ZEC skips seed dormancy makes it advantageous in allowing breeders to massively reduce the generation time of their crops. This was demonstrated by Randolph et al. (14, 15), who made use of the simple fact that excising young embryos and transferring them to nutrient media lead to a direct transition between embryonic and seedling development. In carefully applying tissue culture to iris production, Randolph et al. (14, 15) shortened its breeding cycle from years to months.

Besides its practical applications, ZEC can be an excellent experimental system for pure scientific research. Understanding embryo growth provides a better theoretical understanding of plant growth and development in general, especially during the unique period when tissues, organs, and apical meristems are being established. A surge of information published in recent years concerning zygotic embryo development can attest to growing interest in the field. This information was recently summarized in Vol. 427 of this series, edited by Suarez and Bozhkov (16). In experimental studies on the structural functions, hormonal roles, and the molecular biology of embryo development, embryo culture complements in ovulo studies of zygotic embryogenesis (17). The culture of young embryos of *Phaseolus coccineus* with and without a suspensor has been used to examine this organ's role during early development. These studies have shown that the suspensor plays an important nutritional role in embryo development (18). Auxin has been of special interest in the study of the hormone physiology of embryo development; auxin has a key regulatory function and is essential to axis establishment at the proembryo stage (19). Alteration of auxin transport using auxin transport inhibitors changes embryo symmetry from bilateral to radial (20). Also, interfering with polar auxin movement can lead to shoot apical meristem abortion (21, 22). In vitro culture of embryo mutants enables us to gain better insight to additional molecular mechanisms of embryo development (17). Hence, ZEC is an integral part of many varied research programmes concerning embryo development. The knowledge gained through these methods will also have a significant impact on practical applications.

The purpose of the present chapter is to provide the reader with a brief overview of standard procedures involved in ZEC and how one can modify these methods to suit one's own needs. We begin by emphasizing some key factors that are important for ZEC. Table 1 provides selected references to literature in which species- and need-specific protocols can be found. Next, the model organism *Capsella bursa-pastoris* is used as an example of how general protocols can be modified, in this instance to fulfil the stringent requirements of proembryo culture.

2. General Approaches and Methodology

2.1. Understanding Your System and Establishing a Stage of Interest

Prior to culturing embryos, it is important to establish what the developmental stage of interest is, which depends on the aim of the study. With embryo rescue, for example, it is important to know when the embryos begin to abort so that embryo dissections are carried out prior to abortion. To determine when

Table 1
Selected references to culture methods

Organism	Stage cultured	Method	Reference
Arabidopsis thaliana	Fertilized ovule	Plant transfer through a series of simple media suiting the developing embryo's needs	(60, 61)
	Proembryo	Plant transfer through a series of simple media in Petri plates, multiple-well culture plates, to initiate multiple shoot growth	(62)
	Various	Multiple, simple media tested on different individuals, for mutant rescue and investigation of knockout gene function	(63)
Brassica juncaea	Globular and heart-stage	Multiple, auxin supplemented-media tested on different individuals	(64)
Carica papaya	Mature embryo	Multiple, simple media tested on different individuals for mutant screening	(65)
Centaurea tchihatcheffi	Immature embryos	Simple medium for seed dormancy investigation	(66)
Cicer arietinum	Early globular stage	Plant transfer through a series of liquid culture media in polycarbonate tubes, adjusted to the developing embryo's needs for embryo rescue	(8)
Citrus limon	Immature embryos	Multiple, simple media tested on different individuals for embryo rescue	(10)
Citrus spp.	Immature embryos	Plant transfer through a series of simple media	(9)
Cocos nucifera	Mature embryos	Growth in simple media for ZEC/in vivo comparison	(67)
Cucumis spp.	Immature embryos	Simple medium in tubes, for observation of hormone effects	(11)
Gossypium hirsutum	Fertilized ovule	Solid and liquid-stationary/shaken/rotated simple media tested on different plants	(68)
Helianthus spp.	Various stages	Plant transfer between two separate media, for investigation of embryo rescue	(69)
Manihot esculenta	Immature embryos	Simple medium in tubes, with applications in evading seed dormancy	(70)
Musa acuminata	Mature embryos	Simple medium, for investigation of embryo rescue	(6)
Nicotiana tabacum	Various stages	Nurse cell culture	(51)
Oryza sativa	Various stages	Nurse cell culture, for development of a single-cell regeneration system	(50)

(continued)

Table 1 (continued)

Organism	Stage cultured	Method	Reference
Phaseolus vulgaris	Immature embryos; pod culture	Multiple simple solid and liquid media tested on different individuals	(71)
Prunus persica	Immature embryos	Simple medium, for investigation of embryo rescue	(12)
Taxus baccata	Mature, somatic embryos	Multiple simple media on different individuals, with applications in evading seed dormancy	(72)
Taxus brevifolia	Mature embryos	Simple medium, applications in genetic transformation	(73)
Triticum aestivum	Fertilized ovule	Solid and liquid media tested on different individuals	(49)
Vitis vinifera	Fertilized ovule	Multiple media tested on different individuals	(74)
Zea mays	Embryo sac culture	Double layer	(57)

abortion occurs, the simplest approach is to dissect the seed and observe directly. If necessary, histological sections can be used to determine when abnormalities begin to form. Histological sections can also show potential causes of abortion, such as endosperm failure or seed coat abnormalities. Technical information regarding histological methods is readily available in the literature, for example, Yeung and Sexena (23).

2.2. Establishing a Developmental Timetable

For theoretical studies of embryo development, establishing a developmental timetable is extremely useful. Although creating the timetable is time consuming, it greatly facilitates subsequent embryo collection at the desired stage. There are various methods of establishing a timetable. In *P. coccineus* and *Phaseolus vulgaris* embryo development (24), pod and seed lengths, embryo morphological stages, and colour changes provide a quick assessment of embryo stages for various types of studies. In contrast, in the study of *Capsella* embryo development, measuring ovule size can help determine developmental stage while eliminating the need to open the ovule (25).

2.3. General Consideration of Media Components

The most important aspect of embryo culture work is selection of a medium that meets the needs of isolated, growing embryos. Although there are a number of medium formulas in use, many have not been vigorously tested. In general, younger embryos have more complex nutritional requirements, while more mature embryos can be grown in a simpler inorganic salt media (26).

Culture of relatively young embryos requires proper osmotic adjustment of the culture medium, as well as supplementation with vitamins, amino acids, and growth hormones.

Murray (27) documented the nutrition of angiosperm embryos and discussed the construction of media for embryo growth in vitro. His work provides comprehensive information on embryo nutrition. Recently, key chemical components for in vitro culture and their properties were discussed in detail in a book edited by George et al. (28). Examination of different media additives has shown that reduced nitrogen strongly influences embryo growth in culture. Although ionic ammonium is a ready source of reduced nitrogen and is essential to embryo culture (27), at too high a concentration it can be toxic to cell and embryo cultures (29). Amino acids are readily absorbed and can be used directly as a source of nitrogen. The addition of amino acid mixtures, such as casein hydrolysate, or specific amino acids, such as glutamine, can thus be highly beneficial. The positive effects of glutamine were demonstrated by Rijven (30), and studies continue to indicate that glutamine and other amino acids have a positive influence on embryo culture. One example is provided by Emershad et al. (31), using fertilized grape ovule culture. However, embryonic responses to different nitrogen sources change greatly over the course of development (32). Optimizing the source and concentration of nitrogen in nutrient media can be highly profitable for ZEC studies.

2.4. The Osmotic Environment

The physical and chemical environments surrounding the zygotic embryo in ovulo are very complex (17, 33). For successful in vitro culture of small proembryos, the best approach is to simulate an environment as close to in ovulo conditions as possible. Zygotic embryos develop in an environment with highly negative water potential (34); the beneficial effect of high osmolarity on embryo growth was established more than 50 years ago. It was found that the addition of higher concentration of sucrose improved growth of the zygotic embryo in vitro (35). It is important to note that the majority of media components also contribute to the total water potential of the medium. Further decreases in water potential can be made by addition of carbohydrates such as sucrose, permeable osmotica such as mannitol, or non-permeable osmotica such as polyethylene glycol. The selection and concentrations used must be tested and optimized.

The negative osmotic environment may also have a morphogenetic role and appears to regulate precocious germination of maturing embryos (17, 36). Selective use of different osmotic compounds can affect many growth characteristics of the explants. A dramatic demonstration of this is shown in work by Ilic-Grubor et al. (37, 38) on canola microspore-derived embryos (MDEs). Culture of MDEs using 20% (wt/vol) non-permeable polyethylene glycol 4000 instead of 13% sucrose as an osmoticum produced embryos similar to their zygotic counterparts, i.e., having similar morphology and storage products. The embryos also had a 100% conversion upon germination. On the other hand, MDEs developed from medium containing 13% sucrose as an osmoticum tended to grow larger and had a lower percentage of conversion relative to their zygotic counterparts (38, 39).

2.5. The Nutrient Environment for the Culture of Proembryos

2.5.1. The Role of the Suspensor in Embryonic Development

Zygotic embryos develop within a complex environment in vivo (33); however, details of their nutritional needs are not clearly understood. The first stage of zygotic embryo development is characterized by slow growth of the embryo proper and precocious development of the suspensor. Experimental studies indicate that the suspensor is necessary for development of young embryos (18, 40). The suspensor connects the embryo proper to the seed coat. Its precise role is poorly understood, although its development is tightly linked to the embryo's changing nutritional requirements. The suspensor cells in *P. coccineus* have several unique features that hint at organ function; polyteny, well-developed organelles, and wall ingrowths (41, 42). Putative functions include serving as a conduit channelling nutrients from the maternal tissues and endosperm into the developing embryo proper and regulation of embryo nourishment, growth, and differentiation (43, 44). In vitro studies have shown that, as the embryo proper can be readily cultured on simple medium and germinated precociously post-histogenesis, the suspensor is no longer necessary after this stage in vivo (18). These studies indicate that the proembryo is heterotrophic; suspensor cells may well function as "feeder" cells only during early embryo development.

Because the suspensor is cleaved during dissection in preparation for ZEC, Monnier (32) has suggested that observed lower survival rates of proembryos relative to mature embryos in culture are caused by leaching of indispensable cellular substances from the suspensor into the medium. Exceptionally, the presence or absence of the suspensor does not appear to play an important role in the survival of *Brassica campestris* proembryos (20, 45). This discrepancy relative to other studies could be explained by the fact that many zygotic embryos used for proembryo culture studies are large relative to *Brassica* embryos and may have higher nutrient demands. It is also possible that despite having been severed from the embryo, a few remaining suspensor cells attached to the embryo proper are sufficient for its development.

2.5.2. The Role of the Endosperm in Embryonic Development

Embryonic development is closely tied to that of the endosperm; failure in endosperm development often leads to embryo abortion. The successful inclusion of coconut water (endosperm from coconut palm) to culture medium demonstrates the importance of endosperm nutrients in ZEC (46). Inclusion of liquid endosperm, which has a complex chemical composition (47), in ZEC often leads to increased embryonic growth and/or survival. This has led many researchers to speculate that there is at least one nutrient or growth factor usually supplied to the developing embryo by the endosperm that we are as yet unaware of, and that is therefore not included in medium recipes presently in use (48).

2.5.3. Feeder Cell Systems and Double-Layered Media

Feeder cell and double medium methods have been successfully used in the culture of zygotes and proembryos, respectively. These procedures are designed based on general scientific understanding of the embryonic environment in vivo and of the nutritional requirements of the embryo.

Feeder cell cultures can be useful for increasing the survival rate of zygotes after fertilization and for allowing continuous development into an embryo or embryogenic cell mass. In this method, selected cells such as microspore suspensions, living ovules, mesophyll protoplasts, or suspension cultures are used as feeder cells (see Table 1). A sterile mini-dish insert is placed onto medium containing the feeder cells, and the isolated zygotes are placed onto the membrane of the mini-dish. The use of feeder cells may provide necessary nutrients for growth of the zygote and early proembryo, substituting the functions of the suspensor and endosperm. Kumelehn et al. (49) discuss feeder cell culture in wheat, Zhang et al. (50) describe at length a similar system for rice, and He et al. (51) provide an excellent reference for ZEC of tobacco, as well as an in-depth discussion of culture methods using a somatic cell protoplast feeder system to support development. One interesting finding from feeder cell culture research is that the "unknown factor" that is putatively released from endosperm that improves embryo culture is not universally effective. Furthermore, it is not always effective on its own embryo. This has led many scientists to conclude that the "embryo factor" is neither species specific nor universal (52).

As proembryos develop in vitro, their osmotic and nutritional requirements change. In general, a more positive water potential is favoured as the embryos mature, and their nutritional requirements become less stringent relative to those of the proembryo. In order to avoid transfers of cultured embryos that may cause damage or microbial infection, "double" medium systems can be used. A complex medium with more negative water potential is placed as a well insert within or as a layer atop a simpler medium with more positive water potential. Over time, diffusion of components will take place, and the complex medium gradually becomes dilute and more suitable for the continued development of the maturing embryo in culture. Monnier (32) devised a double medium by a "well" insert method to study the in vitro culture of proembryos of *Capsella*.

Capsella bursa-pastoris (shepherd's purse) is an annual, invasive, widespread, and ruderal member of the Brassicaceae (Cruciferae) family. It is a popular choice for ZEC studies for several reasons. First, its embryos are easy to extract from the ovules because the endosperm stays liquid for a long time, and its long racemes provide a basipetal sequence of developing fruits from the same genetic background. Furthermore, the developmental pattern of *Capsella* embryos is quite consistent between individuals. Important work with *C. bursa-pastoris* ZEC has been conducted by Rijven (35), Raghavan and Torrey (53), and Monnier (32, 54–56).

As mentioned earlier, for very young globular embryos, design of the incubation vessel may require some engineering to increase the embryo's chances of survival. Problems in this instance arise from the fact that young embryos have different nutritional requirements relative to older embryos. Young embryos require a high concentration of sucrose as an osmoticum to prevent precocious germination, as well as higher calcium concentration, which has been observed to have a role in protecting embryos during development. Young embryos also require low concentrations of selected minerals that can be toxic at higher concentrations, as they can be especially sensitive to their negative effects. They also require a high concentration of amino acids, as they lack enzymes necessary for nitrate catabolism. Conversely, media designed for older embryos are generally characterized by lower concentrations of sucrose and amino acids and higher concentrations of nitrates and mineral salts, especially iron. To accommodate these needs without resorting to the often tedious and damaging transfer of embryos between media, Monnier contrived a system consisting of two concentric rings of solid media. Medium suitable for more mature embryos surrounds a central section of young embryo medium in the plate where young embryos are cultured. Monnier chose to use glass Petri plates for his experiment and sterilized glass cylinders as moulds for the central well; however, there are many other approaches that could be taken to creating a central well. One can use a multi-well plate such as the six-well multi-well culture plates where each well can serve as the outer vessel, and the central well can be created using a sterile insert cup.

Recent procedures tend to favour the use of a double-layered medium for the culture of zygote and proembryo, for an example, see the *Zea mays* study by Mòl et al. (57). The media used by Monnier is detailed in Table 2 as a general reference. Readers are encouraged to consult the detailed protocol on *Capsella* ZEC published in Vol. 6 of this series (55).

Table 2
Composition of the two media used in different parts of the culture dish to obtain uninterrupted growth of globular stage *Capsella* embryos to maturity

Nutrient element	Concentration (mg/L) in inner medium	Concentration (mg/L) in outer medium	Nutrient element	Concentration (mg/L) in inner medium	Concentration (mg/L) in outer medium
KNO_3	1990	1990	H_3BO_3	12.4	12.4
$CaCl_2 \cdot 2H_2O$	1320	484	$MnSO_4 \cdot H_2O$	33.6	33.6
NH_4NO_3	825	990	$ZnSO_4 \cdot 7H_2O$	21.0	21.0
$MgSO_4 \cdot 7H_2O$	370	407	KI	1.66	1.66
KCl	350	420	$Na_2MoO_4 \cdot 2H_2O$	0.5	0.5
KH_2PO_4	170	187	$CuSO_4 \cdot 5H_2O$	0.05	0.05
Na_2EDTA	0	37.3	$CoCl_2 \cdot 6H_2O$	0.05	0.05
$FeSO_4 \cdot 7H_2O$	0	27.8	Sucrose	180,000	0
Glutamine	600	0	Vitamin B1	0.1	0.1
Vitamin B6	0.1	0.1	Difco Agar	7,000	7,000

3. Prospect

In vitro culture of zygotic embryo provides a useful experimental tool to study many aspects of the developmental events important to embryo development. In recent years, tremendous advances have been made in our understanding of zygotic embryo development through molecular and genetic studies (16, 17, 58). These combined approaches as emphasized by Wetmore and Wardlaw in their review (59) will continue to advance our understanding of embryo development in plants. The knowledge gained will also have direct practical implication in improvements to horticulture and agriculture.

Acknowledgments

This work was supported by a Discovery Grant from the Natural Sciences and Engineering Research Council of Canada to E.C.Y.

References

1. Wardlaw CW (1955) Embryogenesis in plants. Methuen, London
2. Raghavan V (2003) One hundred years of zygotic embryo culture investigations. In Vitro Cell Dev Biol – Plant 39:137–142
3. Hannig E (1904) Zur Physiologie pflanzlicher Embryonen. I. Ueber die Cultur von Cruciferen-Embryonen ausserhalb des Embryosacks. Bot Ztg 62:45–80
4. Raghavan V, Srivastava PS (1982) Embryo culture. In: Johri BM (ed) Experimental embryology of vascular plants. Springer, Berlin, pp 195–230
5. Laibach F (1925) Das Taubwerden von Bastardsamen und die künstliche Aufzucht früh absterbender Bastard-Embryonen. Z Bot 17:417–459
6. Asif MJ, Mak C, Othman RY (2001) In vitro zygotic embryo culture of wild *Musa acuminata* ssp. *Malaccensis* and factors affecting germination and seedling growth. Plant Cell Tissue Organ Cult 67:267–270
7. Bakry F (2008) Zygotic embryo rescue in bananas. Fruits 63:111–115
8. Clarke HJ, Wilson JG, Kuo I, Lülsdorf MM, Mallikarjuna N, Kuo J, Siddique KHM (2006) Embryo rescue and plant regeneration in vitro of selfed chickpea (*Cicer arietinum* L.) and its wild annual relatives. Plant Cell Tissue Organ Cult 85:197–204
9. Viloria Z, Grosser JW, Bracho B (2005) Immature embryo rescue, culture and seedling development of acid citrus fruit derived from interploid hybridization. Plant Cell Tissue Organ Cult 82:159–167
10. Pérez-Tornero O, Porras I (2008) Assessment of polyembryony in lemon: rescue and in vitro culture of immature embryos. Plant Cell Tissue Organ Cult 93:173–180
11. Ondrej V, Navratilova B (2000) Influence of genotype and medium on culture of immature zygotic embryos of *Cucumis sativus* L. and *Cucumis melo* L. Acta Biol Crac Ser Bot 42:79–81
12. Tukey HB (1933) Artificial culture of sweet cherry embryos. J Hered 24:7–12
13. Fischer-Iglesias C, Neuhaus G (2001) Zygotic embryogenesis: hormonal control of embryo development. In: Bhojwani SS, Soh WY (eds) Current trends in the embryology of angiosperms. Kluwer, Dordrecht, pp 223–247
14. Randoph LF (1945) Embryo culture of *Iris* seed. Bull Am Iris Soc 97:33–45
15. Randolph LF (1943) Factors influencing the germination of iris seed and the relation of inhibiting substances to embryo dormancy. Proc Am Soc Hort Sci 43:284–300
16. Suarez MF, Bozhkov PV (eds) (2008) Plant embryogenesis, methods in molecular biology, vol 427. Humana, Totowa, N.J
17. Yeung, E. C., Nickle, T. C., and Meinke, D. (2001) Embryology of flowering: an overview. *Phytomorphology* (Golden Jubilee Issue) 289–304
18. Yeung EC, Sussex IM (1979) Embryogeny of *Phaseolus coccineus*: The suspensor and the growth of the embryo – proper in vitro. Z Pflanzenphysiol 91:423–433
19. Hamann T (2001) The role of auxin in apical-basal pattern formation during *Arabidopsis* embryogenesis. J Plant Growth Regul 20:292–299
20. Liu CM, Xu ZH, Chua NH (1993) Auxin polar transport is essential for the establishment of bilateral symmetry during early plant embryogenesis. Plant Cell 5:621–660
21. Ramesar-Fortner NS, Yeung EC (2001) Tri-iodobenzoic acid affects shoot apical meristem formation and function in zygotic embryos of *Brassica napus* cv. Topas. Can J Bot 79:265–273
22. Ramesar-Fortner NS, Yeung EC (2006) Physiological influences in the development and function of the shoot apical meristem of microspore-derived embryos of *Brassica napus* cv. Topas. Can J Bot 84:371–383
23. Yeung EC, Sexena P (2005) Histological techniques in the study of somatic embryogenesis. In: Jain SM, Gupta P (eds) Protocols of somatic embryogenesis – woody plants. Kluwer, Dorchrect, pp 517–538
24. Walbot V, Clutter M, Sussex IM (1972) Reproductive development and embryogeny in *Phaseolus*. Phytomorphology 22:59–68
25. Lagriffol J, Monnier M (1983) E'tude de divers paramèters en vue de la culture in vitro des ovule de *Capsella bursa-pastoris*. Can J Bot 61:3471–3477
26. Yeung EC, Thorpe TA, Jensen CJ (1981) *In vitro* fertilization and embryo culture. In: Thorpe TA (ed) Plant tissue culture - methods and applications in agriculture. Academic, New York, pp 253–271
27. Murray DR (1988) Nutrition of the angiosperm embryo. Wiley, New York
28. George EF, Hall MA, De Klerk GJ (2008) Plant propagation by tissue culture, 3rd edn. Springer, Dordrecht
29. George EF, de Klerk GJ (2008) The components of plant tissue culture media I: macro- and

micro- nutrients. In: George EF, Hall MA, De Klerk GJ (eds) Plant propagation by tissue culture, 3rd edn. Springer, Dordrecht, pp 65–114
30. Rijven AHGG (1956) Glutamine and asparagines as nitrogen sources for the growth of plant embryos in vitro: a comparative study of 12 species. Aust J Biol Sci 9:511–527
31. Emershad RL, Ramming DW, Serpe MD (1989) In ovulo embryo development and plant formation from stenospermic genotypes of *Vitis vinifera*. Am J Bot 76:397–402
32. Monnier M (1978) Culture of zygotic embryos. In: Thorpe TA (ed) Frontiers of plant tissue culture. University of Calgary Press, Calgary, pp 277–286
33. Steeves TA, Sussex IM (1989) Patterns in plant development, 2nd edn. Cambridge University Press, Cambridge
34. Yeung EC, Brown DCW (1982) Embryogeny of *Phaseolus*: the osmotic environment of developing embryos. Z Pflanzenphysiol 106:149–156
35. Rijven AHGG (1952) In vitro studies on the embryo of *Capsella bursa-pastoris*. Acta Bot Neerl 1:157–200
36. Thorpe TA, Stasolla C, Yeung EC, de Klerk GJ, Roberts A, George EF (2008) The components of plant tissue culture media II: organic additions, osmotic and pH effects, and support systems. In: George EF, Hall MA, De Klerk GJ (eds) Plant propagation by tissue culture, 3rd edn. Springer, Dordrecht, pp 115–173
37. Ilic-Grubor K, Attree SM, Fowke LC (1998) Induction of microspore-derived embryos of *Brassica napus* L. with polyethylene glycol (PEG), as osmoticum in a low sucrose medium. Plant Cell Rep 17:329–333
38. Ilic-Grubor K, Attree SM, Fowke LC (1998) Comparative morphological studies of zygotic and microspore-derived embryos of *Brassica napus* L. as revealed by scanning electron microscopy. Ann Bot 82:157–165
39. Yeung EC, Rahman MH, Thorpe TA (1996) Comparative development of zygotic and microspore-derived embryos in *Brassica napus* L. cv. Topas. I. Histodifferentiation. Int J Plant Sci 157:27–39
40. Yeung EC (1980) Embryogeny of *Phaseolus*: the role of the suspensor. Z Pflanzenphysiol 96:17–28
41. Yeung EC, Clutter ME (1978) Embryogeny of *Phaseolus coccineus*: growth and microanatomy. Protoplasma 94:19–40
42. Yeung EC, Clutter ME (1979) Embryogeny of *Phaseolus coccineus*: the ultrastructure and development of the suspensor. Can J Bot 57:120–136
43. Yeung EC, Meinke DW (1993) Embryogenesis in angiosperms: development of the suspensor. Plant Cell 5:1371–1381
44. Czapik R, Izmaitov R (2001) Zygotic embryogenesis: structural Aspects. In: Bhojwani SS, Soh WY (eds) Current trends in the embryology of angiosperms. Kluwer, Dordrecht, pp 197–222
45. Liu CM, Xu ZH, Chua NH (1993) Proembryo culture: in vitro development of early globular stage zygotic embryos from *Brassica juncea*. Plant J 3:291–300
46. Van Overbeek J, Conklin ME, Blakeslee AF (1941) Factors in coconut milk essential for growth and development of very young *Datura* embryos. Science 94:350–351
47. Smith JG (1973) Embryo development in *Phaseolus vulgaris*. II. Analysis of selected inorganic ions, ammonia, organic acids, amino acids, and sugar in the endosperm liquid. Plant Physiol 51:454–458
48. Srivastava PS (1982) Endosperm culture. In: Johri BM (ed) Experimental embryology of vascular plants. Springer, Berlin, pp 175–193
49. Kumlehn J, Lörz H, Kranz E (1998) Differentiation of isolated wheat zygotes into embryos and normal plants. Planta 205:327–333
50. Zhang J, Dong WH, Galli A, Potrykus I (1999) Regeneration of fertile plants from isolated zygotes of rice (*Oryza sativa*). Plant Cell Rep 19:128–132
51. He Y-C, He Y-Q, Qu L-H, Sun M-X, Yang HY (2007) Tobacco zygotic embryogenesis in vitro: the original cell wall of the zygote is essential for maintenance of cell polarity, the apical-basal axis and typical suspensor formation. Plant J 49:515–527
52. Hu C, Wang P (1986) Embryo culture: technique and applications. In: Evans DA, Sharp WR, Ammirato PV (eds) Handbook of plant cell culture: techniques and applications, vol 4. Macmillan, New York, pp 43–96
53. Raghavan V, Torrey JG (1963) Growth and morphogenesis of globular and older embryos of *Capsella* in culture. Am J Bot 50:540–551
54. Monnier M (1984) Survival of young immature *Capsella* embryos cultured in vitro. J Plant Physiol 115:105–113
55. Monnier M (1990) Culture of zygotic embryos of higher plants. Meth Mol Biol 6:129–139
56. Monnier M (1995) Culture of zygotic embryos. In: Thorpe TA (ed) In vitro embryogenesis in plants. Kluwer, Dordrecht, pp 117–153
57. Mòl R, Matthys-Rochon E, Dumas C (1995) Embryogenesis and plant regeneration from

maize zygotes by in vitro culture of fertilized embryo sacs. Plant Cell Rep 14:743–747
58. Park S, Harada JJ (2008) *Arabidopsis* embryogenesis. In: Suarez MF, Bozhkov PV (eds) Methods in molecular biology, vol 427. Humana, Totowa, pp 3–16
59. Wetmore RH, Wardlaw CW (1951) Experimental morphogenesis in vascular plants. Annu Rev Plant Physiol 2:269–292
60. Sauer M, Friml J (2004) In vitro culture of *Arabidopsis* embryos within their ovules. Plant J 40:835–843
61. Sauer M, Friml J (2008) *In vitro* culture of *Arabidopsis* embryos. In: Suarez MF, Bozhkov PV (eds) Methods in molecular biology, vol 427. Humana, Totowa, pp 71–76
62. Kost B, Potrykus I, Neuhaus G (1992) Regeneration of fertile plants from excised immature zygotic embryos of *Arabidopsis thaliana*. Plant Cell Rep 12:50–54
63. Baus AD, Franzmann L, Meinke DW (1986) Growth in vitro of arrested embryos from lethal mutants of *Arabidopsis thaliana*. Theor Appl Genet 72:577–586
64. Hadfi K, Speth V, Neuhaus G (1998) Auxin induced developmental patterns in *Brassica juncea* embryos. Development 125:879–887
65. Sharma NK, Bedi S (1990) Effects of *Phytophthora palmivora* on zygotic embryos of papaya in vitro. Ann Bot 66:597–603
66. Ozel CA, Khawar KM, Mirici S, Ozcan S, Arslan O (2006) Factors affecting in vitro plant regeneration of the critically engendered Mediterranean knapweed (*Centaurea tchihatcheffii* Fisch et. Mey). Naturwissenschaften 93:511–517
67. López-Villalobos A, Dodds PF, Hornung R (2001) Changes in fatty acid composition during development of tissues of coconut (*Cocos nucifera* L.) embryos in the intact nut and in vitro. J Exp Bot 52:933–942
68. Beasley CA (1971) In vitro culture of fertilized cotton ovules. Bioscience 21: 906–907
69. Sukno S, Ruso J, Jan CC, Melero-Vara JM, Fernández-Martinez JM (1999) Interspecific hybridization between sunflower and wild perennial *Helianthus* species via embryo rescue. Euphytica 106:69–78
70. Fregene M, Ospina JA, Roca W (1999) Recovery of cassava (*Manihot esculenta* Crantz) plants from culture if immature zygotic embryos. Plant Cell Tissue Organ Cult 55:39–43
71. Geerts P, Sassi K, Mergeal G, Baudoin J-P (2000) Development of an in vitro pod culture technique for young pods of *Phaseolus vulgaris* L. In Vitro Cell Dev Biol Plant 36:481–487
72. Hoseini-Nasr SM, Modanloo S, Jalilvand H, Mofidabadi AJ (2007) Seed dormancy breakage of recalcitrant yew species (*Taxus baccata* L.) using embryo culture. J Biol Sci 7:781–785
73. Luan EC, Allen SE, Bolyard MG (1996) Transient GUS expression in zygotic embryos of *Taxus brevifolia*. In Vitro Cell Dev Biol Plant 32:81–85
74. Cain DW, Emershad RL, Tarailo RE (1983) In-ovulo embryo culture and seedling development of seeded and seedless grapes (*Vitis vinifera* L.). Vitis 22:9–14

Chapter 2

In Vitro Fertilization with Rice Gametes: Production of Zygotes and Zygote and Embryo Culture

Takashi Okamoto

Abstract

In vitro fertilization (IVF) systems using isolated male and female gametes have been utilized to dissect fertilization-induced events in angiosperms, such as egg activation, zygote development, and early embryogenesis, since the female gametophytes of plants are deeply embedded within ovaries. A rice IVF system was established to take advantage of the abundant resources stemming from rice research for investigations into the mechanisms of fertilization and early embryogenesis. Fusion of gametes can be performed using electrofusion and the fusion product, a zygote, forms a cell wall and an additional nucleolus. The zygote divides into an asymmetric two-celled embryo and develops into an early globular embryo, as in planta. The embryo further develops into irregularly shaped cell masses and fertile plants can be regenerated from the cell masses. This rice IVF system is a powerful tool for studying the molecular mechanisms involved in the early embryogenesis of angiosperms and for making new cultivars.

Key words: Egg cell, Globular embryo, In vitro fertilization, Regeneration, Sperm cell, Two-celled embryo, Zygote

1. Introduction

In angiosperms, the sporophytic generation is initiated by double fertilization, resulting in the formation of seeds (reviewed in (1)). During fertilization, a sperm cell from a pollen grain fuses with an egg cell and the resultant zygote develops into an embryo. The central cell fuses with a second sperm cell and develops into the endosperm (reviewed in (2)). In many plant species, for example, cotton (3) and grasses such as maize (4), variant cell division occurs during embryogenesis, although in some dicot plants, such as the crucifers: *Arabidopsis*, *Brassica napus*, and *Capsella bursa-pastoris*, the pattern of cell division during early embryogenesis is fixed and

Trevor A. Thorpe and Edward C. Yeung (eds.), *Plant Embryo Culture: Methods and Protocols*, Methods in Molecular Biology, vol. 710, DOI 10.1007/978-1-61737-988-8_2,

the cell fate is traceable (5–7). Despite such variable patterns in embryogenesis, zygotic cell division is asymmetric in most angiosperms and the apical cell of the two-celled embryo develops into the embryo proper, while the basal cell develops into the suspensor and hypophysis (4–6, 8, 9) (reviewed in (10)). In addition to such cytological observations, it has been reported that two putative homeotic genes, *WUSCHEL HOMEOBOX2* (*WOX2*) and *WOX8*, are specifically expressed in the apical and basal cells of the *Arabidopsis* two-celled embryo, respectively, suggesting that the two daughter cells from a zygote possess different transcriptional profiles (11). Moreover, temporal accumulation of a phytohormone auxin via PIN7, an auxin efflux carrier protein, and the YODA-dependent MAPKKK signaling pathway are thought to be crucial for cell fate specification of the apical and basal cells of the *Arabidopsis* two-celled embryo, respectively (12, 13). Based on these cytological and genetic analyses, an asymmetric cell division of the zygote appears to be the first step for formation of the apical–basal axis of plants and is a fundamental feature of early embryogenesis and morphogenesis in angiosperms.

After asymmetrical zygote division, the formation of a globular embryo is a general event during early embryogenesis. Morphogenetic events for organ differentiation occur after the globular embryo stage (3–7) and it has been proposed that a globular embryo can be divided into domains, demarcated by gene expression patterns, with distinct developmental fates (14, 15). For example, the homeobox gene *WUSCHEL*, which regulates stem cell fate in the *Arabidopsis* shoot meristem, is first expressed in the apical subepidermal cells at the 16-cell stage of embryogenesis (16), although the tunica-corpus structure, a characteristic of shoot apical meristems, becomes evident in the late heart or torpedo stage embryo. In addition, *OSH1*, a KNOX-family homeobox gene, is expressed in the ventral region of rice globular embryos, where the shoot apex will differentiate later (17). Therefore, investigations into early embryogenesis from the zygote to the globular embryo stage will be of great importance to understand how the subdomains of globular embryos are specified and/or zonated.

In contrast to animals and lower plants, which use naturally free-living gametes, in angiosperms, the fertilization and subsequent events such as gamete fusion, embryogenesis, and endosperm development occur in the embryo sac deeply embedded in ovular tissue. Difficulties associated with research directly addressing the biology of the female gamete, zygote, and early embryo have impeded investigations into the molecular mechanisms of fertilization and embryogenesis. Therefore, such investigations have been conducted predominantly through mutant analyses using *Arabidopsis* models. However, for a decade, in vitro fertilization (IVF) has been utilized as a tool in angiosperms to observe and analyze fertilization and postfertilization processes directly (reviewed in (18)). The IVF system used for angiosperms includes a combination of three basic microtechniques (a) the isolation and selection of male and female gametes, (b) the fusion of pairs of gametes, and (c) single cell culture (19). Procedures for the isolation of viable gametes have been reported for a wide range of plant species including maize, wheat, tobacco, rape, rice, barley, *Plumbago zeylanica*, and *Alstroemeria* (20–28). The isolated gametes can be fused electrically (19, 29) or chemically using calcium (30–32), polyethylene glycol (PEG) (33, 34), or bovine serum albumin (35), as the gametes are generally protoplasts. Analyses with calcium-based fusion of maize gametes showed that an influx of calcium is triggered by gamete fusion and that calcium influx induces cell wall formation, an event known as egg activation (36). In addition, fusion behavior and gamete interaction have been traced by video-enhanced microscopy using PEG-mediated gamete fusion (37). However, tobacco zygotes produced by calcium- or PEG-fusion became arrested in development (34, 37) and maize zygotes produced by calcium-fusion did not fully develop (31), suggesting that the procedure of calcium-mediated gamete fusion needs to be optimized to obtain sufficient zygotes for studies of embryogenesis. A complete IVF system was developed by Kranz and Lörz (29) using maize gametes and electrical fusion. A maize zygote produced in vitro by the electrical fusion of an egg cell with a sperm cell developed into an asymmetrical two-celled embryo, a proembryo, and a transition phase embryo via zygotic embryogenesis in a similar manner to that in planta (38). Moreover, the IVF-produced embryo continued to develop and grow into a fertile plant (29). This maize IVF system has been successfully used to observe and analyze postfertilization events including karyogamy in zygotes (39), zygote development (38), decondensation of paternal chromatin in zygotes (40), changes in the microtubular architecture in zygotes (41), and identification of fertilization-induced/suppressed genes (42).

Rice (*Oryza sativa* L. cv. Nipponbare) is an excellent model plant among the monocot crop species as it has a relatively small genome of about 440 Mb. The whole genome sequence (43), Tos17 retrotransposon insertion plants (20,000 independent loci; (44)) and over 28,000 full-length cDNA clones (45) are available; these databases and resources have been released for academic use. A rice IVF system using viable isolated egg and sperm cells from rice flowers has been established to apply these resources for investigation of the mechanisms involved in fertilization and early embryogenesis (27, 46). The rice zygote produced by IVF divides asymmetrically into a two-celled embryo consisting of small apical and large basal cells and then this divides into the early globular embryo through several rounds of egg cleavage-like cell divisions, as seen in planta. Rice embryogenesis can be separated into ten stages, defined as Em1–10 (47). Em1 represents the zygote stage.

The globular embryo stage is divided into three stages: Em2, the early globular stage; Em3, the middle globular stage, and Em4, the late globular stage. The rice IVF system can be considered to reproduce zygotic embryogenesis within embryo sac at least during the Em1 and Em2 stages. It can thus provide zygotes and early embryos as starting materials for molecular, biochemical, and cytological investigations of egg activation, zygote development, and early embryogenic events. The globular embryo produced by the IVF system does not follow normal embryogenesis within the embryo sac, but can be regenerated into fertile plants with complete seed sets through callus-derived shoot regeneration. The rice IVF system described here might become an important technique for generating new cultivars with desirable characters.

2. Materials

2.1. Isolation and Transfer of Gametes

1. Environmental chamber (K30-7248, Koito Industries Ltd, Yokohama, Japan) (see Note 1).
2. Laminar flow box.
3. Inverted microscope.
4. Nontreated plastic dishes with diameter of 3.5 cm.
5. Coverslips (24 × 40 mm), siliconized at the edges with 5% dichloromethylsilane in 1,1,1-trichloroethane (see Note 2).
6. Mineral oil (embryo culture-tested grade, Sigma-Aldrich, St Louis, MO, USA).
7. Mannitol solution adjusted to 370 mosmol/kg H_2O and autoclaved.
8. Sliding stage for the insertion of a coverslip and a plastic dish.
9. Glass capillaries made from 50-μL aspirator tubes (Drummond Scientific Co., Broomall, PA, USA), tip openings 150–250 μm (drawn by hand).
10. Glass needles with fine tips.
11. Cell transfer systems: computer-controlled dispenser/dilutor (NanoSpuit, Ikeda Rika, Tokyo, Japan), or manual handling injector (UJI-B, ST Science, Tsukumi-gun, Kanagawa, Japan).

2.2. Fusion of Gametes

1. Mannitol solution adjusted to 450 mosmol/kg H_2O and autoclaved.
2. Mannitol solution adjusted to 520 mosmol/kg H_2O and autoclaved.
3. Electrofusion apparatus (PA-4000, Cyto Pulse Sciences Inc., Glen Burnie, MD, USA).

4. Manipulator (UMMT-3FC, Narishige Scientific Instrument Lab., Tokyo, Japan) with a double pipette holder (HD-21, Narishige).
5. Electrodes (platinum–iridium wire, diameter 150 μm) fixed to the pipette holder (see Note 3).

2.3. Culture of Zygotes, Embryogenesis, and Regeneration

1. 3.5-cm plastic dishes (nontreated).
2. Millicell-CM inserts, diameter 12 mm (Millipore, Madison, WI, USA).
3. Feeder cells: rice suspension cell culture (Line Oc, provided by RIKEN Bio-Resource Center, Tsukuba, Japan) (see Note 4).
4. Medium for zygote culture: N6Z-medium (48) with modifications: 2 g/L CHU (N6) basal salt mixture (Sigma-Aldrich), 0.025 mg/L $Na_2MoO_4 \cdot 2H_2O$, 0.025 mg/L $CoCl_2 \cdot 6H_2O$, 0.025 mg/L $CuSO_4 \cdot 5H_2O$, 0.01 mg/L retinol, 0.01 mg/L calciferol, 0.01 mg/L biotin, 1 mg/L thiamine·HCl, 1 mg/L nicotinic acid, 1 mg/L pyridoxine·HCl, 1 mg/L cholin chloride, 1 mg/L Ca-pantothene, 0.2 mg/L riboflavin, 0.2 mg/L 2,4-D, 0.02 mg/L cobalamin, 0.02 mg/L *p*-aminobenzoic acid, 0.4 mg/L folic acid, 2 mg/L ascorbic acid, 40 mg/L malic acid, 40 mg/L citric acid, 40 mg/L fumaric acid, 20 mg/L Na-pyruvate, 1,000 mg/L glutamine, and 250 mg/L casein hydrolysate, 100 mg/L myo-inositol. Osmolality, 450 mosmol/kg H_2O adjusted with glucose. pH 5.7 and filter sterilized.
5. Regeneration medium: solidified MS medium with some modifications (49). MS salt, MS vitamin, 100 mg/L myo-inositol, 2 g/L casamino acid, 30 g/L sucrose, 30 g/L sorbitol, 0.2 mg/L 1-naphthaleneacetic acid (NAA), 1 mg/L kinetin, and 0.3% Gelrite.
6. Rooting medium: the same as the regeneration media, but omitting kinetin and NAA.

3. Methods

3.1. Isolation of Gametes

1. Collect panicles in which some flowers have already opened and others remain unflowered. Pick up the unflowered ones from the panicles and dissect them. Isolate ovaries and anthers, and transfer them separately into 3.5-cm plastic dishes filled with 3 mL of mannitol solution (370 mosmol/kg H_2O) for isolating egg and sperm cells, respectively.
2. For egg cell isolation, remove the stigmas from ovaries and transfer them into new 3.5-cm plastic dishes filled with 3 mL of

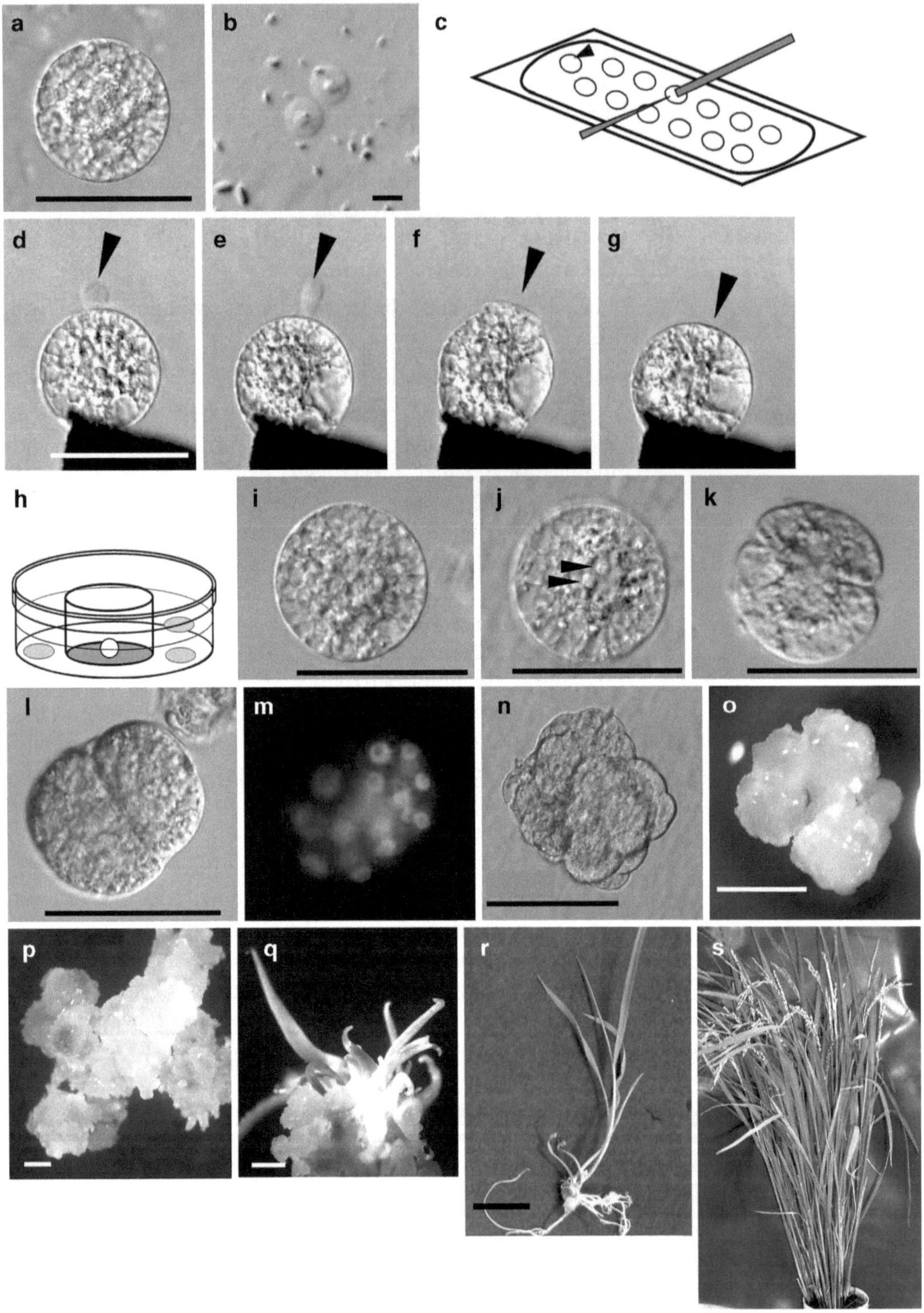

Fig. 1. In vitro fusion of rice gametes (**a**–**g**), early development of a zygote produced by in vitro fertilization (IVF) into a globular embryo (**h**–**m**) and development and regeneration of the globular embryos (**n**–**s**). (**a**) An isolated rice egg cell. (**b**) Rice sperm cells released from pollen grain. (**c**) An illustration of the fusion droplets on a coverslip covered with mineral oil. An *arrowhead* indicates a mannitol droplet for tentative storage of isolated egg cells for subsequent fusion. The *gray bar* and *thin triangle* indicate electrodes. (**d**) Alignment of an egg cell with a sperm cell (*arrowhead*) on one of the electrodes under an alternating current (AC) field in a fusion droplet. (**e**) Aligned egg and sperm cells after the addition of mannitol solution

the mannitol solution (see Note 5). Sink the ovaries to the bottom of the dishes and cut them transversely with a razor blade at the middle (see Note 6). Approximately, 15 egg cells released from the lower parts of the cut ovaries are then transferred into a mannitol droplet on coverslips using the cell transfer system under an inverted microscope (see Fig. 1a, c; Note 7).

3. For sperm cell isolation, roughly break anthers in mannitol solution with forceps to free the pollen grains. Use the sperm cells released from the burst pollen grains for electrofusion (see Fig. 1b; Note 8).

3.2. Fusion of Gametes

1. Overlay the siliconized coverslip with 0.3 mL mineral oil. For tentative storage of isolated egg cells, make one or two 2 μL mannitol droplets (370 mosmol/kg H_2O) using a microcapillary and a micropump. In addition, inject 2 μL mannitol droplets (370 mosmol/kg H_2O) in two rows, each with six droplets (see Fig. 1c). Take care that the droplets do not spread over the glass surface, but are located inside the oil and have no access to the air.
2. Set up fusion apparatus and adjust the position of electrodes.
3. Transfer one egg cell to each of the six mannitol droplets (see Note 9), then transfer one or two sperm cells to each droplet.
4. Align and fix the two gametes at one electrode under an alternating current (AC) field (1 MHz, 0.4 kV/cm). By moving the microscope stage, first fix an egg cell to the electrode. Using the same procedure, fix a sperm cell to the female gamete (see Fig. 1d). Adjust the final distance of the electrodes to approximately twice the sum of the diameters of the cells.
5. Add 0.5–1.0 μL of mannitol solution (520 mosmol/kg H_2O) gently to the fusion droplet using a thin glass capillary (see Fig. 1e; Note 10).

Fig. 1. (continued) with a higher osmolality to the fusion drop. The sperm cell becomes oblong (*arrowhead*). (**f**) Fusion of gametes following a negative direct current (DC) pulse. An *arrowhead* indicates fusion point. (**g**) A zygote 10 s after fusion. The *arrowhead* indicates the fusion point. (**h**) An illustration of zygote culture. A *white circle* in the Millicell insert indicates a zygote. *Gray oblong circles* represent aggregates of feeder cells. (**i**) A zygote 1 h after fusion. (**j**) A zygote 4 h after fusion. Two nucleoli are indicated by *arrowheads*. (**k**) An asymmetric two-celled embryo 18 h after fusion. (**l, m**) Nuclear staining of an embryo 48 h after fusion, visualized by brightfield and fluorescence microscopy, respectively. (**n**) A cell mass 5 days after fusion, which developed from the globular-like embryo. (**o**) A white cell colony 18 days after fusion. (**p**) A developed cell colony 4 days after transferring the white cell colony (panel **o**) into regeneration medium (22 days after fusion). Green spots are visible in/on the cell colony. (**q**) Regenerated shoots. Generation of shoots can be observed after 8 days of subculturing the white cell colony (26 days after fusion). (**r**) A plantlet after 12 days of subculturing a regenerated shoot in hormone-free medium (43 days after fusion). (**s**) A regenerated plant with seed sets (100 days after fusion). Scale bars indicate 50 μm in **a**, **d**, **i–l**, and **n**; 10 μm in **b**; 1 mm in **o–q**; and 1 cm in **r** (originally published by ref. (47), with permission of Springer).

6. Induce cell fusion by applying a single negative direct current (DC) pulse (50 μs, 14–15 kV/cm) (see Fig. 1f, g; Note 11).
7. Remove the fusion products from the electrode by gently moving the sliding stage. Move the electrodes out of the droplet and conduct the next gamete fusion (see Subheading 3.2, Step 4).

3.3. Zygote and Embryo Culture and Plant Regeneration

1. Place 0.2-mL zygote culture medium in a Millicell-CM insert and put it into a 3.5-cm plastic dish containing 2 mL of the medium. Add 40–60 μL of a rice suspension cell culture into the dish as feeder cells.
2. After sterilization of the microcapillary by washing with absolute ethanol and sterilized water, transfer IVF-produced zygotes into fresh mannitol droplets (450 mosmol/kg H_2O) twice and then transfer them onto the membranes of a Millicell-CM insert (see Fig. 1h; Note 12).
3. After overnight culture of zygotes at 26°C in the dark without shaking, continue culture with gentle shaking (40 rpm) (see Fig. 1i–m; Notes 13 and 14).
4. Five days after fusion, remove feeder cells by transferring the Millicell dishes containing the embryos into new 35-mm diameter dishes filled with 2 mL of fresh zygote culture medium (see Fig. 1n; Note 15). Continue culturing as above.
5. After 18 days in culture, subculture cell colonies developed from the IVF-produced zygotes onto a regeneration medium by use of a sterilized Pasteur pipette. Incubate under continuous light at 30°C for 12–30 days (see Fig. 1o; Note 16).
6. Transfer the differentiated shoots into a rooting medium and culture them under a 13 h/11 h light/dark cycle at 28°C for 11–13 days (see Fig. 1p, q).
7. Transfer the resulting plantlets to soil pods and grow in environmental chambers as described in Note 1 (see Fig. 1r). If needed, harvest seeds from the regenerated plants and germinate them (see Fig. 1s).

4. Notes

1. Rice plants (*Oryza sativa* L. cv Nipponbare) were grown in environmental chambers under conditions of 26°C in a 13/11 h light/dark cycle with a photosynthetic photon flux density of 150–300 μmol/m^2/s. Under these growth conditions, flowers can be obtained throughout all seasons.
2. Coverslips should be noncoated, as using coated coverslips will result in attachment of the cells to the surface of the coverslip. Coverslips supplied from Fisher Scientific (No. 125485J) are recommended.

3. Tips of two wires are patted with small hammer to make their thickness to 10–20 µm. Each tip of the flattened wires is trimmed into square or thin-triangle shape, and set to the pipette holder.
4. Rice suspension cells, Line Oc, were subcultured once weekly according to instructions from RIKEN Bio-Resource Center. No difference in feeder effects between freshly subcultured cells and 1-week-cultured cells has been observed.
5. Without removing the stigmas, ovaries always float on the mannitol solution. To isolate egg cells, sinking ovaries into the mannitol solution is essential. Usually, 15–25 ovaries are put into a dish.
6. Usually, three to eight egg cells are automatically released from approximately 20 cut ovaries. Gentle pushing of the basal portion of the lower part of the cut ovary with a glass needle will produce additional egg cells.
7. Egg cells can be kept in the mannitol droplet until 6 h after isolation for conducting IVF without decreasing fusion efficiency.
8. Sperm cells should be used for IVF within 1 h after isolation. Otherwise, sperm cells appear to degenerate and cannot be fused with egg cells.
9. At each round of fusion procedures, five to six sets of gamete fusions are recommended.
10. The addition of mannitol solution with a higher osmolality changes the shape of the sperm cell to oblong and makes the attachment of the egg cell to the electrode more stable (see Fig. 1d, e). Without this treatment, egg cells are often released from the electrode upon fusion induced by a DC pulse and fusion efficiency is greatly reduced.
11. If no cell fusion occurs, reduce the distance between the two electrodes and pulse again. Alternatively, transfer the egg cells into mannitol solution (370 mosmol/kg H_2O) and then reuse for IVF.
12. The efficiency of successful electrofusion is approximately 85% under optimal conditions. A total of 20–50 egg cells can be isolated from 100 processed ovaries, and 20–30 egg cells can be fused with sperm cells by one experimenter in a day.
13. Gamete fusion occurs within 1 s and the shape of the zygote on the electrode recovers to a spherical shape at about 10 s after fusion (see Fig. 1f, g). The rice zygotes produced by IVF start to form cell walls (see Fig. 1i) and two nucleoli can be observed in a zygote at least 4 h after fusion (see Fig. 1j). At around 12 h after fusion, well-developed granular organelles, probably starch granules, are visible in the zygotes and the first asymmetric cell division of the zygotes is observed at 17–22 h after fusion (see Fig. 1k). After the first division, the

two-celled embryos continue to develop into early embryos at 40–50 h after fusion (see Fig. 1l, m).

14. Approximately 90% IVF-produced zygotes divide into two-celled embryo, and 90% IVF-produced two-celled embryos develop into globular embryos.
15. After 5 days culture of the IVF-produced zygotes, cocultivation with feeder cells is not needed.
16. Normally, after 4 days of subculture of the cell colony on a solidified-regeneration medium (22 days after fusion), green spots become visible and the emergence of multiple shoots is observed after 8 days of subculturing (26 days after fusion).

References

1. Raghavan V (2003) Some reflections on double fertilization, from its discovery to the present. New Phytol 159:565–583
2. Russell SD (1992) Double fertilization. Int Rev Cytol 40:357–390
3. Pollock EG, Jensen WA (1964) Cell development during early embryogenesis in *Capsella* and *Gossypium*. Am J Bot 51:915–921
4. Schel JHN, Kieft H, van Lammeren AAM (1984) Interactions between embryo and endosperm during early developmental stage of maize caryopses (*Zea mays*). Can J Bot 62:2842–2853
5. Mansfield SG, Briarty LG (1991) Early embryogenesis in *Arabidopsis thaliana*. II. The developing embryo. Can J Bot 69:461–476
6. Schulz R, Jensen WA (1968) *Capsella* embryogenesis: the egg, zygote and young embryo. Am J Bot 55:807–819
7. Tykarska T (1979) Rape embryogenesis: II. Development of embryo proper. Acta Soc Bot Pol 48:391–421
8. Pritchard NH (1964) A cytochemical study of embryo development in Stellaria media. Am J Bot 51:472–479
9. Tykarska T (1976) Rape embryogenesis: I. The proembryo development. Acta Soc Bot Pol 45:3–16
10. Lindsey K, Topping JE (1993) Embryogenesis: a question of pattern. J Exp Bot 259:359–374
11. Haecker A, Gross-Hardt R, Geiges B, Sarkar A, Breuninger H, Herrmann M, Laux T (2004) Expression dynamics of WOX genes mark cell fate decisions during early embryonic patterning in *Arabidopsis thaliana*. Development 131: 657–668
12. Friml J, Vieten A, Sauer M, Weijers D, Schwarz H, Hamann T, Offringa R, Jürgens G (2003) Efflux-dependent auxin gradients establish the apical–basal axis of *Arabidopsis*. Nature 426: 147–153
13. Lukowitz W, Roeder A, Parmenter D, Somerville C (2004) A MAPKK kinase gene regulates extra-embryonic cell fate in *Arabidopsis*. Cell 116:109–119
14. Bowman JL, Eshed Y (2000) Formation and maintenance of the shoot apical meristem. Trends Plant Sci 5:110–115
15. Laux T, Wurschum T, Breuninger H (2004) Genetic regulation of embryonic pattern formation. Plant Cell 16:S190–S202
16. Mayer KF, Schoof H, Haecker A, Lenhard M, Jürgens G, Laux T (1998) Role of *WUSCHEL* in regulating stem cell fate in the *Arabidopsis* shoot meristem. Cell 95: 805–815
17. Sato Y, Hong SK, Tagiri A, Kitano H, Yamamoto N, Nagato Y, Matsuoka M (1996) A rice homeobox gene, *OSH1*, is expressed before organ differentiation in a specific region during early embryogenesis. Proc Natl Acad Sci USA 93:8117–8122
18. Wang YY, Kuang A, Russell SD, Tian HQ (2006) In vitro fertilization as a tool for investigating sexual reproduction of angiosperm. Sex Plant Reprod 19:103–115
19. Kranz E (1999) In vitro fertilization with isolated single gametes. Methods Mol Biol 111:259–267
20. Dupuis I, Roeckel P, Matthys-Rochon E, Dumas C (1987) Procedure to isolate viable sperm cells from corn (*Zea mays* L.) pollen grains. Plant Physiol 85:876–878
21. Kranz E, Bautor J, Lörz H (1991) In vitro fertilization of single, isolated gametes of maize mediated by electrofusion. Sex Plant Reprod 4:12–16
22. Holm PB, Knudsen S, Mouritzen P, Negri D, Olsen FL, Roué C (1994) Regeneration of fertile barley plants from mechanically isolated protoplasts of the fertilized egg cell. Plant Cell 6:531–543

23. Kovács M, Barnabás B, Kranz E (1994) The isolation of viable egg cells of wheat (*Triticum aestivum* L.). Sex Plant Reprod 7:311–312
24. Tian HQ, Russell SD (1997) Micromanipulation of male and female gametes of *Nicotiana tabacum*: I. Isolation of gametes. Plant Cell Rep 16:555–560
25. Katoh N, Lörz H, Kranz E (1997) Isolation of viable egg cells of rape (*Brassica napus* L.). Zygote 5:31–33
26. Cao Y, Russell SD (1997) Mechanical isolation and ultrastructural characterization of viable egg cells in *Plumbago zeylanica*. Sex Plant Reprod 10:36–73
27. Uchiumi T, Komatsu S, Koshiba T, Okamoto T (2006) Isolation of gametes and central cells from *Oryza sative* L. Sex Plant Reprod 19:37–45
28. Hoshino Y, Murata N, Shinoda K (2006) Isolation of individual egg cells and zygotes in *Alstroemeria* followed by manual selection with a microcapillary-connected micropump. Ann Bot 97:1139–1144
29. Kranz E, Lörz H (1993) In vitro fertilization with isolated, single gametes results in zygotic embryogenesis and fertile maize plants. Plant Cell 5:739–746
30. Faure J-E, Digonnet C, Dumas C (1994) An in vitro system for adhesion and fusion of maize gametes. Science 263:1598–1600
31. Kranz E, Lörz H (1994) In vitro fertilization of maize by single egg and sperm cell protoplast fusion mediated by high calcium and high pH. Zygote 2:125–128
32. Khalequzzaman M, Haq N (2005) Isolation and in vitro fusion of egg and sperm cells in *Oryza sativa*. Plant Physiol Biochem 43:69–75
33. Sun M, Yang H, Zhou C, Koop H-U (1995) Single-pair fusion of various combinations between female gametoplasts and other protoplasts in *Nicotiana tabacum*. Acta Bot Sin 37:1–6
34. Tian HQ, Russell SD (1997) Micromanipulation of male and female gametes of *Nicotiana tabacum*: II. Preliminary attempts for in vitro fertilization and egg cell culture. Plant Cell Rep 16:657–661
35. Peng XB, Sun MX, Yang HY (2005) A novel in vitro system for gamete fusion in maize. Cell Res 15:734–738
36. Antoine AF, Faure J-E, Dumas C, Feijo JA (2001) Differential contribution of cytoplasmic Ca^{2+} and Ca^{2+} influx to gamete fusion and egg activation in maize. Nat Cell Biol 3:1120–1123
37. Sun M-X, Moscatelli A, Yang H-Y, Cresti M (2002) In vitro double fertilization in *Nicotiana tabacum* (L.): polygamy compared with selected single pair somatic protoplast and chloroplast fusions. Sex Plant Reprod 13:113–117
38. Kranz E, von Wiegen P, Lörz H (1995) Early cytological events after induction of cell division in egg cells and zygote development following in vitro fertilization with angiosperm gametes. Plant J 8:9–23
39. Faure J-E, Mogensen HL, Dumas C, Lörz H, Kranz E (1993) Karyogamy after electrofusion of single egg and sperm cell protoplasts from maize: cytological evidence and time course. Plant Cell 5:747–755
40. Scholten S, Lörz H, Kranz E (2002) Paternal mRNA and protein synthesis coincides with male chromatin decondensation in maize zygotes. Plant J 32:221–231
41. Hoshino Y, Scholten S, von Wiegen P, Lörz H, Kranz E (2004) Fertilization induced changes in the microtubular architecture of the maize egg cell and zygote – an immunocytochemical approach adapted to single cells. Sex Plant Reprod 17:89–95
42. Okamoto T, Scholten S, Lörz H, Kranz E (2005) Identification of genes that are up- or down-regulated in the apical or basal cell of maize two-celled embryos and monitoring their expression during zygote development by a cell manipulation- and PCR-based approach. Plant Cell Physiol 46:332–338
43. Ito Y, Arikawa K, Antonio BA et al (2005) Rice annotation database (RAD): a contig-oriented database for map-based rice genomics. Nucleic Acids Res 33:D651–D655
44. Miyao A, Tanaka K, Murata K, Sawaki H, Takeda S, Abe K, Shinozuka Y, Onosato K, Hirochika H (2003) Target site specificity of the Tos17 retrotransposon shows a preference for insertion within genes and against insertion in retrotransposon-rich regions of the genome. Plant Cell 15:1771–1780
45. Kikuchi S, Satoh K, Nagata T et al (2003) Collection, mapping and annotation of over 28, 000 cDNA clones from japonica rice. Science 301:376–379
46. Uchiumi T, Uemura I, Okamoto T (2007) Establishment of an in vitro fertilization system in rice (*Oryza sativa* L.). Planta 226:581–589
47. Itoh J, Nonomura K, Ikeda K, Yamaki S, Inukai Y, Yamagishi H, Kitano H, Nagato Y (2005) Rice plant development: from zygote to spiklet. Plant Cell Physiol 46:23–47
48. Kumlehn J, Lörz H, Kranz E (1998) Differentiation of isolated wheat zygotes into embryos and normal plants. Planta 205:327–333
49. Hiei Y, Ohta S, Komari T, Kumashiro T (1994) Efficient transformation of rice (*Oryza sativa* L.) mediated by *Agrobacterium* and sequence analysis of the boundaries of the T-DNA. Plant J 6:271–282

Chapter 3

Canola Zygotic Embryo Culture

Nicole S. Ramesar-Fortner and Edward C. Yeung

Abstract

To further understand the events occurring during embryogenesis, it is imperative to have an experimental system. While somatic embryo is the system of choice due to the prolific number and success of protocols available, it is necessary to compare in vitro results with those in vivo. This process is often difficult due to the inaccessibility of the developing embryo and the complications of manipulating the embryo in vivo. The development of protocols that allow for manipulation and comparison of both somatic and zygotic embryos is key to elucidating the differences between the two types of embryos and determining if results observed using one system can be applied to the other. This chapter details a simple protocol for the culture of zygotic embryos of canola that allows for the processing of large numbers of embryos with little physical damage. Furthermore, this protocol allows for the experimental manipulation of zygotic embryos in vitro and comparisons to be made with a well-established microspore-derived embryo system.

Key words: Apical meristem, *Arabidopsis*, *Brassica napus*, Canola, Conversion

1. Introduction

Brassica spp. include many important agricultural crops such as the oilseed, canola (*Brassica napus*), and other vegetables such as broccoli, cauliflower, cabbages, and kales (different species of *B. olearacea*). Canola is becoming an increasingly important target for research because of its economic and nutritional values. There is a continuous effort to improve the yield and nutritional value of this oilseed which is now primarily performed through functional genomics. The model plant, *Arabidopsis thaliana*, that is indispensable to plant biotechnology, is closely related to *Brassica* spp. and has been the key to genetically engineering many desirable traits of canola. Unfortunately, the size of *A. thaliana* often hinders experimental manipulation and its larger relative,

Trevor A. Thorpe and Edward C. Yeung (eds.), *Plant Embryo Culture: Methods and Protocols*,
Methods in Molecular Biology, vol. 710, DOI 10.1007/978-1-61737-988-8_3,

canola, becomes the system of choice for experimental studies. The information gained from either plant can be easily applied to the other.

One area of research that is essential to genetic improvements is experimental embryogeny. Embryogenesis, in plants, is an important process that serves to establish the body plan and primary tissues that will eventually form the basis of the mature plant. One of the most fundamental events to occur is the formation of the two apical meristems. The proper formation of these two generating centers is critical for the post-embryonic development of plants. Hence, a proper understanding of the formative process related to meristem initiation and development is of theoretical interest and has practical implications. Tremendous advances have been made in recent years in the understanding of the genetic mechanism related to embryo and seed development in *Arabidopsis*. Numerous reviews have been published on this topic, e.g., Suarez and Bozhkov (1). Because of the larger size of the *Brassica* embryos and the ability of some lines of *Brassica* species to form microspore-derived embryos (2, 3), additional information on embryo development has been gained through the use of *Brassica* species as experimental systems.

Through the use of zygotic embryo culture, Liu et al. (4) were the first to show that treatment of globular embryos with auxin transport inhibitors prevents proper cotyledon formation. Instead of having two separate cotyledons, only a single "trumpet" shape cotyledon is formed. This results in the change of the embryo symmetry from bilateral to radial. A subsequent study by Hadfi et al. (5) using a defined medium confirmed and extended the work of Liu et al. (4) and clearly demonstrated that auxin has an important morphogenetic role during embryo development. A similar result was also obtained in the study of wheat zygotic embryo development (6). Using zygotic embryo culture (7) and subsequently through the use of the microspore-derived embryo (8), we demonstrated that the process of meristem formation is sensitive to changes in auxin distribution and levels. Treatment of the globular embryos with triiodobenzoic acid prevents shoot apical meristem formation. However, it has no effect once the shoot apical meristem has formed by the heart-shape stage of development. Our work indicates that the shoot apical meristem forms within a narrow window of time. Perturbation of auxin levels prevents meristem formation. However, once formed, it is a fully determined structure. All these studies clearly indicate the importance of auxin during embryo development.

Experimental embryogenesis augments current studies on zygotic embryo development. The successful in vitro culture of zygotic embryos has been a process that has required diligent experimentation and improvement over the years. The ability to culture earlier and earlier stages of zygotic embryos has allowed for physiological manipulation of the embryo environment that furthers our understanding of embryogenesis.

Different embryo culture methods have been devised for the culture of zygotic embryos in Brassicaceae, e.g., *Brassica juncea* (9) and *B. napus* (7), *Capsella* (10–14), and *Arabidopsis* (15–17). Liu et al. (9) developed a system to culture proembryos of *B. juncea* as small as 35 μm. They optimized the embryo culture medium by including sterile nonautoclaved coconut water, varying the sugars and organic acid composition, and improving the method by which the embryos were cultured. The method involves first dissecting the proembryos in a 9% (w/v) glucose solution in order to prevent osmotic shock. After isolation, the embryos were rinsed once with the same glucose solution before transferring to the culture medium. A two-layer system was used with the top layer having a high osmotic pressure due to the addition of 6% sucrose (w/v) to the culture medium. A low temperature gelling agarose was used as the gelling agent. The culture medium was gently heated using a microwave oven. Once the agarose was melted, the entire medium was sterilized using a disposable filter unit with the aid of a vacuum. Twenty four-well culture plates were used for the culture of proembryos. A small volume (300 μL) of the bottom-layer medium was added to each well. Embryos were then dissected and placed onto the surface of the medium. The embryos were then overlaid with 100 μL of the top medium. The top medium was maintained at 38°C to prevent it from gelling prior to applying it onto the embryos. The embryos were sandwiched between the two media. This procedure allowed for the growth of proembryos that are similar to their in vivo counterparts. The success of this system allowed Liu et al. (4) to investigate the physiological control of embryo development. The ability to culture proembryos successfully provides an experimental system that is critical to further understand and dissect fundamental processes in embryogenesis. Details of media components and techniques of embryo isolation and culture can be found in Liu et al. (9).

In *Arabidopsis*, due to the small size of the developing proembryo, Sauer and Friml (15–17) cultured fertilized ovules instead of culturing isolated embryos. The culture of ovules enables the study of in vitro embryo development after a successful fertilization event. Excision of intact ovules is rapid and a large number of fertilized ovules can be cultured at one time. Furthermore, the nutrient requirement may not be as stringent as for the isolated embryos. The maternal tissues, i.e., the developing seed coat, may provide additional factors for proper embryo development, as well as initial physical protection to the developing embryo. The ovules were first cultured using a high osmotic pressure medium (10% sucrose in a half-strength Murashige and Skoog (18) (MS) medium) supplemented with 400 mg/L glutamine (15). In order to allow further development of the embryo, the ovules were then transferred to a low osmotic pressure medium (1% sucrose in ½ MS) 5 days after the initial culture. Ovules had to be placed on the surface of the medium to allow for proper development of the embryo. This method was successful in rearing young zygotic embryos as judged by the normal progression of embryo development within the fertilized ovule. Details of the culture protocol can be found in Sauer and Friml (15–17).

In our study of *B. napus* cv Topas, we used a half seed culture procedure instead of excising intact proembryos to increase the number of embryos sampled and avoid physical damage to the embryo during dissection (7). Although culturing intact fertilized ovules as described for *Arabidopsis* is a faster culturing procedure, seed coat development usually lags behind embryo growth in vitro and this can hinder embryo development. Hence, culturing of zygotic embryos with half of the seed coat removed enables a large number of embryos to be cultured and easy assessment of embryo development in vitro. In our studies, the proembryo with a small portion of the seed coat attached is placed directly onto the culture medium. The process of dissection is relatively fast with no physical damage to the small zygotic embryo. A single medium is used instead of the double layered method as detailed in Liu et al. (9). The medium composition for the culturing of the embryos is similar to that reported by Liu et al. (9). However, in order to assess the normality of embryo development, i.e., the functional integrity of the apical meristems, after initial culture, a conversion test needs to be performed. For the conversion test, the developing embryos have to be transferred to half MS medium with a low percentage (2%) of sucrose. The following protocol details the method of culture. It is important to note that we are culturing slightly older embryos when compared with those used by Liu et al. (9). The culture requirement for the globular embryo is not as stringent as the very young proembryos as indicated in Liu et al. (9).

2. Materials

1. Maturation medium: the components of the maturation are detailed in Table 1. Prepare stock solutions for the "organics," "sugar mixture" (minus sucrose and glucose), and "organic acids," dispatch into small volumes and frozen.
2. Coconut water (Sigma C-5915).
3. Difco Noble agar (Fisher Scientific).
4. Conversion medium: half-strength MS medium (18) with 2% sucrose and 0.8% agar.

Table 1
Composition of maturation medium for culture of zygotic embryos of *Brassica napus*

Component	Maturation medium (mg/L)
Macronutrients	
NH_4NO_3	200
KNO_3	1,500
$CaCl_2 \cdot 5H_2O$	622
$MgSO_4 \cdot 7H_2O$	400
$KH_2PO_4 \cdot 2H_2O$	79
$Na_2EDTA \cdot 2H_2O$	33
$FeSO_4 \cdot 7H_2O$	28
Microelements	
KI	0.75
H_3BO_3	3
$MnSO_4 \cdot H_20$	13.2
$ZnSO_4 \cdot 7H_2O$	2
$NaMoO_4 \cdot 2H_2O$	0.25
$CuSO_4 \cdot 5H_2O$	0.025
$CoCl \cdot 6H_2O$	0.025
Organics	
Myo-inositol	500
Glutamine	200
Thiamin · HCl	1
Nicotinic acid	0.1
Pyrodoxine · HCl	0.1
d-Biotin	0.01
Organic supplements	
Casein hydrolysate	100
Coconut water	100 mL/L
Sugar mixture	
Sucrose	40 g
Glucose	20 g
Fructose	100
Ribose	100
Xylose	100
Mannose	100
Rhamnose	100
Cellobiose	100
Sorbitol	100
Mannitol	100

(continued)

Table 1
(continued)

Component	Maturation medium (mg/L)
Organic acids	
Sodium pyruvate	20
Citric acid	40
Malic acid	40
Fumaric acid	40
Agar	10 g

5. Sterilization solution: a 30% bleach solution prepared by diluting commercial Javex® bleach which can be obtained locally from stores (see Note 1) with sterile water and 0.1% (v/v) Tween 20 (Sigma) as a wetting agent.
6. Equipment and supplies: a stereomicroscope, forceps, scalpels, Petri dishes, and Parafilm.

3. Methods

3.1. Plant Growth and Maintenance

1. Grow plants of *B. napus* cv Topas in a greenhouse mix soil in a growth chamber maintained at 25°C days and 16°C nights with a 16 h photoperiod. In order to sustain the growth of plants until seed set, large pots about 1 gal in size should be used per plant and the plants should be fertilized regularly with a complete fertilizer.
2. The plants start to flower approximately 6–7 weeks after the initiation of germination (see Note 2).
3. Hand-pollinate the flowers and tag on the day of anthesis (see Note 3).

3.2. Maturation Medium

1. Prepare maturation medium as detailed in Table 1 in preparation of autoclaving without the coconut water.
2. After autoclaving, allow the media to cool, and then add the coconut water (see Note 4). Depending on the volume, use sterile pipettes to distribute the coconut water to the media flasks.
3. After the medium is poured into Petri plates, store at 4°C.

3.3. Preparing Half Seeds for In Vitro Culture

1. Pick developing siliques and surface sterilize in the sterilization solution for 15 min and subsequently wash three times with sterile water (see Note 5).
2. The embryos are at the globular stage of development at approximately 6 days after pollination and at the heart stage at approximately 8 days after pollination. Within one silique, the seeds could contain embryos at slightly different stages of development, but these could be easily differentiated under the dissecting microscope (see Note 6).
3. Split siliques open along the replum with the tip of a fine forceps while holding the base of the silique with another pair of forceps. The two halves of the wall (the valves) can then be separated by gently pulling them exposing the partition (the replum) and the seeds. Since a majority of seeds are still attached to the replum, exposed seeds can be easily cut with a sharp double-edge razor blade or using a scalpel with a no. 11 sterile stainless steel surgical blade (see Note 7). All attached seeds can be cut quickly one at a time since they are "immobilized" and attached to the replum through a funiculus (see Note 8). Quickly remove the half attached seeds using a fine forceps and place onto the maturation medium. Using fine forceps partially submerge the half seed with the cut surface facing up at the same level as the medium. Do not submerge the half seeds into the medium.
4. Place approximately 20–25 embryos in each Petri plate, seal with Parafilm, and place in the dark for 2 days and then under light conditions (photon flux density of 90–95 μmol/m^2/s, PAR) with a 16 h photoperiod for a further 12 days.

3.4. Transfer to Conversion Medium to Evaluate Normality of Development

1. At the end of the maturation period, transfer the half seeds to a conversion medium and maintain under similar conditions. The embryo should have grown large enough to see readily with an unaided eye or through a stereomicroscope.
2. Embryos that develop normally with properly developed apical meristems will start to grow giving rise to a shoot with new leaves and a root within days. These embryos usually have a bilateral symmetry.
3. Embryos that fail to develop apical meristems will become arrested. The cotyledons often fuse to form a trumpet.

4. Notes

1. It is important to ensure that the commercial bleak solution is still effective and not to use the bleach solution beyond the

expiration date indicated on the bottle. The cap of the bottle needs to be tightly secure after each use.

2. The canola plants when matured are relatively large with active transpiration. It is imperative the plants are not allowed to dry out at the time of flowering as this can cause flower abortion or poor seed set.
3. Canola is a prolific plant that continues to flower and set seeds for several weeks. Based on the tagging of flowers, we find that the first set of seeds goes through defined changes in morphological stages in a reproducible manner. For those flowers that form later in the flowering period, embryo development tends to be slower. This is most likely due to competition for nutrients among the numerous siliques present. Hence, it is preferable to use the first crop of seeds for experimental purposes.
4. Coconut water can be purchased readily from various commercial sources. Since the solution is already sterile, it can be added directly to the warm culture medium. If a large volume of coconut water is added such as 10%, it is important to dissolve proper amounts of the major and minor salts and additives in a reduced volume of water allowing for the added volume of coconut water used.
5. In order to ensure proper embryo staging, it is important to check the stage of embryo development by careful dissection of a few seeds under a stereomicroscope before sampling. This is especially important when siliques are from older plants.
6. Filter sterilization of media components is preferred for components other than the major and minor elements. In a number of studies, it has been recommended that a shorter autoclave time be used, i.e., 10 min (15).
7. For excising plant tissues, often a scalpel with a no. 11 pointed blade is used. However, the blade tends to be a bit thick and does not give a clean cut. We prefer breakable carbon steel double-edge blades as our cutting knives. A special blade holder is used to clamp and break off a small piece of the steel blade and which can then be used as a knife. Since the blades are thin, a sharp clean cut can be made. Such instruments can be obtained from Fine Science Tools (catalog number 10050-00 and 10052-11).
8. It is essential to know the location of the micropylar end of a seed. When excising the seed half, locate the micropylar end of the seed as the embryo is sitting next to the micropyle to the inside. Gently hold onto the micropylar end of a seed with a pair of fine forceps and make the incision at the opposite end. If the seeds are still attached to the silique, excision of seed halves can be done simultaneously as the seeds have the same orientation.

Acknowledgments

This work was supported by a Discovery Grant from the Natural Sciences and Engineering Research Council of Canada to E.C.Y.

References

1. Suarez MF, Bozhkov PV (2008) Plant embryogenesis. Methods in molecular biology, vol 427. Humana Press, Totowa
2. Yeung EC (2002) The canola microspore-derived embryo as a model system to study developmental processes in plants. J Plant Biol 45:119–133
3. Ilic-Grubor K, Attree SM, Fowke LC (1998) Comparative morphological study of zygotic and microspore-derived embryos of *Brassica napus* L. as revealed by scanning electron microscopy. Ann Bot 82:157–165
4. Liu C-M, Xu Z-H, Chua N-H (1993) Auxin polar transport is essential for the establishment of bilateral symmetry during early plant embryogenesis. Plant Cell 5:621–630
5. Hadfi K, Speth V, Neuhaus G (1998) Auxin induced developmental patterns in *Brassica juncea* embryos. Development 125: 879–887
6. Fischer C, Neuhaus G (1996) Influence of auxin on the establishment of bilateral symmetry in monocots. Plant J 9:659–669
7. Ramesar-Fortner NS, Yeung EC (2001) Triiodobenzoic acid affects shoot apical meristem formation and function in zygotic embryos of *Brassica napus* cv Topas. Can J Bot 79:265–273
8. Ramesar-Fortner NS, Yeung EC (2006) Physiological influences in the development and function of the shoot apical meristem of microspore-derived embryos of *Brassica napus* cv. Topas. Can J Bot 84:371–383
9. Liu C-M, Xu Z-H, Chua N-H (1993) Proembryo culture: in vitro development of early globular-stage zygotic embryos from *Brassica juncea*. Plant J 3:291–300
10. Monnier M (1976) Culture in vitro del'embryon immature de *Capsella bursa-pasforis* Moench. Rev. Cyt. Biol. Văg. 39, 1–120. Thesis
11. Monnier M (1978) Culture of zygotic embryos. In: Thorpe TA (ed) Frontiers of plant tissue culture. University of Calgary Press, Calgary, pp 277–286
12. Monnier M (1984) Survival of young immature *Capsella* embryos cultured in vitro. J Plant Physiol 115:105–113
13. Monnier M (1990) Culture of zygotic embryos of higher plants. Methods Mol Biol 6:129–139
14. Monnier M (1995) Culture of zygotic embryos. In: Thorpe TA (ed) In vitro embryogenesis in plants. Kluwer Academic Publishers, Dordrecht, pp 117–153
15. Sauer M, Friml J (2004) In vitro culture of *Arabidopsis* embryos within their ovules. Plant J 40:835–843
16. Sauer M, Friml J (2005) In vitro culture of *Arabidopsis* embryos. In: Mujib A, Samaj J (eds) Somatic embryogenesis. Springer, Berlin, pp 343–354
17. Sauer M, Friml J (2008) In vitro culture of *Arabidopsis* embryos. In: Suarez MF, Bozhkov PV (eds) Plant embryogenesis, methods in molecular biology, vol 427. Humana Press, Totowa, pp 71–76
18. Murashige T, Skoog F (1962) A revised medium for rapid growth and bioassays with tobacco tissue cultures. Physiol Plant 15:897–911

Chapter 4

Immature Seeds and Embryos of *Medicago truncatula* Cultured In Vitro

Sergio J. Ochatt

Abstract

Legumes are an important source of proteins and lipids for food and feed. In addition, they are environmentally friendly because of their capacity to fix nitrogen through a symbiosis with *Rhizobium* that permits them to produce abundant proteins even in the absence of nitrogen fertilization. Seed development in plants follows three chronological steps (1) seed coat differentiation, embryo morphogenesis and endosperm development; (2) embryo maturation with storage accumulation and (3) dehydration and the acquisition of desiccation tolerance. Finally, germination occurs when the environmental conditions become favourable. Working with the model legume *Medicago truncatula*, an in vitro protocol was developed for the culture of immature embryos that permits their development in a way comparable to that observed in plants.

In this chapter, the usefulness of this system for investigating embryo development in legumes is outlined.

Key words: Abscisic acid, Embryo morphogenesis, Flow cytometry, Gibberellin, *Medicago truncatula*, Nitrogen, Storage proteins, Sulphur

1. Introduction

Seed development in plants follows three chronological steps. The first step corresponds to seed coat differentiation, embryo morphogenesis, and endosperm development. Following this step, embryo maturation with storage accumulation occurs. The last step is dehydration and the acquisition of desiccation tolerance. Germination may then occur when the environmental conditions become favourable (1, 2). The immature seeds of many species have been cultured in vitro to study the regulation of storage product accumulation, but this frequently resulted in callus development from the embryo axis or the cut surfaces of explanted tissues (3, 4). In addition, even if explants expanded through water uptake, cell

Trevor A. Thorpe and Edward C. Yeung (eds.), *Plant Embryo Culture: Methods and Protocols*,
Methods in Molecular Biology, vol. 710, DOI 10.1007/978-1-61737-988-8_4,

division slows down and storage product accumulation either stops or strongly reduces as compared with in vivo rates. Hence, a strategy to sustain in vitro development comparable with that observed in vivo is of great value to investigate the effects of plant growth factors (nutrients and hormones) on storage product accumulation, its metabolic regulation and the interactions between different seed tissues and with the mother plant (5). Non-viable mutants affected during seed development may also be analyzed if such embryo rescue in vitro is possible.

Legumes are an important source of proteins and lipids for food and feed. In addition, they are also environmentally friendly as a result of their capacity to fix nitrogen through a symbiosis with *Rhizobium* that permits them to produce abundant proteins even in the absence of nitrogen fertilization. *Medicago truncatula* is an annual diploid ($2n = 16$) autogamous legume species that originated in the Mediterranean basin with a small genome size 500 Mb/1C (6). It is designed as a model species in the context of studies on plant-*Sinorhizobium meliloti* symbiosis (7) and has been used extensively since then in research on mutagenesis, genomics and also at the plant physiology level. *M. truncatula* is phylogenetically close to the most widely cultivated legumes in Europe, pea and field bean (8, 9), and its seeds have been shown to consist mainly of the embryo at maturity. However, they also possess persistent endosperm that contributes about 10% of the final seed mass. *M. truncatula* seed protein content, at 35–45%, is significantly higher than that of pea or faba bean and most of the remaining carbon stored in the mature seed is in the form of oil with starch content being less than 1% (10).

The in vitro protocol for seed filling with immature seeds of *M. truncatula* described in this chapter was modified from a method previously used to shorten generation cycles with various protein legumes (4). It was used to analyse the effects of variations in the nitrogen content of the medium, demonstrating that the accumulation of storage proteins in embryos cultured in vitro on an MS medium (5) corresponded to that observed in planta (11). Once established, the same protocol was used to determine the time of competence for in vitro culture to compare the performance of isolated embryos with that of entire grain explants, and to analyse the response of both variations in sulphur and hormonal nutrition. In this respect, grain legumes are generally penalized as feed when contrasted with cereals because of the poor sulphur content in their protein. Modification of the exogenous sulphur supply appeared to be an interesting way of studying its effects on the accumulation of storage proteins richer in this element. Likewise, it was tempting to assess the effects of some growth regulators on embryo development and seed filling as a method to improve the stability of this trait in protein legumes. Naturally, the choices of gibberellin and abscisic acid were made. The use of flow cytometry (12, 13) then appeared as a tool to analyse the mitotic index and cell cycle of such cultured seeds and embryos.

2. Materials

2.1. Plant Material and Culture Conditions of Donor Plants

1. *M. truncatula* seeds, harvested at maturity, scarified (H_2SO_4, 3 min) and sown to give the donor plants.
2. N:P:K (20:20:20) fertilizer solution (PlantProd; Quebec, Canada, or any other brand with the same composition).
3. 2-L pots.
4. A 1:1 soil:pozzolane potting mixture.
5. Glasshouse/growth chamber facilities for controlled environmental conditions of donor plants.

2.2. In Vitro Culture

1. Stock solutions for preparation of Murashige and Skoog (14) basal medium (MS) as follows (see Table 1 for details): macroelements concentrated 10× (kept in the fridge until use or for

Table 1
Composition of the basal Murashige and Skoog (14) medium used for culture of immature seeds and embryos that supported embryo development and seed filling as observed in planta

Components		Concentration (mg/L)
Macroelements	NH_4NO_3	1,650
	KNO_3	1,900
	$CaCl_2 \cdot H_2O$	440
	$MgSO_4 \cdot 7H_2O$	370
	KH_2PO_4	170
Microelements	H_3BO_3	6.22
	$MnSO_4 \cdot 4H_2O$	22.3
	$ZnSO_4 \cdot 7H_2O$	10.6
	KI	0.83
	$Na_2MoO_4 \cdot 2H_2O$	0.25
	$CuSO_4 \cdot 5H_2O$	0.025
	$CoCl_2 \cdot 6H_2O$	0.025
Fe EDTA	$FeSO_4 \cdot 7H_2O$	27.85
	Na_2EDTA	37.23
Vitamins/organics	Nicotinic acid	0.5
	Pyridoxine-HCl	0.5
	Thiamine-HCl	0.1
	Glycine	2
	myo-Inositol	100
Sugars	Sucrose	130,000
Agar		6,000

maximum 1 month), microelements concentrated 1,000× (kept in the fridge until use or for maximum 1 year), FeEDTA concentrated 200× (kept in fridge in a brown bottle until use or up to 1 year) and vitamins/organics (altogether) are concentrated 500× and aliquoted before use (frozen stock may be kept for up to 1 year while in the fridge for a few months only).

2. Sucrose.
3. Agar.
4. KOH and HCl 1 N to adjust pH.
5. Ultrafiltration equipment to filter-sterilize hormones to be added to media as required.
6. Standard tissue culture equipment (autoclave, dissection tools, laminar flow hood, etc.).

2.3. Flow Cytometry

1. Flow cytometer equipped with UV excitation lamp.
2. Dissection tools (for chopping of material).
3. Petri dishes for chopping of tissues.
4. Buffers for extraction and nuclei and their staining with DAPI (4′,6-diamidino-2-phenylindole): two-step procedure consisting of nuclei extraction buffer followed by staining buffer, or single-step procedure with a unique buffer.
5. Plastic or nylon meshed sieves (50-μm pore size) for filtering of the nuclei suspension prior to measurements.

2.4. FDA Assessments of Viability

1. Fluorescein diacetate (FDA) solution prepared by diluting 60 μL of a stock solution of 5 mg FDA/mL acetone in 8 mL mannitol (9%, w/v). The stock can be kept frozen for a very long time.
2. Photonic epi-fluorescence microscope.

3. Methods

3.1. Plant Material and Culture Conditions of Donor Plants

M. truncatula genotype A17, derived from cultivar Jemalong (provided by P. Guy, Plant Breeding Station, INRA Lusignan, France), was used throughout.

1. Culture plants in 2-L pots with soil mixed with pozzolane (1:1), with fertilization (N:P:K, 20:20:20; PlantProd; Quebec, Canada) through drop watering during the first week and once weekly thereafter.
2. Maintain plants in the glasshouse, under a photoperiod of 16 h, with warm fluorescent lamps at an intensity of 220 μE/m^2/s, a temperature of 19°C (day) and 22°C (night) and a relative humidity of 60–70%.

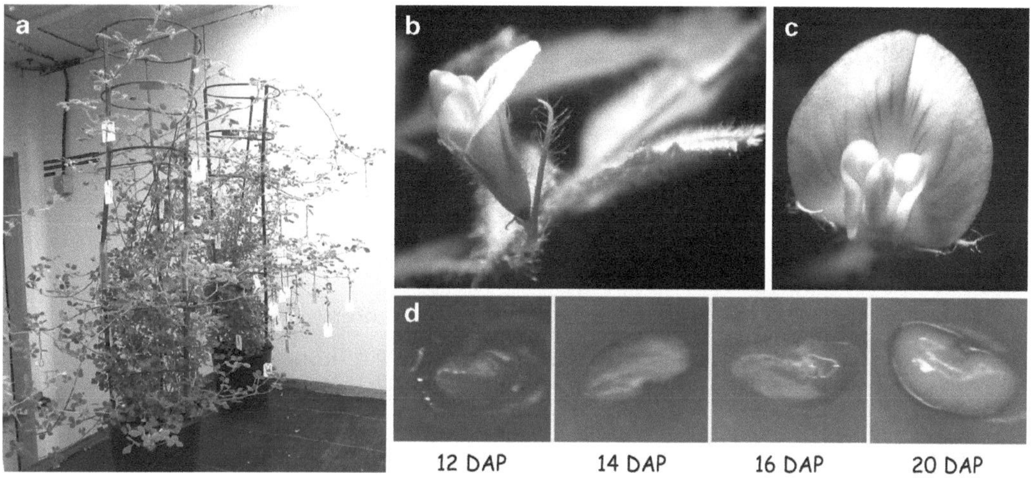

Fig. 1. Developmental stages of harvest of immature pods of *Medicago truncatula*. (**a**) Plant with flowers labelled with the date of pollination; (**b**, **c**) front and side view of a flower at the pollination stage, used to calculate pod age subsequently; (**d**) immature seeds at 12, 14, 16 and 20 DAP as used for culture.

3. For harvest of immature seeds and embryos, identify flowers morphologically on the day following pollination (Fig. 1a–c; see Note 1) and label with the date for the determination of the number of days after pollination (DAP) (see Note 2).
4. Immature seeds (Fig. 1d; see Note 3) and embryos (see Note 4) are harvested from 8 DAP onwards depending on the kind of studies to be performed, with material of the respective ages (Fig. 1d) harvested in plants used as controls. The optimum age to compare in vitro and in planta seeds and embryos is 12 DAP.

3.2. In Vitro Culture

1. All media are based on Murashige and Skoog (14) basal medium (MS) supplemented with 6-g/L agar and 130-g/L sucrose (MS130) (Table 1; see Note 5). The pH of all media is adjusted to 5.6 with 1 N KOH before autoclaving, for 30 min at 112°C.
2. No addition of growth regulators is needed to support development and seed filling comparable to that observed in planta. Thus, growth regulators are added only when aimed at assessing their role on embryo development and seed filling (see Note 6). Similarly, for studies on the nitrogen or sulphur content of the medium, modifications are performed from this basal formulation (Table 1).
3. Dispense media as 2-mL aliquots into 5 × 5 multi-well plastic dishes and store at 4°C in the dark until used.
4. For culture, seal dishes with Parafilm and keep them under a photoperiodic regime of 16-h light/8-h dark from warm fluorescent tubes at an intensity of 90 mE/m^2/s and under a thermoperiod of 22/20°C.

5. At least 15 embryos and 15 seeds per age should be cultured on each tested medium and all experiments should be repeated three times.
6. Immature pods are disinfected for 1 min in 70% ethanol and 15 min in calcium hypochlorite (37 g/L) then rinsed four times in sterile water.
7. Aseptically open sterilized pods and excise grains or embryos for culture.
8. Seeds are examined at 24-h interval (see Note 7).
9. This protocol, without major modifications, should also work with field pea, grass pea and other grain legumes.

3.3. Flow Cytometry

Flow cytometry assessments were carried out on material at 12-, 16,- and 20-DAP harvested directly from the mother plants for the controls (Fig. 1c) and on seeds and embryos that had been cultured for different periods of time (i.e. typically for 4 days: 12→16 and 16→20, or 8 days: 12→20) on the various media assessed depending on the studies carried out (MS0, ABA10 and ABA50).

The typical stepwise methodology for flow cytometry analyses of *M. truncatula* (12) does not differ from that used for other species (13) by our team, and is as the following:

1. All assessments are performed using a Partec PAS II flow cytometer, equipped with a mercury HBO 100-W lamp, a dichroic mirror (TK 420), and a built-in programme for the treatment of data (Flomax, Partec GmbH).
2. All plant tissues (leaves of mother plant, seeds, and embryos) are prepared following methods as described (12, 13) and are stained with DAPI (4′,6-diamidino-2-phenylindole) (see Note 8). Briefly, a small amount of tissue is chopped in 400 μL nuclei extraction buffer, diluted with 1,600 μL of staining buffer [both from Partec GmbH (Germany), Cystain UV Precise T kit] or in a one-step procedure (Cystain UV Precise one-step kit from Partec GmbH, Germany), where 1 mL of buffer is used to chop tissues for simultaneous extraction and staining of nuclei. Following extraction and staining, the nuclei suspension is sieved through a 50-μm mesh (Celltrics from Partec GmbH, Germany, or home-made) and recovered into vessels (generally 2.5-mL haemolysis tubes) adapted to the flow cytometer employed.
3. For each measurement, a minimum of 2,500 nuclei are counted and the results presented as a linear scale on a real-time graph with the size of nuclei (intensity of the epifluorescence emitted) in the abscissa and the number of nuclei counted in the ordinates. The parameters for flow cytometry readings were as described previously for *M. truncatula* (12, 13). Two instrument settings are very important to

obtain exploitable data from analyzed samples. The first is the adjustment of the lower threshold (LL) to avoid acquisition of small and unwanted "noise" signals below this emission value and to allow the system to have more time for the particles of interest. So, measurements must be performed with a LL high enough to remove all the small "noise" signals in the histograms but low enough to retain all signals from the particles of interest. The second one is the speed since highly accurate measurements require a low speed (~2 μL/s). If the speed is too high, peaks in the histograms tend to become wider and hence accuracy decreases. However, care should be taken not to use a speed too low, because in that case, particle sedimentation effects can influence a counting result.

4. For peak and cell cycle analysis (Fig. 2), the studied tissues are compared with leaves taken from mother plants of *M. truncatula* A17 cultured on MS0 medium (blank).
5. The cytometer is calibrated prior to the analyses with leaves of the pea cultivar Victor (*Pisum sativum* L.) as a standard.
6. The mitotic index is calculated and analyzed as described (12, 13) (see Notes 9 and 10).

3.4. FDA Assessments of Viability

1. Viability of tissues is evaluated with FDA (15) under UV, using a OMRB Leica photonic microscope.
2. The FDA solution employed is prepared by diluting 60 μL of a stock solution of 5 mg FDA/mL acetone in 8 mL mannitol (9%, w/v).
3. Then, two droplets (~50 μL) of this solution are added to each slide. Once excited with UV, living cells fluoresce yellow–green, confirming membrane integrity and the conservation of enzymatic activity (Fig. 3).
4. Viability is expressed as a percentage (of the number of fluorescing cells counted divided by the total number of cells in a sample and multiplied by 100).
5. At least 100 cells per microscopic fields are counted and three or more independent counts are performed for each sample analyzed.

3.5. Comments

By the end of the cell division phase, the cell number in the cotyledons is established and, in grain legumes, it determines the storage capacity of the organ (16). The onset of endoreduplication is a progressive phenomenon in storage accumulating organs during the transition between cell division and maturation phases (17) whose control is therefore of considerable agronomic interest. In plants with a small genome, this alternative to a cell cycle is a means of increasing genetic and metabolic capacities.

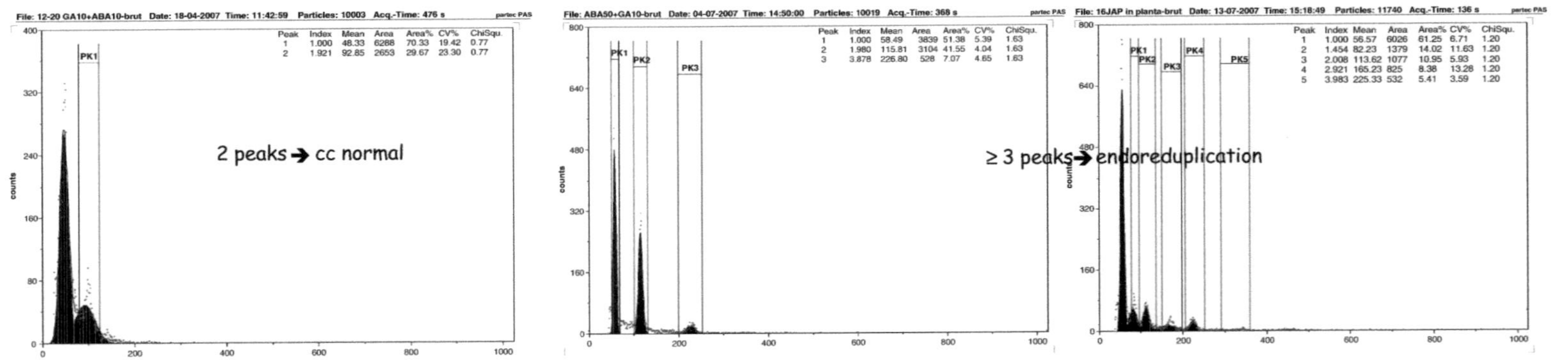

Fig. 2. Typical flow cytometry profiles of immature seeds of *M. truncatula*. From *left to right*: a normal cell cycle with two peaks, a profile showing onset of endoreduplication (three peaks), and a profile with established endoreduplication (more than three peaks) as observed in planta with 16 DAP seeds and embryos.

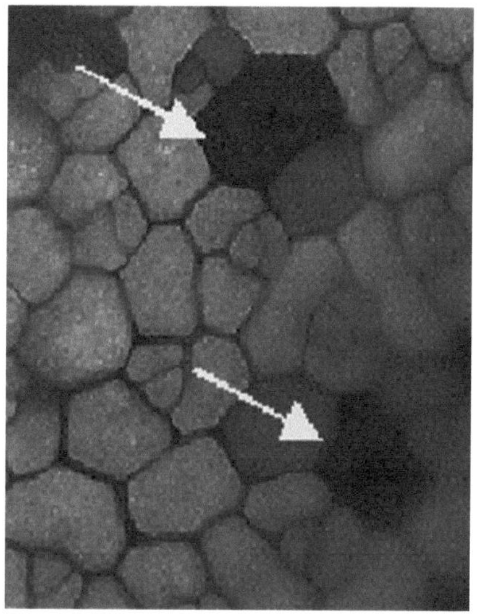

Fig. 3. An immature *M. truncatula* embryo stained with FDA observed under UV light. Two dead, non-fluorescing cells are *arrowed*.

With an exogenous nitrogen supply, both 12-DAP seeds and embryos developed with storage protein synthesis comparable to that observed in vivo. Conversely, in the absence of added nitrogen, seeds during initial stages of embryo development exhibited a remobilisation of endogenous nitrogen from tissues surrounding the embryo thereby ensuring initial storage protein accumulation, whereas isolated embryos rapidly ceased synthesizing de novo proteins and their development appeared arrested, presumably reflecting a shortage of nitrogen.

When studying sulphur exogenous supply, it was observed that the sulphur content of the medium affects development of embryos and seeds as well as their maturation except in media Sh and Sh + Nd, where an increased sulphur content seemed to compensate for the nitrogen deficiency. Conversely, sulphur deficiency dramatically altered development by demonstrating anthocyanin accumulation in the tissues (possibly a stress-response) and a significantly reduced albumin synthesis that were both resolved by adding sulphur to the medium. This system is hence useful to investigate the embryo's response to nitrogen and sulphur.

Using this strategy for the analysis of ABA and GA_3 helped to provide evidence for their actions in several important processes during seed development and stressed that 12-DAP is a key stage corresponding to the end of embryo morphogenesis and the beginning of the maturation phase and storage protein accumulation (11). Beyond 12-DAP, development and morphogenesis are finished, the embryo is formed, and the storage accumulation phase starts (2, 5, 18).

Seeds germinated in the presence of 10-mM ABA showed that only root elongation was delayed. While cell division did occur, it was slowed down. Flow cytometry analysis at 12-DAP gave profiles with a number of peaks of decreasing intensity, reflecting endoreduplication (13). There is therefore an arrest of divisions; phases G1, S, and G2 do take place but with very few mitoses, and most nuclei in the second peak are in phase G2 rather than M of the cell cycle. Conversely, on 50-mM ABA, germination was completely blocked (the lack of a second peak in flow cytometry profiles indicates that nuclei were blocked at G1 and/or S phase) and, even if they remained viable (Fig. 3), seeds looked withered, which is probably correlated with the role of ABA intolerance to desiccation (18, 19). Thus, embryos and immature seeds at 12-DAP are able to germinate within 48 h on MS0 medium, while, later, only excised embryos remain capable of rapid germination that supports the hypothesized presence of exogenous ABA in the seed coat (20). In other words, the combination of endogenous and exogenous ABA determines the kinetics of storage protein accumulation and modulates the onset of germination.

On the other hand, studies of GA_3 alone or combined with ABA proved that on GA10 and GA50, cell divisions increased and the cell cycle typically showed endoreduplication, while on media with both GA and ABA, the cell cycles were normal (Fig. 2). Likewise, the cell cycle of control in planta in embryos from 16 DAP showed a strong endoreduplication (Fig. 2), but, when the time course of the onset of endoreduplication was observed between 12- and 20-DAP, it also appeared that these would not be fixed phases but more a progression of events instead, as while division frequency is reduced the accumulation of storage compounds commences (i.e. at 12 DAP cell division continues despite an even stronger endoreduplication than at

Table 2
Relative DNA content/nucleus of various tissues of *M. truncatula* A17 in planta

	2C	4C	8C	16C	32C/64C/128C
Tissue	**PgDNA/nucleus**	**PgDNA/nucleus**	**PgDNA/nucleus**	**PgDNA/nucleus**	**PgDNA/nucleus**
Leaf	0.483	0.977			
Embryo 12DAP	0.418	0.801	1.326	2.02	5.43
Embryo 16DAP	0.485	0.8455	1.483	1.92	
Embryo 20DAP	0.45	1.028	1.438		

16 DAP; Table 2). Hence, it seems that there are transition periods rather than distinct phases, and the use of this strategy was instrumental to demonstrate it.

This strategy also showed that GA not only stimulates cell division, but it also compensates for the inhibitory effect of ABA and delays the onset of storage protein accumulation in seeds.

4. Notes

1. When the petal is at its largest, the pistil will not be apparent.
2. *M. truncatula* pods are collected from the mother plants before the accumulation of storage proteins (8 and 12 DAP) and during early stages of storage protein accumulation (14 and 16 DAP), as in intact plants storage proteins begin to accumulate by 14 DAP (5). The authentic in vitro development of grains on MS130 medium prompted a comparison with isolated embryos on this same medium. Then, both seeds and embryos developed and yielded protein profiles corresponding to that of material grown in vivo. However, the arrest in cotyledon growth and lack of anthocyanin accumulation coupled with a reduced chlorophyll pigmentation for embryos cultured on medium-N may reflect metabolic defects and a possible hormonal imbalance for such tissues, such as increased endogenous ABA levels (21, 22), which encouraged further studies on the effects of this growth regulator on seed filling using this same strategy. In addition, a study of other environmental factors such as the sulphur content in the medium and the temperature during culture is also of interest, having affected protein synthesis in soybean (23) and pea (24).
3. Seeds and embryos were harvested from 8 to 12 DAP for studies with nitrogen and from 12 to 20 DAP for those with hormones or sulphur. The choice of testing the in vitro response of seeds isolated at 8 and 12 DAP is because 12 DAP immediately precedes the onset of accumulation of storage proteins at 14 DAP (5) and 8 DAP is the earliest stage at which seeds can be easily detached without damage. In this respect, it should be mentioned that the minimum number of DAP from which seeds and/or embryos may be harvested for culture will vary with different species, but for protein legumes, in general, will remain at around 8–12 DAP.
4. To assess the autonomy of metabolic processes in place during embryo development, embryos were cultured in parallel to seeds under the same conditions. With nitrogen, cotyledons enlarge significantly, root tips extend and within 6 days embryos doubled in size, coupled with an accumulation of anthocyanins as seen for the seeds. On MS130-N, cotyledon

size did not increase significantly although the root continued to grow, and there was a progressive loss of chlorophyll pigmentation, suggesting a loss of photosynthetic capacity. In contrast to embryos on MS130+N, there was no anthocyanin or carotenoid accumulation in embryos cultured in the absence of nitrogen.

5. For assessments on the effects of sulphur, seeds and embryos at different stages of maturity were cultured on 12 different MS media (14) of a modified composition to render sulphur either deficient or excessive and to combine it with accordingly deficient or excessive nitrogen contents as follows: Sn, Sd, Sh, Nd, Nh, Nh+, Sd + Nd, Sd + Nh, Sd + Nh+, Sh + Nd, Sh + Nh and Sh + Nh+. Thus, media were based on MS formula (14) with 130-g/L sucrose (see Table 1), but were modified to supply deficient (Sd; no S; $MgSO_4$ in MS formula replaced by $MgCl_2$), normal (Sn; 1.5-mM S as in MS formula) or high (Sh; 4.5 mM S supplied as Na_2SO_4 because, at such concentration, $MgSO_4$ would precipitate) sulphur contents and nitrogen at deficient (Nd; 544-mg N_2/L), normal (Nn; 824.1-mg N_2/L), high (Nh; 1384.1-mg N_2/L) or very high (Nh+; 1944.1-mg N_2/L) concentrations.
6. When appropriate, the growth regulators ABA (*cis*-trans racemic mixture, Sigma) at 0, 1, 5, 10 or 50 μM (i.e. 0, 0.264, 1.321, 2.643 or 13.215 mg/L) or GA_3 (gibberellic acid) at 0, 10 or 50 μM (i.e. 0, 3.46 or 17.3 mg/L) were added to autoclaved media. Stock solutions of the growth regulators tested were filter-sterilized. To avoid imposing an osmotic stress to the cultured tissues, the sucrose content in the medium used in these studies was reduced from 130 g/L used in our previous work with nitrogen (5) and sulphur (see above) to 30 g/L, as used routinely in vitro (4, 14).
7. At 12 DAP in planta, a seed measured 3 mm and an embryo 1.5 mm in length, while at 20 DAP, both measured 5 mm stressing the role of the endosperm, whose size decreases as the embryo grows.
8. DAPI specifically binds to the adenine and thymine bases of DNA is excited under UV (at 372 nm) and emits (at 456 nm) fluorescence that is proportional to the relative DNA content per nucleus.
9. For normal tissue, the flow cytometry profile will exhibit two peaks corresponding to the nuclei in the G1 (2C) and G2/M (4C) phases of mitosis (12, 13), respectively.
10. Similarly, such a profile will include around 80% of nuclei in G1, 10% in S and 10% in G2/M of the mitotic cycle. From this set of data, it is equally possible to analyse the cell cycle and the division frequency expressed as the mitotic index (MI):

$$MI = 4 \times 4C / \Sigma 2C + 4C.$$

Where 2C and 4C correspond to the mean intensity of the first (nuclei in phase G1) and second peak (nuclei in phase G2/M) in the profile obtained, respectively. For a normal cell cycle, $MI = 2 \pm 0.15$. A slight variation from this figure indicates a problem with cell division and the cell cycle itself (12).

The onset of endoreduplication in *M. truncatula* is indicative of the transition between cell division and seed filling. This is reflected by the start of the storage protein accumulation phase during the embryo maturation cycle concomitant with the appearance of endoreduplication peaks in the flow cytometry profiles from the developing embryos (see Table 2; (12, 13)).

References

1. Borisjuk L, Weber H, Panitz R, Manteuffel R, Wobus U (1995) Embryogenesis of *V. faba* histodifferentiation in relation to starch and storage protein synthesis. J Plant Physiol 147:203–218
2. Weber H, Borisjuk L, Wobus U (2005) Molecular physiology of legume seed development. Ann Rev Plant Biol 56:253–279
3. Ochatt SJ, Pontécaille C, Rancillac M (2000) The growth regulators used for bud regeneration and shoot rooting affect the competence for flowering and seed set in regenerated plants of protein peas. In Vitro Cell Dev Biol Plant 36:188–193
4. Ochatt SJ, Sangwan RS, Marget P, Assoumou Ndong Y, Rancillac M, Perney P (2002) New approaches towards the shortening of generation cycles for faster breeding of protein legumes. Plant Breed 121:436–440
5. Gallardo K, Kurt C, Thompson R, Ochatt S (2006) In vitro culture of immature *M. truncatula* grains under conditions permitting embryo development comparable to that observed in vivo. Plant Sci 170:1052–1058
6. Blondon F, Marie D, Brown S, Kondorosi A (1994) Genome size and base composition in *Medicago sativa* and *M. truncatula* species. Genome 37:264–270
7. Barker DG, Bianchi S, Blondon F, Dattée Y, Duc G, Essad S, Flament P, Gallusci P, Génier G, Guy G, Muel X, Tourneur J, Dénarié J, Huguet T (1990) *Medicago truncatula*, a model plant for studying the molecular genetics of the *Rhizobium*-legume symbiosis. Plant Mol Biol Rep 8:40–49
8. Aubert G, Morin J, Jacquin F, Loridon K, Quillet MC, Petit A, Rameau C, Lejeune-Hénaut I, Huguet T, Burstin J (2006) Functional mapping in pea, as an aid to the candidate gene approach and for investigating the synteny with the model species *Medicago truncatula*. Theor Appl Genet 112:1024–1041
9. Kaló P, Seres A, Taylor SA, Jakab J, Kevei Z, Kereszt A, Endre G, Ellis THN, Kiss GB (2004) Comparative mapping between *Medicago sativa* and *Pisum sativum*. Mol Gen Genomics 272:235–246
10. Djemel N, Guedon D, Lechevalier A, Salon C, Miquel M, Prosperi JM, Rochat C, Boutin JP (2005) Development and composition of the seeds of nine genotypes of the *Medicago truncatula* species complex. Plant Physiol Biochem 43:557–566
11. Gallardo K, Le Signor C, Vandekerckhove J, Thompson R, Burstin J (2003) Proteomics of *Medicago truncatula* seed development establishes the time frame of diverse metabolic processes related to reserve accumulation. Plant Physiol 133:664–682
12. Ochatt SJ (2006) Flow cytometry (ploidy determination, cell cycle analysis, DNA content per nucleus). In: Medicago truncatula handbook, Chapter 2.2.7. Available at: http://www.noble.org/MedicagoHandbook/, November 2006. ISBN number 0-9754303-1-9
13. Ochatt SJ (2008) Flow cytometry in plant breeding. Cytometry A 73:581–598
14. Murashige T, Skoog F (1962) A revised medium for rapid growth and bioassays with tobacco tissue cultures. Physiol Plant 15:473–497
15. Widholm JM (1972) The use of fluorescein diacetate and phenosaphranine for determining the viability of cultured cells. Stain Technol 47:189–194
16. Munier-Jolain NG, Ney B (1998) Seed growth rate in grain legumes. J Exp Bot 49:1963–1969
17. Kowles RV, Srienc F, Phillips RL (1990) Endoreduplication of nuclear DNA in the

developing maize endosperm. Dev Genet 11: 125–132

18. Reidt W, Wohlfarth T, Ellestrom M, Czihal A, Tewes A, Ezcurra I, Rask L, Baumlein H (2000) Gene regulation during late embryogenesis: the RY motif of maturation-specific gene promoters is direct target of the FUS3 gene product. Plant J 21:401–408
19. Quatrano RS, Bartels D, Ho THD, Pages M (1997) New insights into ABA-mediated processes. Plant Cell 9:470–475
20. Wang H, Qi Q, Schorr P, Cutler AJ, Crosby WL, Fowke LC (1998) ICK1, a cyclin-dependent protein kinase inhibitor from *Arabidopsis thaliana* interacts with both Cdc2a and CycD3, and its expression is induced by abscisic acid. Plant J 15:501–510
21. Xu NF, Bewley JD (1995) The role of abscisic acid in germination, storage protein synthesis and desiccation tolerance in alfalfa (*Medicago sativa* L.) seeds, as shown by inhibition of its synthesis by fluridone during development. J Exp Bot 46:687–694
22. Xu NF, Bewley JD (1995) Temporal and nutritional factors modulate responses to abscisic acid and osmoticum in their regulation of storage protein synthesis in developing seeds of alfalfa (*Medicago sativa* L.). J Exp Bot 46:675–686
23. Ohtake N, Kawachi T, Okuyama I, Fujikake H, Sueyoshi K, Ohyama T (2001) Effect of short-term application of nitrogen on the accumulation of beta-subunit of beta-conglycinin in nitrogen-starved soybean (*Glycine max* L.) developing seeds. Soil Sci Plant Nutr 48:31–41
24. Karjalainen R, Kortet S (1987) Environmental and genetic variation in protein content of peas under northern growing conditions and breeding implications. J Agric Sci Finland 59:1–9

Chapter 5

In Vitro Culture and Germination of Terrestrial Asian Orchid Seeds

Yung-I. Lee

Abstract

Orchidaceae is one of the largest families of flowering plants, and many of its species are highly valuable as herbal medicines and to the horticultural industry. To meet commercial requirements and to conserve natural resources, in vitro seed germination has been utilized to produce large quantities of uniform seedlings. In general, terrestrial orchid seeds are more difficult to germinate and grow than epiphytic orchids. Terrestrial orchid seeds have a hardened seed coat and more stringent requirements for germination in vitro. In this chapter, we document the timing of seed collection and pretreatments for improving in vitro germination of some terrestrial Asian orchids. The process of in vitro germination is demonstrated, including (1) the culture of immature seeds; (2) the culture of mature seeds; and (3) subsequent seedling development. For immature seed culture, optimal timing of seed harvest is key to maximizing germination; for mature seed culture, selection of adequate pretreatment conditions (i.e., the duration and concentrations of pretreatment solutions) is essential to improve germination.

Key words: Asymbiotic germination, *Calanthe*, Cuticular material, *Cypripedium*, Seed coat, Zygotic embryos

1. Introduction

Orchidaceae is one of the largest families of flowering plants, consisting of more than 22,000 species (1). Some orchid genera, such as *Phalaenopsis*, *Dendrobium*, *Oncidium*, *Paphiopedilum*, and *Cymbidium*, have become economically important as potted plants and cut flowers. According to a United States Department of Agriculture report, the total annual wholesale value of potted orchids for 2007 was 126 million USD in the United States alone. Some species of *Dendrobium*, *Gastrodia*, and *Vanilla* are also valuable herbs and flavoring agents (2). Many orchids that remain unstudied have potential use in natural medicine and as ornamentals. However, the

Trevor A. Thorpe and Edward C. Yeung (eds.), *Plant Embryo Culture: Methods and Protocols*,
Methods in Molecular Biology, vol. 710, DOI 10.1007/978-1-61737-988-8_5,

majority of plants currently sold by nursery growers are collected from wild strains without regard for their scarcity. Therefore, practical and efficient methods of propagation are necessary to prevent depletion of natural populations and to supply commercial production.

The numerous and tiny seeds of orchids are unique in several ways (3). As orchid capsules mature, they bear seeds that consist of a globular embryo covered by a thin seed coat. The seeds are without an endosperm and are regarded as precociously released proembryos (4). Under natural conditions, successful germination of orchid seeds is dependent on the formation of mutualistic associations with specific mycorrhizal fungi (5). However, Knudson (6) discovered in 1922 that orchid seeds are able to germinate on some synthetic media without the assistance of mycorrhizal fungi. That is, they can germinate asymbiotically. Asymbiotic culture of orchid seeds (undifferentiated embryos) is similar to embryo rescue of other crops and has become a practical tool for most orchid propagation systems (7). Generally, the process of seed germination of terrestrial orchids is more intricate than that of epiphytic orchids (8). It has been suggested that difficulties encountered in the former may be due to impermeability of the seed coat. In *Phalaenopsis amabilis* var. *formosa*, an easy-to-germinate epiphytic species, cuticular materials, form a discontinuous layer around the embryo proper; this intermittency enables the embryo to access water and nutrients and reduces physical constraints on germination (9). In contrast, histochemical studies of developing seeds of the terrestrial orchid *Cypripedium* have found that the embryo is enclosed tightly and compactly by the carapace (the inner seed coat) and that various chemicals, such as cuticular substances (10), lignin (11), and suberin (12), accumulate in the carapace and seed coat as the seeds mature. In another terrestrial orchid, *Calanthe tricarinata*, cuticular substances and phenolic compounds are present in the seed coat (13). The deposition of these hydrophobic compounds plays a key role in the impermeability of mature seeds. Moreover, high levels of endogenous abscisic acid (ABA) have been found to accumulate in the mature seeds of hard-to-germinate terrestrial orchids (13–15). These findings suggest that the accumulation of ABA in mature terrestrial orchid seeds may be responsible for their low germination percentages.

Immature temperate terrestrial orchid seeds are easier to germinate in vitro than they are at maturity (16–18). At the optimal inoculation time, seeds are usually yellowish-white in color, moist, and remain partially attached to the placenta. Culturing immature seeds may avoid full development of the impermeable seed coat (10, 19) and the accumulation of substances inhibitory to germination (13, 15, 20). For immature seed culture, optimized timing of seed harvest is crucial to maximize germination rates. For example, in *Cypripedium formosanum*, seeds collected from 90 to 105 DAP are the most suitable for culture in vitro, while after 105 DAP, seed germination decreases abruptly (10).

For mature seed culture, pretreating seeds with hypochlorite (20, 21), ultrasound (22), and/or chilling (23, 24) are often good strategies to maximize germination percentages. In *C. tricarinata*, soaking mature seeds into 1 N NaOH and 1% NaOCl solution not only scarifies the seed coat but also demolishes endogenous ABA (13). Thus, pretreatment increases the permeability of the seeds by scarification and releases substances linked to seed dormancy.

Though orchid seeds are tiny and simple, many characteristics of the seed coat (such as cell layers of the seed coat and chemical constituents of the cell wall) vary widely among species. Therefore, the choice of an adequate pretreating condition (i.e., the duration and concentrations of the pretreating solution) is important to improve germination of mature seeds. In this chapter, the optimum timing of seed collection and the seed pretreatments for improving in vitro germination of some Asian terrestrial orchids are documented.

2. Materials

1. Culture media for immature and mature seed culture: 1/4–1/10 strength macroelements of Murashige and Skoog (MS) basal salts with full strength microelements (25–28), or Thomale GD basal salts (29), as detailed in Table 1. Basal media are supplemented with 20 g/L sucrose (Sigma Chemical Co., St. Louis, Mo), 1 g/L tryptone (Merck KGaA, Darmstadt, Germany), and 100 mL/L coconut water (obtained from fresh green coconut fruit) and solidified with 2.2 g/L Phytagel ((Sigma Chemical Co.). See Table 1 for composition.
2. Banana extract: obtained bananas just as they are turning yellow.
3. Potato extract: obtain fresh potatoes from local market.
4. Capsule sterilizing solution: 1% sodium hypochlorite solution containing two drops of a wetting agent (Tween-20; Sigma).
5. Seed scarifying solution: NaOH (1 N) solution with two drops of Tween-20.
6. Hyponex protocorm medium: Hyponex No. 1; 7N-6P-19K (Hyponex Corp., Marysville, Ohio) with 20 g/L sucrose, 1 g/L tryptone, 1 g/L activated charcoal, 20 g/L potato homogenate, and 30 g/L banana homogenate, solidified with 2.2 g/L Phytagel.
7. An ultrasonicator (200 W, 44 kV, Branson 8210; Branson Ultrasonic Corp., Danbury, Conn.) is used for seed scarification.
8. Other laboratory equipment and supplies: a stereomicroscope, laminar flow hood, sterile droppers, scalpels, and forceps.

Table 1
Composition of MS (25) and Thomale GD (29) media

Component	MS medium (mg/L)	Thomale GD medium (mg/L)
Macronutrients		
KNO_3	1,900	400
NH_4NO_3	1,650	370
$(NH_4)_2SO_4$	–	60
KH_2PO_4	170	300
$CaCl_2 \cdot 2H_2O$	440	
$MgSO_4 \cdot 7H_2O$	370	
$Mg(NO_3)_2 \cdot 6H_2O$		110
Micronutrients		
$MnSO_4 \cdot 4H_2O$	22.3	
$ZnSO_4 \cdot 7H_2O$	8.6	
$CuSO_4 \cdot 5H_2O$	0.025	
KI	0.83	
$CoCl_2 \cdot 6H_2O$	0.025	
H_3BO_3	6.2	
$NaMoO_4 \cdot 2H_2O$	0.25	
$FeSO_4 \cdot 7H_2O$	27.84	20
$Na_2 \cdot EDTA$	37.3	
Organics		
Myo-inositol	100	
Nicotinic acid	50	
Pyridoxine-HCl	50	
Thiamine-HCl	10	
Glycine	200	

3. Methods

3.1. Flower Pollination

1. When the flowers are fully open, hand pollinate the flowers to ensure good capsule sets and seed quality. Use a clean toothpick to touch the sticky fluid of the stigma and then touch the pollinia. The pollinia will stick onto the toothpick. Carefully place the pollinia on the surface of the stigma.
2. Capsules will develop after successful pollination and fertilization. For immature seed culture, the timing of seed collection plays an important role in further development of the embryo and in

Table 2
The optimum time for seed collection of some Asian terrestrial species

Taxon	The optimum time (DAP)	References
Calanthe tricarinata	150	Lee et al. (13)
Cymbidium ensifolium var. *misericors*	190	Lu et al. (26)
Cymbidium sinense	150	Lee (27)
Cypripedium formosanum	90–105	Lee et al. (10)
Paphiopedilum armeniacum	120	Unpublished data
Paphiopedilum bellatulum	120–150	Lee (28)
Paphiopedilum delenatii	150	Lee et al. (30)
Paphiopedilum godefroyae	90–120	Lee (28)
Paphiopedilum niveum	120–150	Lee (28)
Paphiopedilum helenae	150	Lee (28)
Paphiopedilum henryanum	120–180	Lee (28)
Paphiopedilum spicerianum	120–180	Lee (28)

This table lists the optimum time for seed collection of Asian terrestrial species; for the European and North American terrestrial species, see Rasmussen (19)

protocorm formation. Table 2 lists the optimal time for seed collection of some species (see Note 1). Prior to seed culture, it is important to check seed quality. Examine the seeds under a microscope to make sure embryos are present and the seed coat has not yet acquired moisture-repellency (19, 30).

3.2. Preparation of Media for the In Vitro Culture of Immature Seeds

Different media can be used for culturing immature embryos from different species; preliminary testing is required to determine the best medium for a specific project. Immature embryos can be successfully cultured using 1/4–1/10 strength macroelements from MS basal salts (25) or Thomale GD basal salts (29) with full strength microelements (Table 1) supplemented with 20 g/L sucrose, 1 g/L tryptone, and 100 mL/L coconut water, solidified with 2.2 g/L Phytagel. The pH of media is adjusted to 5.7 before autoclaving at 121°C. Ten mL of medium is poured into each 25 × 100 mm culture tube.

3.3. Immature Seed Germination

1. Harvest the green capsule at the optimum time and determine whether the seeds hold embryos (see Note 2).
2. Wash the capsules for 3 min in tap water and sterilize for 30 min in capsule sterilizing solution. Wash 3 times with sterile distilled water.
3. Cut and open the capsules in a laminar flow hood. Scoop out the seeds and distribute onto the medium (see Note 3). Add a few drops of sterile distilled water to the immature seeds if necessary (see Note 4).
4. Place the culture tubes in a darkened growth room at 25 ± 2°C (see Note 5).
5. After protocorm formation, transfer the culture tubes to lit conditions with a 12 h photoperiod at 30 μmol/m^2/s (see Note 6).

3.4. Germination from Mature Seeds

1. Harvest the mature seeds just prior to splitting of mature capsules (see Note 7). Place the seeds in a test tube (20 × 100-mm) and surface sterilize for 20–60 min in capsule sterilizing solution (see Note 8). The capsule sterilizing solution is not only used for surface sterilization but also serves to partially erode the seed coat (Fig. 1). This improves in vitro seed germination. Rinse the seeds 3 times with sterile distilled water before sowing.
2. Soak hardened seeds in seed scarifying solution for 10–30 min (see Note 9). Nonhardened seeds do not require this treatment. Rinse the seeds 3 times with sterile distilled water, as in step 1.

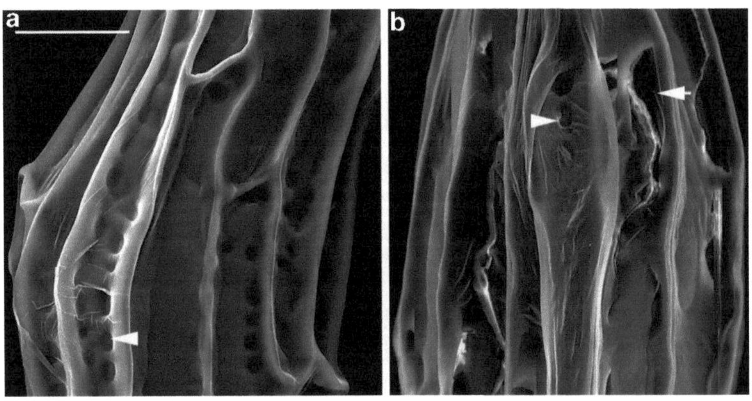

Fig. 1. SEM micrographs of seedcoat of *Paphiopedilum armeniacum* after soaking in sodium hypochlorite solution. (**a**) The surface of seed coat is pitted (*arrowhead*) because of the unevenly deposited cell wall constituents. Scale bar = 50 μm. (**b**) After soaking in 1% sodium hypochlorite solution for 60 min, the surface of seed coat is eroded (*arrowhead*), and some cracks (*arrow*) are formed. Scale bar = 50 μm.

3. Ultrasound treatment is another way to scarify hardened seeds. Scoop the seeds into tubes with sterile distilled water and treat with an ultrasonicator for 8–30 min (see Note 10). Rinse the seeds 3 times with sterile distilled water, as in step 1.
4. Transfer the seeds to 25 × 100-mm test tubes containing culture media as indicated in Subheading 3.2. Maintain the culture tubes in the dark.
5. Liquid culture is another way to hasten and synchronize seed germination. After surface sterilization or ultrasound treatment, place the seeds into liquid medium in a 125 mL flask (1/4 MS or Thomale GD without gelling agent). Place the flasks in the dark on a rotary shaker at approx. 100 rpm. Transfer to gelled medium as the embryos emerge from their seed coats (see Note 11).
6. Transfer the culture tubes to light conditions with a 12 h photoperiod at 30 μmol/m^2/s photon flux density after protocorm formation.

3.5. Seedling Development

1. Once the protocorms have turned green and their first leaves become visible, place them on Hyponex protocorm medium. For the banana extract, peel bananas and cut them into cubes (about 1 cm^3). Boil fresh materials with water (20 g fresh material per 100 mL water) for 10 min and homogenize them with a kitchen blender. For the potato extract, peel potatoes and process them as in the banana extract procedure. Add the homogenate to the basal medium prior to pH adjustment. The pH of the medium is adjusted to 5.7 before autoclaving. Pour 100 mL medium into a 500-mL flask (see Notes 12 and 13).
2. Incubate in the light (30 μmol/m^2/s photon flux density) under a 12 h photoperiod until the seedlings develop roots approximately 1–2 cm in length.
3. Subculture the seedlings to fresh Hyponex medium every 2 months if the species of interest secretes inhibitory chemicals into its environment.
4. In a laminar flow hood, pick 15–20 seedlings of the same size and transplant to the 500 mL flask of Hyponex medium. Adjust the banana homogenate to 50 g/L and the sucrose to 10–15 g/L.
5. Incubate in the light (30 μmol/m^2/s photon flux density) under a 12 h photoperiod for approximately 4–8 weeks and transfer to a greenhouse for further growth (see Note 14).

4. Notes

1. Temperature affects embryo development substantially. Adjust the harvest time according to your local climate. Sample the capsules and examine embryos under a microscope before sowing to confirm that you have collected them at the correct developmental stage. Figure 2 shows the developing embryos of *C. tricarinata* and *C. formosanum* at the optimum time for their inoculation. The plants of *C. tricarinata* and *C. formosanum* are maintained in the greenhouses at Mei-Fong high-land farm (lat. 24° 5′N, long. 121° 11′E, 2,100 m above sea level), Taiwan; the plants of *Paphiopedilum* species are grown in pad-and-fan system greenhouses in flat, low elevation areas of Taiwan.
2. If many of the embryo sacs are not successfully fertilized, the capsule will contain many "empty seeds." Before sowing, quickly examine the seeds under a microscope to ensure that they contain embryos.
3. The germinating seeds of some orchid species secrete inhibitory chemicals into their environment. If the seeds are sown too close together, the germinating seeds/protocorms will necrotize quickly.

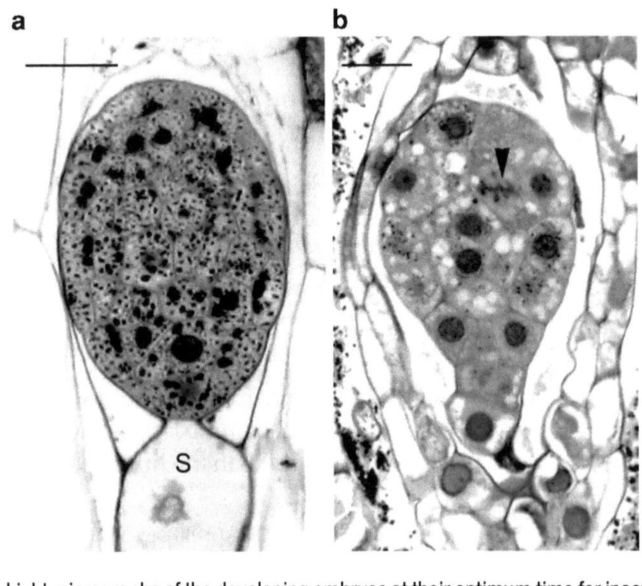

Fig. 2. Light micrographs of the developing embryos at their optimum time for inoculation. (**a**) The globular embryo with a single-celled suspensor (S) of *Calanthe tricarinata* at 150 days after pollination. Scale bar = 50 μm. (**b**) The early globular embryo of *Cypripedium formosanum* at 90 days after pollination. At this stage, cell division (*arrowhead*) is occurring. Scale bar = 20 μm.

4. If the surface of the gelled medium appears dry, add a few drops of sterile distilled water to moisten the seeds after sowing.
5. For some temperate species, incubation of the seeds at 4°C for about 2 weeks can stimulate germination (23, 24).
6. The span of the germination and protocorm developing phases are variable among species, ranging from 1 month to 2 years.
7. Harvest the capsule and place on a clean bench before splitting. Sow the seeds as quickly as possible.
8. Since the character of the seed coat is diverse among different species, the duration and concentration of optimal seed treatments differ. One indicator of how easily a seed will germinate is the color of its seed coat. If the seed coat is darker in color, soak the seeds until the dark color disappears.
9. In the case of *C. tricarinata* (13), NaOH solution may scarify the seed coat, allowing the efflux of endogenous ABA from mature seeds. Use sterile distilled water to prepare NaOH solution.
10. To allow for variations among different seed coats and ultrasonicator models, test different durations before regular ultrasound treatment. A good indicator can be the seeds' buoyancy following ultrasound treatment. If most of the seeds do not float on water, they may not have been scarified enough.
11. Use a sterile dropper with a wide opening to suck up the seeds.
12. Peptone and yeast extract are also good organic sources for orchid seedling development.
13. In many cases, banana homogenate promotes growth of protocorms and seedlings. However, some species are sensitive to banana homogenate, and protocorms will turn brown if they receive this treatment.
14. Acclimatization conditions are important for seedling survival. For many terrestrial orchids, the required light quantity for seedlings is less (about 1/10–1/20) than full sun light. We usually use sphagnum moss to grow seedlings under high humidity and low light conditions.

References

1. Arditti J (1992) Fundamentals of orchid biology. Wiley, New York
2. Hew CS, Arditti J, Lin WS (1997) Three orchids used as herbal medicines in China: an attempt to reconcile Chinese and Western pharmacology. In: Arditti J, Pridgeon AM (eds) Orchid biology: review and perspectives, VII. Kluwer Academic Publishers, Dordrecht, pp 213–283
3. Dressler RL (1981) The orchids natural history and classification. Harvard University Press, Cambridge

4. Harvais G (1982) An improved culture medium for growing the orchid *Cypripedium reginae* axenically. Can J Bot 60:2547–2555
5. Dearnaley JDW (2007) Further advances in orchid mycorrhizal research. Mycorrhiza 17:475–486
6. Knudson L (1922) Nonsymbiotic germination of orchid seeds. Bot Gaz 73:1–25
7. Arditti J (1967) Factors affecting the germination of orchid seeds. Bot Rev 33:1–97
8. Kano K (1968) Acceleration of the germination of so-called hard-to-germinate orchid seeds. Amer Orchid Soc Bull 37:690–698
9. Lee Y-I, Yeung EC, Lee N, Chung MC (2008) Embryology of *Phalaenopsis amabilis* var. *formosa*: embryo development. Bot Study 49:139–146
10. Lee Y-I, Lee N, Yeung EC, Chung MC (2005) Embryo development of *Cypripedium formosanum* in relation to seed germination in vitro. J Am Soc Hortic Sci 130:747–753
11. Carlson MC (1940) Formation of the seed of *Cypripedium parviflorum*. Bot Gaz 102:295–301
12. Harvais G (1980) Scientific notes on a *Cypripedium reginae* of North-western Ontario. Amer Orchid Soc Bull 49:237–244
13. Lee Y-I, Lu CF, Chung MC, Yeung EC, Lee N (2007) Developmental changes in endogenous abscisic acid concentrations and asymbiotic seed germination of a terrestrial orchid, *Calanthe tricarinata* Lindl. J Am Soc Hortic Sci 132:246–252
14. Lee Y-I (2003) Growth periodicity, changes of abscisic acid during embryogenesis, and in vitro propagation of *Cypripedium formosanum* Hay. Ph.D. Thesis, National Taiwan University, Taiwan, ROC
15. Van der Kinderen G (1987) Abscisic acid in terrestrial orchid seeds: a possible impact on their germination. Lindleyana 2:84–87
16. De Pauw MA, Remphrey WR (1993) In vitro germination of three *Cypripedium* species in relation to time of seed collection, media and cold treatment. Can J Bot 71:879–885
17. Withner CL (1953) Germination of "Cyps". Orch J 2:473–477
18. Yamazaki J, Miyoshi K (2006) In vitro asymbiotic germination of immature seed and formation of protocorm by *Cephalanthera falcata* (Orchidaceae). Ann Bot 98:1197–1206
19. Rasmussen HN (1995) Terrestrial orchids from seed to mycotrophic plants. Cambridge University Press, Cambridge
20. Van Waes JM, Debergh PC (1986) In vitro germination of some Western European orchids. Physiol Plant 67:253–261
21. Miyoshi K, Mii M (1995) Enhancement of seed germination and protocorm formation in *Calanthe discolor* (Orchidaceae) by NaOCl and polyphenol absorbent treatments. Plant Tissue Cult Lett 12:267–272
22. Miyoshi K, Mii M (1988) Ultrasonic treatment for enhancing seed germination of terrestrial orchid, *Calanthe discolor*, in asymbiotic culture. Sci Hortic 35:127–130
23. Chu CC, Mudge KW (1994) Effects of prechilling and liquid suspension culture on seed germination of the yellow lady's slipper orchid (*Cypripedium calceolus* var. *pubescens*). Lindleyana 9:153–159
24. Shimura H, Koda Y (2005) Enhanced symbiotic seed germination of *Cypripedium macranthos* var. rebunense following inoculation after cold treatment. Physiol Plant 123:281–287
25. Murashige T, Skoog F (1962) A revised medium for rapid growth and bioassays with tobacco tissue cultures. Physiol Plant 15:473–497
26. Lu IL, Lee CJ, Lee N (1992) Effect of medium composition on seed germination in vitro of *Cymbidium ensifolium* var. *misericors*. J Chinese Soc Hortic Sci 38:161–169
27. Lee CJ (1991) Studies on flowering and seed germination in vitro of *Cymbidium sinense* Wild and *Cymbidium ensifolium* var. *misericors*. M.Sc. Thesis, National Taiwan University, Taiwan, ROC
28. Lee Y-I (2007) The asymbiotic seed germination of six *Paphiopedilum* species in relation to the time of seed collection and seed pretreatment. Acta Hortic 755:381–386
29. Thomale H (1957) Die Orchideen. Einfuhrung in die Kultur und Vermehrung tropischen und einheimischen Orchideen, 2nd edn. Eugen Ulmer, Stuttgart
30. Lee Y-I, Yeung EC, Lee N, Chung MC (2006) Development in the lady's slipper orchid, *Paphiopedilum delenatii*, with emphasis on the ultrastructure of the suspensor. Ann Bot 98:1311–1319

Chapter 6

In Vitro Culture of Coconut (*Cocos nucifera* L.) Zygotic Embryos

Florent Engelmann, Bernard Malaurie, and Oulo N'Nan

Abstract

Coconut is a very important crop for millions of people in tropical countries. With coconut, in vitro culture protocols have been developed with two main objectives, viz. the large scale production of particular types of coconuts and the international exchange and conservation of coconut germplasm. The methods described in this chapter have been developed in the framework of collaborative activities between research institutes in Côte d'Ivoire and France. Two coconut embryo in vitro collecting protocols have been established, one consisting of storing the disinfected embryos in a KCl solution until they are brought back to the laboratory, where they are re-disinfected and inoculated in vitro under sterile conditions, and the other including in vitro inoculation of the embryos in the field. For international germplasm exchange, zygotic embryos inoculated in vitro in plastic test tubes or endosperm cylinders containing embryos in plastic bags are used. For in vitro culture, embryos are inoculated on semi-solid medium supplemented with sucrose and activated charcoal and placed in the dark, and then transferred to light conditions with the same (solid or liquid) medium once the first true leaf is visible and the root system has started developing.

Key words: Coconut (*Cocos nucifera* L.), *In vitro* collection, *In vitro* culture, International germplasm exchange, Zygotic embryo

1. Introduction

Coconut (*Cocos nucifera* L.), the "tree of life", plays a very important role in the life and welfare of millions of people in tropical countries. The world coconut production in 2007 was 61 M tons, with the three main producing countries, Indonesia, the Philippines and India, accounting for over 75% of the total production (1). The most important part of the coconut palm is the nut, from which the solid endosperm is used primarily for oil production. However, over 100 products made from the coconut palm have been identified, with almost all parts of the palm having a use (2).

Trevor A. Thorpe and Edward C. Yeung (eds.), *Plant Embryo Culture: Methods and Protocols*,
Methods in Molecular Biology, vol. 710, DOI 10.1007/978-1-61737-988-8_6,

With coconut, in vitro culture protocols have been developed with two main objectives, viz. the large scale production of particular types of coconuts and the international exchange and conservation of coconut germplasm.

Normal coconuts have a hard and crisp endosperm at maturity, but some coconuts (called Makapuno in the Philippines, Dikiri in Sri Lanka, Kopyor in Indonesia, etc.) have a soft, jelly-like endosperm, which fills the nut cavity. Such coconuts are highly priced in the ice cream and pastry industries as well as for preparing sweetened preserves. However, the yield of Makapuno-bearing palms, which are heterozygous for the character, is only 2–20%, and the phenomenon is believed to be governed by a single recessive gene (3). Makapuno embryos are visually and anatomically similar to normal embryos, but they do not germinate because the soft endosperm does not support their germination. However, efficient protocols have been developed, which allow Makapuno embryos to be extracted from the nuts, inoculated and successfully grown in vitro (4–6). They will germinate into normal coconuts, bearing 100% Makapuno nuts if properly isolated from contamination by pollen from other coconut varieties.

The other application of in vitro culture techniques to coconut is for the international exchange and conservation of germplasm. Indeed, within the plant kingdom, coconut is one of the species with seeds of the largest dimensions. Moreover, coconut seeds are highly recalcitrant; there is no dormancy, and germination immediately follows seed maturation (7). These characteristics drastically limit the amount of material, which can be gathered during collecting missions. Germplasm exchange is made even more difficult due to the high risks of introducing pests and diseases in the recipient country if whole nuts are exchanged, and also to the high cost of transporting whole nuts. Exchanging coconut germplasm in the form of embryos would allow both avoiding the phytosanitary problems and reducing costs linked with transportation of whole nuts. Using in vitro techniques for collecting, exchanging and conserving coconut germplasm requires efficient protocols for in vitro germination of embryos, development of embryos into whole plantlets, their acclimatization to in vivo conditions and further development into seedlings, which can be transferred to the field.

As regards to germplasm exchange, the FAO/IBPGR Technical Guidelines for the Safe Movement of Coconut Germplasm recommended as early as 1993 that coconut germplasm be distributed as zygotic embryos in vitro to reduce risks of introducing diseased material into disease-free areas (8). The need for operational coconut in vitro culture techniques has become especially significant with the establishment of the multi-site International Coconut Genebank or ICG (9) and the implementation of the various international breeding and testing programmes coordinated by the International Coconut Genetic Resources Network (COGENT;

http://www.cogentnetwork.org/), which rely heavily on in vitro techniques for collecting and exchanging germplasm (10).

Research teams in different countries in Africa, the Americas, Asia, Australia, the Caribbean, Europe, India, the Pacific and Oceania have worked towards the development of in vitro techniques for collecting and exchanging coconut germplasm, with varying degrees of success (11, 12). In order to improve and standardize the coconut embryo culture technology, COGENT organized a series of workshops and coordinated research projects with a group of coconut researchers worldwide. The first workshop was held in the Philippines in 1997 (13), the second in Mexico in 2002 (14), and, more recently, a third workshop, funded by the Global Crop Diversity Trust (Rome, Italy, www.croptrust.org/) took place in the Philippines in 2008 (15).

In parallel to the development of embryo in vitro culture protocols, research has been performed towards the establishment of medium- and long-term storage methods for coconut embryos. For medium-term storage, Assy-Bah and Engelmann (16) demonstrated that, after proper modification of the culture medium, embryos could be stored in vitro for 1 year in the growth room and successfully germinate and produce plantlets afterwards. For long-term storage, Assy-Bah and Engelmann (17) have developed an efficient cryopreservation protocol based on pre-treatment of coconut embryos with high sugar medium, partial desiccation and rapid freezing in liquid nitrogen. This protocol has been applied to ten different varieties, with 44–100% cryopreserved embryos giving rise to whole in vitro plantlets (18). Other research teams have successfully adapted the original protocol to their local conditions and plant material, thereby demonstrating its efficiency and broad applicability (19, 20).

2. Materials

1. For in vitro culture experiments, embryos collected from mature nuts (11–12 months after fecundation) were used. The fresh weight of embryos varied between 80 and 160 mg, depending on the variety and the maturity stage. Selection of mature nuts in the field was performed based on the date of fecundation, in the case of hand-pollinated nuts or on the colour of the nuts, which changes from shiny to matt at maturity, in the case of open-pollinated nuts (21).
2. Tools and equipment used during field collection: Folding table, hammer or machete, cork borer (3 cm diameter), camping stove, dissection tools, culture tubes (24 × 150 mm).
3. KCl solution for storing disinfected endosperm cylinder: 16.2 g/L.

4. Disinfection solutions: 80% ethanol; commercial bleach solution.
5. Cryotubes (2 mL sterile polypropylene) for storage of embryos.
6. Embryo culture medium: Murashige and Skoog (22) macro- and microelements, Morel and Wetmore (23) vitamins, 100 mg/L sodium ascorbate, 60 g/L sucrose, 2 g/L activated charcoal (Sigma C5386) and 8 g/L agar (Labosi AL 540, France). The pH is adjusted to 5.5 before adding agar and charcoal and autoclaving at 110°C for 20 min.
7. Carbendazin-based fungicide solution: benlate (2 g/L).
8. Nutritive solution for plantlet acclimatization (see Table 1).
9. Fertilizing solution for plantlets: 50 mL of an N:P:K solution (8:11:14 g/L), prepared using urea, phosphate bicalcite and potassium chloride from SPPC Vidri company.
10. Chelated iron solution: 6% solution prepared by mixing 26.1 g EDTA with 24.9 g FeSO·7H_2O in 2 L ultrapure water.

Table 1
Composition of the nutritive solution used for acclimatization of plantlets (mg/L) (30)

KNO_3	274
$Ca(NO_3)_2 \cdot 4H_2O$	1,095
KH_2PO_4	137
$MgSO_4 \cdot 7H_2O$	274
$(NH_4)_2SO_4$	137
KCl	2.74
H_3BO_3	3
$MnSO_4 \cdot H_2O$	1.7
$ZnSO_4 \cdot 7H_2O$	2.74
$(NH_4)_6MO_7O_{24} \cdot 4H_2O$	2.74
H_2SO_4	0.137
$CuSO_4 \cdot 5H_2O$	1.37
EDTA	26.1
$FeSO_4 \cdot 7H_2O$	24.9

EDTA ethylenediaminetetraacetic acid

11. Forest leaf mould mixed with sand (1:1 ratio, leaf mould collected from forest around the plantation, river sand collected from nearby river).

3. Methods

Various protocols have been developed for in vitro collecting, international exchange and in vitro culture of coconut zygotic embryos, which can be found in Engelmann (12, 24), Batugal and Engelmann (13), Engelmann et al. (14) and references therein. The methods described hereafter are those developed in the framework of the long-lasting collaboration between Côte d'Ivoire (CNRA – Centre National de Recherche Agronomique) and France (IRD and CIRAD – Centre de coopération internationale en recherche agronomique pour le développement).

3.1. In Vitro Collecting

Assy-Bah et al. (25) developed two coconut embryo in vitro collecting protocols, one consisting of storing the disinfected embryos in a KCl solution until they are brought back to the laboratory, where they are re-disinfected and inoculated in vitro under sterile conditions, the other including in vitro inoculation of the embryos in the field. These two protocols have been described in details as follows (24).

3.1.1. Protocol 1 (Inoculation of Embryos in the Laboratory)

1. Preliminary operations are performed in the open air, on a folding table that has been washed and disinfected with pure commercial bleach.
2. Select and dehusk mature nuts.
3. Break nuts open with a clean hammer or machete.
4. Use a cork borer to remove a cylinder of solid endosperm containing the embryo, and use forceps to transfer the cylin der to a jar containing 500 mL commercial bleach. Disinfect all instruments with 100% commercial bleach and sterilize in the flame of the gas burner.
5. Immerse batches of 25 cylinders in commercial bleach for 20 min.
6. Immediately after disinfection, transfer endosperm cylinders without rinsing to individual 30 mL containers containing 15 mL KCl solution.
7. In the laboratory, under the laminar airflow cabinet, remove endosperm cylinders from the KCl solution and immerse them in batches of 25 cylinders in 100% commercial bleach for 20 min.
8. Place one cylinder in a sterile Petri dish and dissect out the embryo using forceps and a scalpel. Flame dissecting tools

before manipulating a new embryo to reduce the risk of cross-contamination.

9. Rinse the embryo once in sterile water (using one flask per embryo to reduce the risk of cross-contamination) and transfer it to solid embryo medium in a culture tube.
10. Seal the tube with cling film and place it on a rack for culture in the growth room.

3.1.2. Protocol 2 (Inoculation of Embryos in the Field)

The following operations are performed inside a wooden box, which provides some protection from external contaminants. The inside walls of the box are disinfected with bleach.

Steps 1–5 are the same as in Subheading 3.1.1

6. Place one cylinder in a sterile Petri dish and dissect out the embryo using forceps and a scalpel. Flame dissecting tools before manipulating a new embryo to reduce the risk of cross-contamination.
7. Rinse the embryo once in sterile water (using one flask per embryo to reduce the risk of cross-contamination) and transfer it to solid embryo culture medium in a culture tube.
8. Seal the tube with cling film and place it on a rack for transport to the laboratory (see Note 1).

3.2. International Exchange of Germplasm

Two different methods have been employed for international exchange of coconut germplasm, viz. zygotic embryos inoculated in vitro or endosperm cylinders containing embryos in plastic bags.

3.2.1. Embryos Inoculated In Vitro (26, 27)

Steps 1–9 are the same as in Subheading 3.1.1, except that embryos are inoculated in polypropylene sterile tubes (15 × 100 mm) on solid embryo culture medium devoid of sugar (26), or in 2 mL sterile cryotubes containing 1 mL solid agar medium (0.45% agar + water) (27).

10. Seal the tubes with cling film and place them on polystyrene holding racks.
11. Place racks with tubes in cardboard boxes, fill boxes with polystyrene chips and dispatch by air courier.
12. Upon arrival in the recipient laboratory, tubes are inspected for contamination; non-contaminated embryos are transferred to solid medium in culture tubes (see Subheading 3.3 for medium composition), tubes are sealed with cling film and placed in the growth room (see Note 2).
13. Upon arrival in the recipient laboratory, non-contaminated embryos shipped in cryotubes on solid agar embryo culture medium are disinfected for 5 min with diluted commercial bleach (6% active chlorine) and then rinsed once in sterile water before inoculation on culture medium (see Note 3).

3.2.2. Endosperm Cylinders in Plastic Bags (28)

1. Preliminary operations are performed in the open air, on a folding table that has been washed and disinfected with pure commercial bleach.
2. Select and dehusk mature nuts.
3. Break nuts open with a clean hammer or machete.
4. Use a cork borer to remove a cylinder of solid endosperm containing the embryo, rinse the cylinders under running tap water and then use forceps to transfer the cylinders to a jar containing 500 mL 80% ethanol. Disinfect all instruments with 100% commercial bleach and sterilize in the flame of the gas burner.
5. Immerse batches of 25 cylinders in 80% ethanol for 5 min.
6. Remove ethanol, replace with commercial bleach and immerse cylinders for 30 min.
7. Rinse six times with sterile water.
8. Transfer cylinders by batches of 10 in small plastic bags previously disinfected with 80% ethanol and seal the bags.
9. Place 10 small plastic bags in a larger plastic bag and seal bag.
10. Place larger bags in a polystyrene isothermal box, add sealed plastic bags filled with ice and dispatch by air courier.
11. Upon arrival in the recipient laboratory, disinfect cylinders by batches of 10 in commercial bleach for 20 min and rinse five times with sterile water.
12. In the laboratory, under the laminar airflow cabinet, place one cylinder in a sterile Petri dish and dissect out the embryo using forceps and a scalpel. Flame dissecting tools before manipulating a new embryo to reduce the risk of cross-contamination.
13. Disinfect the embryo for 5 min with commercial bleach, rinse five times in sterile water (using one flask per embryo to reduce the risk of cross-contamination) and transfer it to solid embryo culture medium in a culture tube.
14. Seal the tube with cling film and place it on a rack for culture in the growth room (see Note 4) (Malaurie B, personal communication).

3.3. In Vitro Culture of Zygotic Embryos

The method described is based on the papers published by Assy-Bah (29), Assy-Bah et al. (30) and Verdeil et al. (26).

1. Inoculate embryos in 24 × 150 mm culture tubes containing 20 mL embryo culture medium, seal with cling film and place them in the growth room at 27 ± 1 °C in the dark. Subcultures to fresh medium are performed every 4–6 weeks.
2. When the first true leaf is visible and the root system has started developing (at least one root with ramifications),

transfer plantlets to light conditions (12 h light/12 h dark photoperiod, 45 μmol/m²/s, light intensity, Sylvania Grolux daylight tubes) either on the same solid medium (29, 30) or to 1 L glass bottles containing 100 mL liquid medium (26).

3. Transfer plantlets every 4–6 weeks into large tubes (36 × 200 mm) containing fresh medium. Using liquid medium, generally no haustorium (cotyledon) growth and development is observed as is often the case on solid medium (30). When haustorium development is observed, cut it off after 3 months of embryo culture, when the gemmule is 2–4 cm in height. Haustorium removal will improve survival of plantlets during acclimatization. Plantlets can be acclimatized when they display 3–4 unfolded green leaves. The more advanced plantlets reach the acclimatization stage 6–7 months after initial inoculation.

3.4. Acclimatization of Plantlets

The method described is based on the papers published by Assy-Bah (29), Assy-Bah et al. (30) and Verdeil et al. (26).

1. Remove plantlets from culture tubes, rinse them with distilled water and plunge them for 5 min in a benlate solution to prevent fungal development.
2. Transfer plantlets to the greenhouse and plant in pots filled with sterilized river sand. Plantlets are covered with a plastic bag (acrylic polypropylene) during the first 2 weeks to maintain maximum relative humidity conditions. They are watered daily during the first month, after which a nutritive solution is applied every 2 days.
3. After 2 months, plantlets are transferred to plastic bags filled with forest leaf mould mixed with sand. Fifty millilitre of an N:P:K solution (8:11:14 mL/L) are applied every 2 weeks, and 50 mL of chelated iron (6%) every 2 months.

4. Notes

1. Using Protocol 2, contamination was around 10%, while it was only around 5% with Protocol 1. No differences were noted in germination and development between embryos treated following Protocols 1 and 2. Embryos could be stored for up to 14 days in the KCl solution without any impact on their further development. After direct inoculation in the field (following Protocol 2), embryos could be kept on semi-solid medium under non-controlled environmental conditions for 2 months before being grown in the culture room of a laboratory (31).
2. This method has been used successfully for shipping several thousands embryos between Côte d'Ivoire and France (21).

3. This method has been used successfully for shipping several hundred embryos from Sri Lanka to France (27).
4. This method has been used successfully for shipping over 20,000 embryos from Côte d'Ivoire to France (28) as well as around 1,000 embryos from Sri Lanka to France (27).

References

1. FAO (2009) http://faostat.fao.org/site/339/default.aspx
2. Persley GJ (1992) Replanting the tree of life: towards an international agenda for coconut palm research. CAB-ACIAR, Oxon, UK
3. Zuñiga LC (1953) The probable inheritance of the Makapuno character of coconut. Phil Agri 36:403–414
4. de Guzman EV, del Rosario DA (1964) The growth and development of *Cocos nucifera* L 'Makapuno' embryos in vitro. Phil Agri 48:83–94
5. Rillo EP (1998) PCA's embryo culture technique in the mass production of Makapuno coconuts. In: Batugal PA, Engelmann F (eds) Coconut embryo in vitro culture – Proceedings of the first workshop on embryo culture, 27–31 October 1997, Banao, Guinobatan, Albay, Philippines. IPGRI-APO, Serdang, Malaysia, pp 69–78
6. Weerakoon LK, Vidhanaarachchi VRM, Fernando SC, Fernando A, Gamage CKA (2002) Increasing the efficiency of embryo culture technology to promote coconut germplasm collecting and exchange in Sri Lanka. In: Engelmann F, Batugal P, Oliver JT (eds) Coconut embryo in vitro culture part II. IPGRI-APO, Serdang, Malaysia, pp 27–40
7. Chin HF, Pritchard HW (1988) Recalcitrant seeds, a status report. International Board for Plant Genetic Resources, Rome
8. Frison EA, Putter CAJ, Diekmann M (1993) FAO/IBPGR technical guidelines for the safe movement of coconut germplasm. Food and Agricultural Organisation of the United Nations/International Board for Plant Genetic Resources, Rome
9. Rao RV, Batugal P (1998) Proceedings of the COGENT regional coconut genebank planning workshop, 26–28 February 1998, Pekanbaru, Indonesia. IPGRI-APO, Serdang, Malaysia
10. Engelmann F, Batugal P (2002) Background on the development and implementation of the coconut embryo in vitro culture project. In: Engelmann F, Batugal P, Oliver JT (eds) Coconut embryo in vitro culture part II. IPGRI-APO, Serdang, Malaysia, pp 1–4
11. Engelmann F (1998) Current state of the art and problems with in vitro culture of coconut embryos in coconut embryo in vitro culture. In: Batugal PA, Engelmann F (eds) Proceedings of the first workshop on embryo culture, 27–31 October 1997, Banao, Guinobatan, Albay, Philippines. IPGRI-APO, Serdang, Malaysia, pp 6–11
12. Engelmann F (2002) Coconut. In: Pence VC, Sandoval JA, Villalobos VMA, Engelmann F (eds) In vitro collecting techniques for germplasm conservation, IPGRI technical bulletin No 7. IPGRI, Rome, pp 68–71
13. Batugal PA, Engelmann F (1998) Coconut embryo in vitro culture – Proceedings of the first workshop on embryo culture, 27–31 October 1997, Banao, Guinobatan, Albay, Philippines. IPGRI-APO, Serdang, Malaysia
14. Engelmann F, Batugal P, Oliver JT (2002) Coconut embryo in vitro culture part II. IPGRI-APO, Serdang, Malaysia
15. Engelmann F, George ML Coconut embryo in vitro culture part III. In: Proceedings of a training and research workshop for coconut embryo culture to improve collecting and safe movement of germplasm, 11–15 December 2008, Zamboanga City, Philippines, in press
16. Assy-Bah B, Engelmann F (1993) Medium-term conservation of mature embryos of coconut. Plant Cell Tissue Org Cult 33:19–24
17. Assy-Bah B, Engelmann F (1992) Cryopreservation of mature embryos of coconut (*Cocos nucifera* L.) and subsequent regeneration of plantlets. Cryoletters 13:117–126
18. Engelmann F, Malaurie B, N'Nan O, Borges M (2005) Status of cryopreservation research in coconut. In: Batugal P, Rao VR, Oliver J (eds) Coconut genetic resources. IPGRI-APO, Serdang, Malaysia, pp 142–148
19. Sajini KK, Karun A, Kumaran PM (2006) Cryopreservation of coconut (*Cocos nucifera* L.) zygotic embryos after pre-growth desiccation. J Plant Crops 34:576–581
20. Sisunandar, Samosir YMS, Adkins SW (2006) Cryopreservation of coconut embryos using desiccation procedures. In: Adkins SW, Foale M, Samosir YMS, (eds) Coconut revival: new

possibilities for the "tree of life" – Proceedings of the international coconut forum, 22–24 Nov 2005, Cairns, Australia, ACIAR Proceedings No 125, p 102

21. Assy-Bah B (1992) Utilisation de la culture *in vitro* d'embryons zygotiques pour la collecte et la conservation des resources génétiques du cocotier (*Cocos nucifera* L.). Ph.D. Thesis, University Paris 6, France
22. Murashige T, Skoog F (1962) A revised medium for rapid growth and bioassays with tobacco tissue cultures. Plant Physiol 15:473–497
23. Morel G, Wetmore RM (1951) Fern callus tissue culture. Am J Bot 38:138–140
24. Engelmann F (2005) In vitro collecting of coconut germplasm. In: Batugal P, Ramanatha Rao V, Oliver J (eds) Coconut genetic resources. IPGRI-APO, Serdang, Malaysia, pp 65–74
25. Assy-Bah B, Durand-Gasselin T, Pannetier C (1987) Use of zygotic embryo culture to collect germplasm of coconut (*Cocos nucifera* L.). Plant Genet Res Newsl 71:4–10
26. Verdeil J-L, Hocher V, Triques K, Lyakurwa R, Rival A, Durand-Gasselin T, Engelmann F, Sangaré A, Hamon S (1998) State of research on coconut embryo culture and acclimatization techniques in the IDEFOR (Côte d'Ivoire) and ORSTOM/CIRAD laboratories (France). In: Batugal PA, Engelmann F (eds) Coconut embryo in vitro culture – Proceedings of the first workshop on embryo culture, 27–31 October 1997, Banao, Guinobatan, Albay, Philippines. IPGRI-APO, Serdang, Malaysia, pp 17–28
27. Malaurie B, Bandupriya HDD, Fernando SC, Verdeil JL (2006) Optimisation du procédé de cryoconservation de la plumule de cocotier Les Actes du BRG No 6, Ressources génétiques ressources partagées, 449–468. Available at: http://www.brg.prd.fr/brg/pdf/LaRochelle_Malaurie.pdf
28. N'Nan O (2004) Utilisation des biotechnologies pour les échanges et la conservation des resources génétiques du cocotier (*Cocos nucifera* L.). Ph.D. Thesis, Angers University, France
29. Assy-Bah B (1986) In vitro culture of coconut zygotic embryos. Oléagineux 41:321–328
30. Assy-Bah B, Durand-Gasselin T, Engelmann F, Pannetier C (1989) The in vitro culture of coconut (*Cocos nucifera* L.) zygotic embryos: revised and simplified method for obtaining coconut plantlets suitable for transfer to the field. Oléagineux 44:515–523
31. Engelmann F, Assy-Bah B (1992) Maintenance of coconut genetic resources: *in vitro* techniques for medium and long-term conservation. In: Papers of the IBPGR Workshop on Coconut Genetic Resources, Cipanas, Indonesia, 8–11 October 1991, International Crop Network Series No 8, IBPGR, pp 63–69

Part II

Application of Embryo Culture Techniques: Embryo Rescue

Chapter 7

Immature Embryo Rescue and Culture

Xiuli Shen, Fred G. Gmitter Jr., and Jude W. Grosser

Abstract

Embryo culture techniques have many significant applications in plant breeding, as well as basic studies in physiology and biochemistry. Immature embryo rescue and culture is a particularly attractive technique for recovering plants from sexual crosses where the majority of embryos cannot survive in vivo or become dormant for long periods of time. Overcoming embryo inviability is the most common reason for the application of embryo rescue techniques. Recently, fruit breeding programs have greatly increased the interest in exploiting interploid hybridization to combine desirable genetic traits of complementary parents at the triploid level for the purpose of developing improved seedless fruits. However, the success of this approach has only been reported in limited number of species due to various crossing barriers and embryo abortion at very early stages. Thus, immature embryo rescue provides an alternative means to recover triploid hybrids, which usually fail to completely develop in vivo. This chapter will provide a brief discussion of the utilization of interploid crosses between a monoembryonic diploid female with an allotetraploid male in a citrus cultivar improvement program, featuring a clear and comprehensive illustration of successful protocols for immature embryo rescue and culture. The protocols will cover the complete process from embryo excision to recovered plant in the greenhouse and can easily be adapted to other plant commodities. Factors affecting the success and failure of immature embryo rescue to recover triploid progeny from interploid crosses will be discussed.

Key words: *Citrus*, Culture medium, Embryo abortion, Embryo developmental stage, Genotypes, Interploid hybridization, Shoot tip grafting, Triploid

1. Introduction

1.1. Interploid Hybridization in Crop Improvement

Crosses between two different ploidy levels of the same or different species or genera are referred to as interploid hybridization (1). Like interspecific or intergeneric crosses at the same ploidy level, interploid hybridization has been a useful tool for the introduction of novel genetic traits for new cultivar development and crop improvement (2). One of the advantages of interploid crosses is that some crosses which are not successful at the same ploidy

Trevor A. Thorpe and Edward C. Yeung (eds.), *Plant Embryo Culture: Methods and Protocols*,
Methods in Molecular Biology, vol. 710, DOI 10.1007/978-1-61737-988-8_7,

level can be made possible via interploid crosses by manipulating parental chromosome levels (3). Interploid crosses also provide a means to increase heterozygosity through sexual polyploidization as compared with colchicine-induced asexual polyploidization (4–6). However, interploid hybridization has only been reported in relatively few crops, including citrus (7–13), grapes (14–20), and banana (21–24) for seedless fruit production via triploidy, and in ornamental plants for novel flower and foliage characteristics (2, 25–29). As a general rule, crosses between different ploidy levels are more recalcitrant in yielding plantlets than crosses carried out using parents of the same ploidy level, primarily due to embryo inviability. The difficulty in obtaining interploid hybrids has been attributed to the taxonomic distance between parental species, chromosome imbalances, and endosperm failure (30).

Crossing barriers are frequently encountered during interploid crosses. Sexual barriers limiting the success and efficiency of interploid hybridization have been classified as pre- and postfertilization (31). Prefertilization barriers can be overcome by selection of cross-compatible parents, ploidy manipulation (3, 32, 33), and a variety of techniques such as direct injection of pollen into the ovary, in vitro pollination, cut-style pollination, the use of mentor pollen, or grafting of the style (31, 34).

Nevertheless, postfertilization barriers are more common in interploid crosses due to the failure of normal endosperm development (35). The endosperm plays an important role in the development of the seed because of its physical, physiological, and genetic relationships to the embryo (36). Therefore, normal endosperm development is a prerequisite for normal embryo development. It has been proposed that a 2 maternal:1 paternal ratio of endosperm balance number (EBN) (in the endosperm itself) is necessary for normal endosperm development. Any maternal/paternal ploidy ratio that deviates from the 2:1 ratio may affect embryo development and subsequent seed or fruit formation (37, 38). Failure in normal endosperm development in most interploid crosses is common due to the abnormal maternal and paternal ratio, especially in crosses between a diploid female and a tetraploid male. As a consequence, abortion of hybrid embryos usually occurs prior to fruit maturity (1, 13, 31).

1.2. Immature Embryo Rescue and Culture

The plant tissue culture technique of embryo rescue offers a means to recover starving embryos prior to abortion (39). Immature or mature zygotic embryos can be excised from their natural growing environment and cultured on an artificial medium in vitro, which is a substitute for the endosperm, allowing continued development and subsequent germination to produce a plant (40, 41). Interploid hybrids rarely have been obtained in vivo from interploid crosses (13, 15). With the aid of embryo rescue, interploid hybrids can be produced with various frequencies through in vitro culture of abortive seeds and embryos (42).

The genotype, the developmental stage of the embryo at excision, and composition of the embryo culture media are the three main factors affecting the success and efficiency of embryo rescue (43). Selection of parents is very important for successful interploid crosses (15–17). Some considerations in parental selection include cross compatibility of parents, ploidy level of seed and pollen parents, and crossing direction (44). A cross is expected to be successful when male and female gametes have matching EBN (3, 35). Ploidy level is also important because it can influence male and female fertility, cross fertility, and plant vigor (45). The crossing direction affects EBN, and thus whether or not embryo rescue is required for hybrid recovery. Some interploid crosses are reciprocal, while others are nonreciprocal (36, 38). It has been postulated that maternal excess crosses, e.g., 4x × 2x (2 maternal:1 paternal in endosperm), typically result in less postzygotic lethality than paternal excess (2x × 4x) crosses (2 maternal:2 paternal) (3), and we have found this to be true with limited experience in citrus (Grosser, unpublished data). However, the availability of quality monoembryonic citrus tetraploids that generate zygotic offspring is quite limited.

1.3. Embryo Growth and Development

Zygotic embryo development within seeds has been divided into a series of stages including globular, heart, torpedo, and cotyledonary for dicots. This classification indicates not only the change in embryo shape but also a gradual progression towards independence from surrounding tissues for nutrients (34). It is understandable that an embryo at its latter developmental stage is easier to isolate and has a greater chance for survival in vitro. It has been postulated that the late heart-stage embryo coincides with the change from heterotrophic to autotrophic growth of the embryo. Heterotrophic embryos generally require a more complex media composition for their growth (46). For embryos that abort at a very early stage, it is impossible to isolate the embryo; thus, instead of embryo culture, ovary, ovule, or embryo sac culture can be used to increase the likelihood of embryo survival (47–49).

A wide array of media with numerous modifications in inorganic salts, sugars, vitamins, amino acids, organic adjuvants (yeast extract, malt extract, and coconut water), and plant growth regulators (PGRs) have been used to culture excised embryos (5, 34, 40, 41, 49). In addition to parental genotypes, the age of the embryo at excision requires particular medium compounds for optimal growth. As a general rule, medium composition is more critical for immature than mature embryos for their growth in vitro. Medium complexity increases with decreasing embryo age (41). For mature or near-mature embryos, a simple medium with only mineral salts and low level of sugar is enough to support normal growth. However, for younger embryos, a complex medium including various organic supplements is generally required (49, 50). Globular or early heart-stage embryos may fail to develop and germinate even on a complex medium, and a section of endosperm from mature seeds, intact suspensor, or "nurse endosperm" may help their growth (42).

1.4. Protocol for Immature Embryo Rescue and Culture

Although embryo rescue techniques have been widely used as an aid in interspecific, intergeneric, and interploid crosses for over 100 years (49), few publications have provided a detailed protocol. To illustrate the process of embryo rescue in a step-by-step manner, this chapter will feature our routine citrus protocol applied to recover triploid hybrids, from an interploid cross between a monoembryonic diploid seed parent ("LB8-10" tangelo ("Clementine" × "Minneola")) and an allotetraploid pollen parent (somatic hybrid of "Valencia" sweet orange + "Murcott" tangor), to generate a large number of progeny toward the selection of high-quality seedless fruits. To date, this protocol has been used successfully to recover more than 12,000 triploid citrus hybrids in our breeding program.

In 2x × 4x crosses, a 2 maternal:2 paternal ratio of endosperm is expected in hybrid seeds, instead of a normal endosperm ratio of 2:1. Therefore, embryo abortion occurs at a certain point of growth, and undeveloped, wrinkled seeds are produced with a size of 1/6 to 1/3 of fully developed seeds (7, 9, 10). Such undeveloped seeds usually contain triploid zygotic embryos at different developmental stages. Thus, properly staged undeveloped seeds serve for triploid hybrid recovery via embryo rescue (12, 51).

2. Materials

1. General tissue culture equipment and tools: culture room with controlled temperature and photoperiod, laminar flow hood, burner, sharp surgical scalpel, forceps, pliers, Petri dishes (100 × 15 mm) (Fisher Scientific, Canada), Magenta box GA-7 vessels (Magenta Corporation, Chicago, USA), and self-sealing film (Nescofilm or Parafilm) (Karlan Research Products Corporation, Cottonwood, AZ, USA).
2. Other equipment: Growth chamber, dissecting microscope (Leica Zoom 2000, model Z45L, Leica Inc., Buffalo, NY, USA), autoclave.
3. Chemical solutions: 0.1 N NaOH to adjust pH of media; regular Clorox bleach to surface sterilize fruits; 95% ethanol to disinfect equipment, tools, and hybrid fruits.
4. Media composition: All medium formulations are listed in Table 1. Embryo germination (EG) and shoot elongation

Table 1
Composition of EG, SE, BH3, EME, DBA3, and RMA media

Components	EG(mg/L)	SE(mg/L)	BH3(mg/L)	EME(mg/L)	DBA3(mg/L)	RMA(mg/L)
Inorganics						
KCl			1,500			
KNO_3	1,900	1,900		1,900	1,710	950
NH_4NO_3	1,650	1,650		1,650	1,485	825
$MgSO_4 \cdot 7H_2O$	181	181	370	370	333	185
$KH_2 \cdot PO_4 \cdot 6H_2O$	170	170	170	170	153	85
$H_3 BO_3$	6.2	6.2	6.2	6.2	5.6	3.1
$MnSO_4 \cdot 4H_2O$	16.8	16.8	16.8	16.8	15.1	8.4
$ZnSO_4 \cdot 7H_2O$	8.6	8.6	8.6	8.6	7.7	3.9
KI	0.8	0.8	0.8	0.8	0.8	0.4
$Na_2MoO_4 \cdot 2H_2O$	0.3	0.3	0.3	0.3	0.2	0.1
$CuSO_4 \cdot 5H_2O$	0.025	0.025	2.5	2.5	2.5	2.5
$CaCl_2 \cdot 2H_2O$	332	332	440	440	440	440
$CoCl_2 \cdot 6H_2O$	0.025	0.025	2.5	2.5	2.5	2.5
Na_2 EDTA$\cdot 2H_2O$	37.3	37.3	37.3	37.3	37.3	37.3
$FeSO_4 \cdot 7H_2O$	27.9	27.9	27.9	27.9	27.9	27.9
Organics						
Adenine	25					
myo-Inositol			100	100	90	50
Thiamine–HCl			10	10	9	5
Pyridoxine–HCl			10	10	9	5
Nicotinic acid			5	5	4.5	2.5
Glycine			2	2	1.8	1
Fumaric acid			40			
Citric acid			40			
Malic acid			40			
Pyruvic acid			20			
Ascorbic acid			2			
Calcium pantothenate			1			
Choline chloride			1			
Folic acid			0.4			
Riboflavin			0.2			
p-Aminobenzoic acid			0.02			
Biotin			0.02			
Retinol (vit. A)			0.01			
Cholecalciferol (vit. D3)			0.01			
Vit. B12			0.02			
Glutamine			3,100			

(continued)

Table 1 (continued)

Components	EG(mg/L)	SE(mg/L)	BH3(mg/L)	EME(mg/L)	DBA3(mg/L)	RMA(mg/L)
Casein enzyme hydrolysate			250			
Fructose			250			
Ribose			250			
Xylose			250			
Mannose			250			
Rhamnose			250			
Cellobiose			250			
Galactose			250			
Malt extract	500		1,000	500	1,500	
Coconut water			20 (mL)		20 (mL)	
Mannitol			82,000			
Sucrose	50	50	51,300	50,000	25,000	25,000
Activated charcoal		4				500
Plant growth regulators						
2,4-D					0.01	
BAP					3.0	
NAA						0.02
GA3		3				

(SE) media were composed of MS (Murashige & Skoog) (52) inorganic salts in combination with various organic supplements and PGRs. BH3, EME, DBA3, and RMA media consisted of MT (Murashige & Tucker) basal medium (53) supplemented with a variety of organics and PGRs. PGRs including BAP (6-benzylaminopurine), GA_3 (gibberellic acid), 2,4-D (2,4-dichlorophenoxyacetic acid), and NAA (1-naphthalene acetic acid) can be stored at room temperature with desiccant, but stock solutions made with them (1 mg/mL) must be stored in a refrigerator at 5°C. Pyruvic acid, ribose, calcium pantothenate, *p*-aminobenzoic acid, vitamin B12 also need to be stored in a refrigerator at 5°C. Retinol must be stored in a freezer. Dissolve retinol and cholecalciferol in ethanol prior to use. Maintain all other chemical materials at room temperature. Adjust the pH of media to 5.8 with 0.1 N NaOH prior to the addition of 8 g/L TC agar (Fisher BioReagents, Fair Lawn, NJ, USA). Autoclave media at 1.2 kg/cm^2 for 20 min. Add GA_3 to autoclaved SE medium through filter sterilization (0.22 μm) (Syringe Driven Filter Unit, Millex™, Millipore Corporation, Bedford, MA, USA), once the medium cools down to about 45°C. Dispense the medium either into 100 × 15-mm sterile Petri plates (20 mL) for embryo germination or into GA-7 vessels (50 mL) for shoot elongation.

3. Methods

3.1. Fruit Sterilization, Seed Extraction

1. Parental trees of various ages were grown in the orchard at the Citrus Research and Education Center of University of Florida in Lake Alfred (28°N, 081°W). Hand pollination for this specific cross was performed on 25 March 2008. Maternal inflorescences were emasculated 1 day before expected opening of the petals and immediately pollinated with fresh pollen. The number of flowers pollinated varied according to the availability of flowers on individual trees. Hybrid fruits from the crosses were harvested on 30 July 2008, which was 126 days after pollination (see Note 1).
2. Wash hybrid fruits thoroughly under running water to remove any dirt on the fruit surface.
3. Disinfect hybrid fruits in 50% Clorox® regular bleach (3% sodium hypochlorite, v:v) for 30 min. (Fig. 1a), followed by immersing them in 95% ethanol for 30 s (Fig. 1b), and immediately flaming them with 95% ethanol for about 10 s (Fig. 1c) (see Note 2).
4. Cut fruits with a sharp knife at the equatorial zone, avoiding the core where seeds are embedded (Fig. 1d) (see Note 3). Twist both halves of fruits in opposite directions until they are totally separated (Fig. 1e).
5. Select and remove immature, wrinkled small seeds (Fig. 1f). Label seeds properly (Fig. 1g).

3.2. Embryo Excision, Growth, Germination, Shoot Elongation

1. Excision of embryo: In a laminar flow hood, examine the seeds under a dissecting microscope (Fig. 2a). Hold the antipodal end of the seed (flattened end) with one hand using forceps (Fig. 2b). Place the other forceps near the micropylar end (pointed end) (Fig. 2c). Tear seed coats apart from each other using both hands (Fig. 2d). Expose and excise the hybrid embryo (Fig. 2e) (see Note 4).
2. Record embryo developmental stage of the rescued embryo. Based on the shape of embryos, they are divided into globular, heart, torpedo, and cotyledonary stages (Fig. 3a–d) (see Note 5).

Fig. 1. Fruit surface sterilization and seed extraction. (**a**) Sterilize fruits in 50% Clorox regular bleach for 30 min. (**b**) Disinfect fruits in 95% ethanol for 30 s. (**c**) Flame fruits with 95% ethanol for 10 s. (**d**) Cut fruits with a sharp knife at equatorial zone avoiding the core. (**e**) Twist both halves of fruits in opposite direction. (**f**) Pick immature, undeveloped seeds. (**g**) Collect and label the seeds.

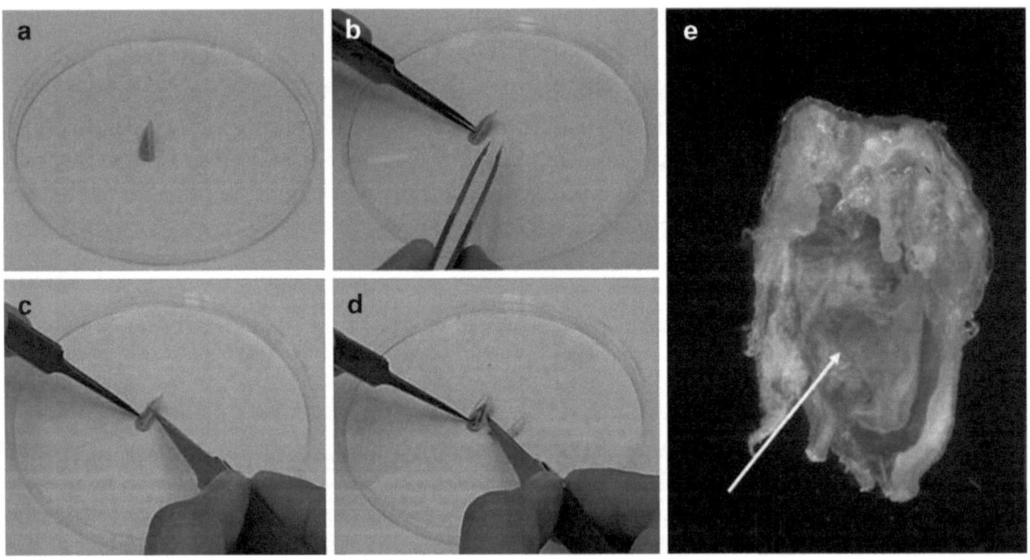

Fig. 2. Excision of the embryo from undeveloped seeds under the dissecting microscope. (**a**) Place an undeveloped seed in a Petri dish. (**b**) Hold the antipodal end of the seed (flattened end) with one hand using forceps. (**c**) Place the other forceps near the micropylar end (pointed end). (**d**) Tear the seed coat apart with both forceps. (**e**) Expose the embryo under the microscope (*arrow*).

3. Immediately culture excised embryos in 100 × 15-mm sterile Petri plates containing 20 mL of EG. Four embryos can be cultured on each plate (see Note 6). Seal plates with Nescofilm. Use only freshly made medium; the use of stored medium is not recommended.

Fig. 3. Developmental stages of the embryo, globular (**a**), torpedo (**b**), heart (**c**), and cotyledonary (**d**).

Additional culture options: Rescued embryos can also be cultured on 0.22-μm cellulose acetate membrane filters placed directly on the surface of the selected embryo culture medium to improve normalization and enlargement (54). Recently we have discovered that the addition of a thin layer of liquid medium with elevated osmoticum over the solid embryo culture medium also enhances embryo survival and growth. This can be achieved by pipetting 2–3 mL of a 1:2 solution (v:v) of BH3 protoplast culture medium: EME liquid suspension culture medium (55) over the selected agar-solidified embryo culture medium. BH3 is a high-osmoticum (0.6 M) medium enriched with additional vitamins and organic acids. We now routinely use the liquid overlay in conjunction with the cellulose acetate membrane filters.

4. Maintain embryo cultures in a culture room at a temperature of 22 ± 3°C with 12/12-h light/dark photoperiod at 40 μmol/m^2/s provided by cool white fluorescent lamps (Lithonia Lighting F40W/SS, Georgia, USA).
5. Observe and record responses of embryos on EG medium (Fig. 4a–c) (see Note 7). Embryos should be transferred to fresh medium every 4 weeks. Normal-sized embryos (usually exhibiting altered morphology) that do not germinate after three passages can be sliced in half longitudinally and cultured on DBA3 medium [56) to induce adventitious shoots, which can take two to three passages (at 4-week intervals).
6. Germinated embryos with good roots and shoots are transferred into GA-7 vessels containing 50 mL of SE medium for shoot elongation. Four germinated embryos can be cultured in each GA-7 vessel (Fig. 4d). Shoots from embryos that produce no roots directly can be removed and cultured in GA-7 vessels containing 50 mL of RMA rooting medium (55) to induce adventitious rooting; this medium also allows for shoot elongation. However, if shoot tip grafting is used (see below) successfully, there may be no need to attempt root induction.

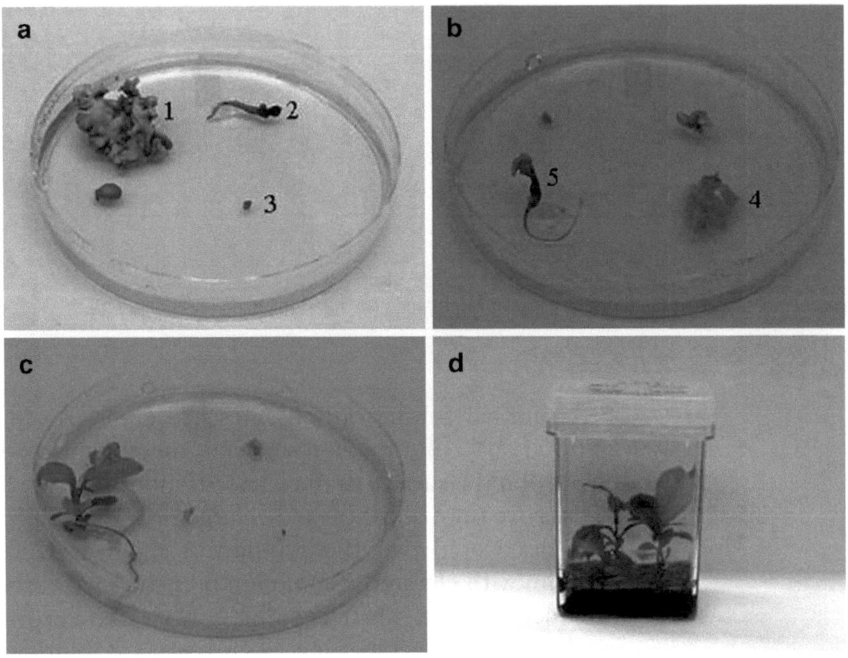

Fig. 4. In vitro responses of embryos rescued from interploid crosses (2x × 4x). Embryos were cultured on MS medium supplemented with 50 g/L sucrose, 0.5 g/l malt extract, 25 mg/L adenine for 8 weeks. (**a**) Formation of leafy structure (*1*), germinated embryo with swollen root and nongrowing apex (*2*), and nonresponse (*3*). (**b**) Callus formation (*4*) and deformed shoots (*5*). (**c**) Normal germinated embryos. (**d**) Developed seedlings from germinated embryos after transfer to the shoot elongation medium composed of MS basal medium, 50 g/L sucrose, 4 g/L activated charcoal, and 8.7 μM GA_3 for another 8 weeks.

3.3. Shoot Tip Grafting to Accelerate Plant Recovery

3.3.1. Preparation of Rootstocks

1. Commercially available Carrizo citrange (*Citrus sinensis* Osb. × *Poncirus trifoliata* L. Raf.) rootstock seeds are germinated to provide rootstock seedlings for grafting (Fig. 5a). Carrizo exhibits trifoliate leaves, so it is easy to identify and remove unwanted rootstock sprouts from the young trees in their early development.
2. Soak seeds in tap water for 2 h. Dry seeds with a paper towel. Remove outer seed coat with pointed forceps (Fig. 5b). Tear inner seed coat apart at the chalazal end but do not completely remove (Fig. 5c) (see Note 8).
3. Autoclave soil at 1.2 kg/cm^2 for 20 min.
4. Sow seeds individually in 38-cell plug trays containing sterilized soil. Seeds are planted about 20 mm deep.
5. Keep trays in the dark at 25°C in a growth chamber. Hand water twice a week (see Note 9).
6. Remove at least 30-day-old germinated rootstock seedlings from the growth chamber and maintain under standard greenhouse conditions (Fig. 5d).

Fig. 5. Rootstock preparation. (**a**) Dry rootstock seeds with paper towel after 2-h soak in water. (**b**). Remove the outer seed coat. (**c**) Tear inner seed coat apart at the chalazal end (flattened end). (**d**) Sow seeds in a plastic tray containing sterilized soil and germinate seeds in the dark at 25°C in a growth chamber.

3.3.2. Preparation of Scions from Recovered Embryo

1. Remove hybrid plantlets from SE medium (Fig. 6a).
2. Excise shoot tips about 10–15 mm long with two or three leaves (Fig. 6b) (see Note 10). Keep excised shoot tips in a moist environment to prevent desiccation.
3. Subculture remaining plants onto fresh SE medium (Fig. 6c); this retains the original hybrid plant for repropagation should the shoot tip grafting procedure fail.

3.3.3. Grafting

1. Decapitate rootstocks to about 90 mm in height and remove all leaves and lateral buds (Fig. 7a, b) (see Note 11).
2. Cut a vertical slit of 2–3 mm in length in the middle of stump (Fig. 7c) (see Note 12).
3. Trim the base of scion to form a "V" 2–3 mm in length (Fig. 6d) (see Note 13).
4. Insert the scion into vertical slit in the rootstock (Fig. 7d) (see Note 14).
5. Cover grafted plants with transparent covers to maintain high humidity (see Note 15).

Fig. 6. Scion preparation. (**a**) Remove hybrid plants from SE medium after 8 weeks culture. (**b**) Excise a shoot tip of 10–15 mm in length with two to three leaves. (**c**) Subculture the remaining part of plants to a fresh SE medium. (**d**) Cut the basal end of the scion to form a "V" shape in a length of 2–3 mm (*arrow*).

6. Acclimatization of the grafted plants must be gradual. Transplant acclimatized plantlets to larger pots as necessary in the greenhouse (Fig. 8a, b) and finally to the field for growth until fruiting and evaluation (Fig. 8c).

4. Notes

1. The timing of fruit harvest varies by cross. Fruits are usually harvested at 120–150 days after pollination (DAP) in citrus. The optimal harvest time (DAP) also varies by species and geographical locations, the same as pollination date and methods.
2. Problems with contamination. Contamination is a big hindrance for any in vitro manipulation. Plant materials are one source of contamination (57, 58). If fruits are used as starting materials and they are not damaged at all, surface sterilization of fruits is sufficient because the internal fruit tissue is sterile. If fruits are cracked or damaged compromising the pulp, seeds must be extracted and surface sterilized to avoid any contamination risk. The embryo is located in the sterile

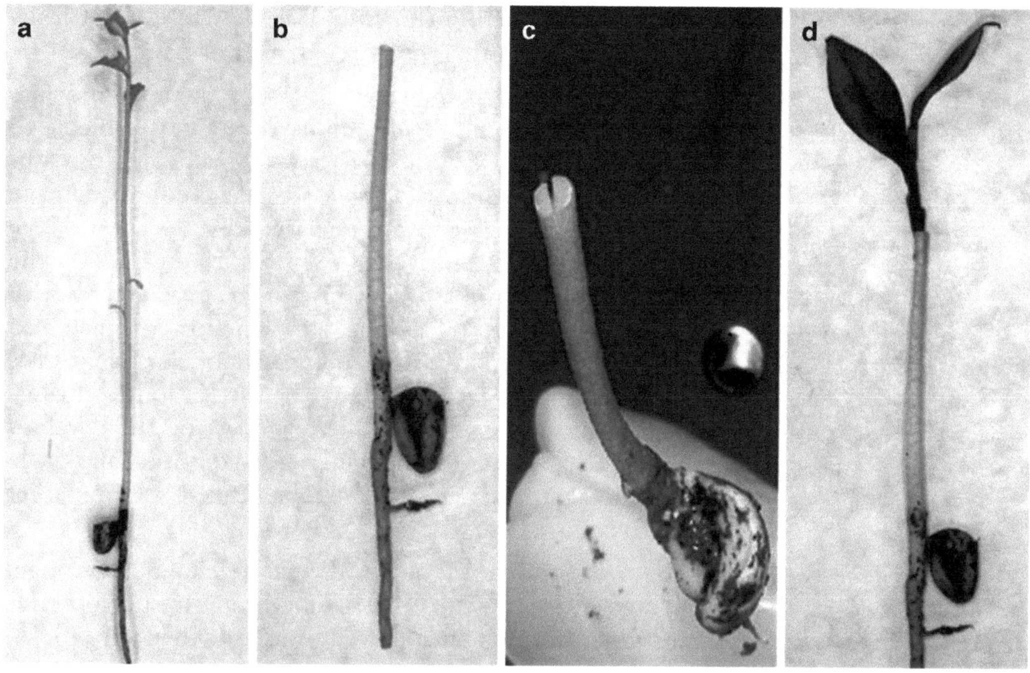

Fig. 7. Grafting technique. (**a**) A germinated rootstock seedling with a pale elongated shoot, small unexpanded leaves of height of about 150 mm after 4 weeks growth in the dark at a temperature of 25°C. (**b**) Decapitate the rootstock to about 90 mm in height and remove all leaves and lateral buds. (**c**) Cut a vertical slit, 2–3 mm long in the middle of the stump. (**d**) Insert the prepared scion into the vertical slit.

Fig. 8. Grafted hybrid plants grown in the greenhouse and the field. (**a**) Grafted hybrid plants were grown in plastic trays 8 weeks after grafting. (**b**) Hybrid plants were transferred to larger pots and grown in the greenhouse to a size acceptable for field transplantation. (**c**) Field evaluation of fruits on triploid hybrid trees.

environment of the seed, and surface sterilization of embryos is usually not necessary (43). However, bacteria and fungi may reside inside the seed, and it is impossible to remove these internal contaminants by surface sterilization. In the situation where seed coats are cracked or if endophytic pathogens exist inside the seed coats, direct disinfection of embryos is needed. Another source of contamination is from the excision operation itself. To minimize this, thoroughly and frequently sterilize tools, and change the operating plates as often as possible.

3. Seeds are embedded in the core of fruits. Do not cut fruits through the core, otherwise seeds may be damaged.
4. The dissection of the embryos to separate them from ovular tissue can be difficult. Large embryos are not difficult to excise. However, it is not easy to isolate small embryos without injury (40). Caution should be taken not to damage embryos when seed coats are removed. The location of the embryo can usually been seen as a darker area under the microscope, so do not touch this darker area directly with forceps. Knowledge of specific seed composition may also help. When dissection of the embryo is not possible, ovary, ovule, and embryo sac culture may be used as long as maternal tissues do not exert an inhibitory effect on the development of the embryo. The medium needed for ovary and ovule culture is also less critical than that needed for immature embryo culture (44).
5. Embryo abortion can occur at any developmental stage. Embryos excised at the optimal stage could result in better embryo germination rates and vigorous growth in vitro (5, 7). Embryos excised earlier than optimal stage are difficult to excise and have a lesser chance of survival in vitro but also require a more complex medium for growth and subsequent germination. Embryos excised too late during fruit development increases the risk of abortion. Fruits or seeds can be harvested at different days after pollination (DAP) and examined to determine the optimal stage of excision. For very immature embryos, care should be taken not to damage the suspensor because the growth of immature embryos can be enhanced in the presence of the suspensor, as the presumed suspensor function is the production of gibberellins and cytokinins that enhance growth (49).
6. It is also important that the excised embryos do not become desiccated. Excised embryos should be immediately placed directly onto culture medium after isolation. Embryos are easily damaged when removed from surrounding protective tissues, so sharp or pointed instruments should not be used to dissect and remove embryos. Embryos usually adhere to the surface of instruments due to their tiny size.
7. Following embryo isolation and culture, variable responses of embryos in vitro are common, including formation of leafy structures, swollen roots, nonresponsive embryos (Fig. 4a), callus formation, or deformed shoot/cotyledon growth (Fig. 4b). In cultures showing normal embryo growth and germination, the radicle of embryos starts to grow down first within 3–4 weeks, followed by the emergence of embryonic leaves in the following weeks (Fig. 4c). Fully developed plantlets are generally obtained within another 8 weeks on SE medium (Fig. 4d).

8. Soaking seeds in water for 2 h helps to soften seed coats and ease its removal. Besides, it stimulates seed germination. Do not remove inner seed coats completely. The micropylar end of the seed, where radicle emerges, is fragile and easily falls apart or can be damaged when sown in the soil. Tearing the inner seed coat apart at the chalazal end is sufficient to allow seed imbibition.
9. The development of plants at a low light level or in the absence of light (in the dark) is referred to as etiolation. Etiolation usually results in pale elongated shoots, small unexpanded leaves, less lignified stem, and enhanced rooting potential (59, 60). All these characteristics are desirable for rootstocks used for micrografting of shoots of recovered germinated embryos. In the dark, rootstock seedlings can grow to a suitable height for grafting in about 3–4 weeks, instead of about 8 weeks in the light.
10. The size and age of scions affect micrografting success (60, 61). Shoots 2–3 months of age exhibiting a size of 10–20 mm in length seem to be optimal in our system. If the size of a scion is too small (<5 mm), it is very difficult to make a "V" cut at the base for grafting; if too large (>20 mm), the scion can easily fall off the rootstock.
11. The height and age of rootstocks influence the success of graft union development (61, 62). All leaves and lateral buds on the rootstock should be removed, otherwise mineral nutrition supplied by the rootstock would go to these organs, instead of the grafted scion, resulting in the eventual death of scion.
12. A clean vertical cut should be made directly in the center of the stump. Since the diameter of the stem of the rootstock is usually less than 2 mm, it is not easy to make the straight cut. Place a sharp scalpel in the middle of the stump, and slowly make the vertical cut keeping the scalpel straight.
13. Place a sharp scalpel at about 2–3 mm above the basal end of the scion, and make one smooth-tapered cut toward the end of the scion. On the opposite side, make a similar symmetric cut. Be sure the basal end of the scion gradually tapers off along both sides and takes the shape of a "V."
14. The key to success in grafting is the intimate contact of the vascular cambium of the scion with that of the rootstock; choosing the scion and the rootstock of about the same diameter maximizes cambial contact between them.
15. Maintaining high relative humidity during the first 2 weeks after grafting is critical for success. During this period, the graft union process has not been completed, and leaves of scion easily dehydrate. Scion shoot flushing generally occurs 3–4 weeks after grafting. Afterwards, humidity can be gradually reduced until plants can grow under natural conditions.

References

1. Bauer MJ (2006) The interploidy hybridization barrier in *Zea mays* L. University of Missouri-Columbia, Ph.D. dissertation
2. Van Huylenbroeck J, Leus L, Van Bockstaele E (2005) Interploidy crosses in Roses: use of triploids. Acta Hort 690:109–112
3. Jansky S (2006) Overcoming hybridization barriers in potato. Plant Breed 125:1–12
4. Yan G, Ferguson AR, McNeilage MA, Murray BG (1997) Numerically unreduced (2n) gametes and sexual polyploidization in *Actinidia*. Euphytica 96:267–272
5. Viloria Z, Grosser JW, Bracho B (2005) Immature embryo rescue, culture and seedling development of acid citrus fruit derived from interploid hybridization. Plant Cell Tiss Org Cult 82:159–167
6. Beuselinck PR, Steiner JJ, Rim YW (2003) Morphological comparison of progeny derived from 4x-2x and 4x-4x hybridization of *Lotus glaber* Mill. and *L. corniculatus* L. Crop Sci 43:1741–1746
7. Jaskani MJ, Khan IA, Khan MM (2005) Fruit set, seed development and embryo germination in interploid crosses of citrus. Sci Hort 107:51–57
8. Fatta Del Bosco S, Siragusa M, Abbate L, Lucretti S, Tusa N (2007) Production and characterization of new triploid seedless progenies for mandarin improvement. Sci Hort 114:258–262
9. Usman M, Ramzan M, Fatima B, Jaskani MJ, Khan MM (2002) Citrus germplasm enhancement by interploid hybridization 1. Reciprocal crosses of Kinnow and Succari. Int J Agri Bio 4:208–210
10. Raza H, Khan MM, Khan AA (2003) Seedlessness in citrus. Int J Agri Biol 5:388–391
11. Jaskani MJ, Khan IA, Khan MM, Abbas H (2007) Frequency of triploids in different interploidal crosses of citrus. Pak J Bot 39:1517–1522
12. Fatima B, Usman M, Ramzan M, Khan MM, Khan IA (2002) Interploid hybridization of Kinnow and Sweet Lime. Pak J Agri Sci 39:132–134
13. Reforgiato Recupero G, Russo G, Recupero S (2005) New promising citrus triploid hybrids selected from crosses between monoembryonic diploid female and tetraploid male parents. HortSci 40:516–520
14. Park SM, Wakana A, Hiramatsu M, Uresino K (2002) A tetraploid hybrid plant from 4x x 2x crosses in *Vitis* and its origin. Euphytica 126:345–353
15. Wakana A, Sarikhani H, Hanada N, Fukudome I, Kajiwara K, Yasukochi K, Hiramatsu M, Sakai K (2007) Characteristics of seedless berries of triploid hybrid grapes (*Vitis vinifera* Complex) derived from eighteen crosses. J Fac Agr Kyushu Univ 52:337–344
16. Wakana A, Hiramatsu M, Park SM, Hanada N, Fufudome I, Xuan B (2002) Degree of abortion and germination rates in triploid seeds from crosses between diploid and tetraploid grapes (*Vitis vinifera* L. and *V.* complex). J Fac Agri Kyushu Univ 46:281–294
17. Wakana A, Fukudome I, Hanada N, Hiramatsu M, Sakai K, Kajiwara K (2008) 'Bea-kei', a new triploid seedless grape cultivar derived from a 'Muscat Bailey A' x 'Kyoho' cross. J Fac Agri Kyushu Univ 53:423–427
18. Wakana A, Park SM, Hiramatsu M, Hanada N, Fukudome I, Yasukochi K (2005) Characteristics of seedless berries of triploid hybrid grapes (*Vitis* complex) from reciprocal crosses between diploid 'Muscat Bailey A' and tetraploid 'Red Pearl'. J Fac Agri Kyushu Univ 50:49–59
19. Yang D, Li W, Li S, Yang X, Wu J, Cao Z (2007) In vitro embryo rescue culture of F1 progenies from crosses between diploid and tetraploid grape varieties. Plant Growth Regul 51:63–71
20. Heo JY, Park KS, Yun HK, Park SM (2007) Degree of abortion and germination percentage in seeds derived from interploid crosses between diploid and tetraploid grape cultivars. Hort Env Biotech 48:115–121
21. Pillay M, Ogundiwin E, Tenkouano A, Dolezel J (2006) Ploidy and genome composition of *Musa* germplasm at the International Institute of Tropical Agriculture (IITA). Afri J Biotech 5:1224–1232
22. Vuylsteke D (2000) Breeding bananas and plantains: from intractability to feasibility. Acta Hort 540:149–156
23. Tomekpe K, Jenny C, Escalant JV (2004) A review of conventional improvement strategies for *Musa*. InfoMusa 13:2–6
24. Jenny C, Tomekpe K, Bakry F, Escalant JV (2003) Conventional breeding of bananas. In: Jacome L, Lepoivre P, Martin D, Ortiz R, Romero R, Escalant JV (eds) Mycosphaerella leaf spot diseases of bananas: present status and outlook. The International Network for the Improvement of Banana and Plantain, Montpellier, France, pp 199–208
25. Olsen RT, Ranney TG (2006) Fertility and inheritance of variegated and purple foliage across a polyploidy series in *Hypericum androsaemum* L. J Am Soc Hort Sci 131:725–730

26. Olsen R (2006) Polyploid breeding in *Hypericum Androsaemum* L. HortSci 41:952
27. Jacob Y, Mastrantuono S, Ferrero F (2007) Interploid crosses in *Anemone coronaria*. Acta Hort 743:55–57
28. Kaori S, Yukio O, Michikazu H (2006) Intrasubgeneric and interploid cross compatibility in evergreen and deciduous azaleas. J Fac Agri Kyushu Univ 51:73–81
29. Sakai K, Ozaki Y, Ureshino K, Miyajima I, Wakana A, Okubo H (2008) Interploid crossing overcomes plastome-nuclear genome incompatibility in intersubgeneric hybridization between evergreen and deciduous azaleas. Sci Hort 115:268–274
30. Hirsch AM, Testolin R, Brown S, Chat J, Fortune D, Bureau JM, De Nay D (2001) Embryo rescue from interspecific crosses in the genus *Actinidia* (Kiwifruit). Plant Cell Rep 20:508–516
31. Van Tuyl JM (1997) Interspecific hybridization of flower bulbs: a review. Acta Hort 430:465–476
32. Tel-Zur N, Abbo S, Mizrahi Y (2005) Cytogenetics of semi-fertile triploid and aneuploid intergeneric vine cacti hybrid. J Hered 96:124–131
33. Datson PM, Murray BG, Hammett KRW (2006) Pollination systems, hybridization barriers and meiotic chromosome behavior in *Nemesia* hybrids. Euphytica 151:173–185
34. Williams EG, Maheswaran G, Hutchinson JF (1987) Embryo and ovule culture in crop improvement. Plant Breed Rev 5:181–236
35. Sokolov VA (2006) Imprinting in plants. Russ J Genet 42:1043–1052
36. Johnston SA, den Nijs TPM, Peloquin SJ, Hanneman RE (1980) The significance of genic balance to endosperm development in interspecific crosses. Theor Appl Genet 57:5–9
37. Johnston SA, Hanneman RE (1995) The genetics of triploid formation and its relationship to endosperm balance number in potato. Genome 38:60–67
38. Burton TL, Husband BC (2000) Fitness differences among diploids, tetraploids, and their triploid progeny in *Chamerion angustifolium*: mechanism of inviability and implications for polyploidy evolution. Evolution 54:1182–1191
39. Navarro L, Olivares-Fuster O, Juarez J, Aleza P, Pina JA, Ballester-Olmos JF (2004) Application of biotechnology to citrus improvement in Spain. Acta Hort 632:221–234
40. Bridgen MP (1994) A review of plant embryo culture. HortSci 29:1243–1246
41. Sharma DR, Kaur R, Kumar K (1996) Embryo rescue in plants – a review. Euphytica 89:325–337
42. Monnier M (1990) Culture of zygotic embryos of higher plants. In: Pollard JW, Walker JM (eds) Methods in molecular biology, vol 6, Plant cell and tissue culture. Humana Press, Totowa, pp 129–139
43. Pierik RLM (1987) In vitro culture of higher plants. Kluwer Academic, Dordrecht, pp 139–148
44. Sharma HC (1999) Embryo rescue following wide crosses. In: Hall RD (ed) Methods in molecular biology, vol 111, Plant cell culture protocols. Humana Press, Totowa, pp 293–307
45. Yao JL, Cohen D (1996) Production of triploid *Zantedeschia* hybrids using embryo rescue. New Zealand J Crop Hort Sci 24:297–301
46. Ramming DW (1990) The use of embryo culture in fruit breeding. HortSci 25:393–398
47. Olsen RT, Ranney TG, Viloria Z (2006) Reproductive behavior of induced allotetraploid x Chitalpa and in vitro embryo culture of polyploidy progeny. J Am Soc Hort Sci 131:716–724
48. Chi HS (2002) The efficiencies of various embryo rescue methods in interspecific crosses of *Lilium*. Bot Bull Acad Sin 43:139–150
49. Raghavan V (2003) One hundred years of zygotic embryo culture investigations. In Vitro Cell Dev Biol Plant 39:437–442
50. Collins GB, Grosser JW (1984) Culture of embryos. In: Vasil IK (ed) Cell Culture and Somatic Cell Genetics of Plants. Academic press, Orlando, Florida, pp 241–257
51. Grosser JW, Gmitter FG Jr, Fleming GH, Chandler JL (2000) Application of biotechnology to citrus cultivar improvement at the Citrus Research and Education Center. Acta Hort 535:213–220
52. Murashige T, Skoog F (1962) A revised medium for rapid growth and bioassay with tobacco tissue culture. Physiol Plant 15:473–495
53. Murashige T, Tucker DPH (1969) Growth factors requirements of citrus tissue culture. Proc First Int Citrus Symp 3:1155–1161
54. Niedz RP, Hyndman SE, Wynn ET, Bausher MG (2002) Normalizing sweet orange [*C. sinensis* (L.) Osbeck) somatic embryogenesis with semi-permeable membranes. In Vitro Cell Dev Biol Plant 38:552–557
55. Grosser JW, Gmitter FG Jr (1990) Protoplast fusion and citrus improvement. Plant Breed Rev 8:339–374
56. Deng XX, Grosser JW, Gmitter FG Jr (1992) Intergeneric somatic hybrid plants from protoplast fusion of *Fortunella crassifolia* cultivar Meiwa with *Citrus sinensis* cultivar Valencia. Sci Hort 49:55–62

57. Kane ME (2000) Culture indexing for bacterial and fungal contaminants. In: Trigiano RN, Gray DJ (eds) Plant tissue culture concepts and laboratory exercise. CRC Press, Boca Raton, pp 427–431
58. Kunisaki JT (1974) Tissue culture of tropical ornamental plants. HortSci 12:141–142
59. Hartmann HT, Kester DE, Davies FT (1990) Plant propagation – principles and practices, 5th edn. Prentice Hall, New Jersey, p 199
60. Muthan SB, Rathore TS, Rai VR (2006) Factors influencing in vivo and in vitro micrografting of sandalwood (*Santalum album* L.): an endangered tree species. J For Res 11:147–151
61. Onay A, Pirinc V, Adiyaman F, Isikalan C, Tilkat E, Basaran D (2003) In vivo and in vitro micrografting of pistachio, *Pistacia vera* L. cv. "Siirt". Turk J Biol 27:95–100
62. Raharjo SHT, Litz RE (2005) Micrografting and ex vitro grafting for somatic embryo rescue and plant recovery in avocado (*Persea americana*). Plant Cell Tiss Org Cult 82:1–9

Chapter 8

Chickpea Hybridization Using In Vitro Techniques

Nalini Mallikarjuna and Fred J. Muehlbauer

Abstract

Tissue culture techniques play an important role in the utilization of wild *Cicer* species for the improvement of cultivated chickpea. Utilization of wild *Cicer* species has become essential as a series of evolutionary bottlenecks have narrowed the genetic base of chickpea, thus making it susceptible to a range of diseases and pests. Crosses with wild *Cicer* can broaden its genetic base and introduce useful traits. Except for two wild species, none of the other *Cicer* species are cross-compatible. To use a range of *Cicer* species for the improvement of chickpea, embryo rescue and tissue culture techniques are necessary. The success of the cross with incompatible species depended on a range of techniques including the application of growth regulators to pollinated pistils and saving aborting embryos in vitro. Further, the chances of successful transfer of hybrid shoots to soil are greater if the hybrid shoots are grafted to chickpea stocks.

Key words: Chickpea, *Cicer* species, Cleft graft, Embryo rescue, Growth regulators, Incompatibility

1. Introduction

Chickpea is one of the Neolithic crops and in cultivation for more than 10,000 years. It is an important source of staple protein source in many Asian, African, and Middle Eastern countries. It has one of the highest nutritional compositions of any dry edible legume. Chickpeas' average nutritional composition is 23% protein, 64% total carbohydrates, 47% starch, 5% fat, 6% crude fiber, 6% soluble sugar, and 3% ash (1). Chickpea protein digestibility is the highest among the dry edible legumes. Like other food legumes, chickpea is rich in the essential amino acid lysine. Chickpeas are unique in moderating the rise in plasma glucose after meals and help control diabetes. Chickpea is high in phosphorus (343 mg/100 g), calcium (186 mg/100 g), magnesium (141 mg/100 g), iron (7 mg/100 g), and zinc (3 mg/100 g) (1). The seeds contain carotenoids such as beta-carotene,

Trevor A. Thorpe and Edward C. Yeung (eds.), *Plant Embryo Culture: Methods and Protocols*,
Methods in Molecular Biology, vol. 710, DOI 10.1007/978-1-61737-988-8_8,

cryptoxanthin, lutein, and zeaxanthin in amounts above the engineered beta-carotene-containing "golden rice" level (2). Nutritional benefits of chickpea could be one of the reasons for the rise of civilization in the Fertile Crescent of Mesopotamia. According to Kerem et al. (3), chickpea contains higher amounts of tryptophan, which is a precursor of the neurotransmitter serotonin, which affects brain function and human behavior. Hence, higher amounts of tryptophan might have played a major role in selecting chickpea during human expansion as tryptophan promotes birth rates and accelerated growth in humans and animals.

Chickpea has undergone a series of evolutionary bottlenecks, which has narrowed its genetic base, and hence is susceptible to more than 47 diseases (4) and 54 insect pests (5). Sources of resistance to the desired level are not present in the cultivated gene pool and this opens up avenues to look for resistant sources in the related wild species germplasm. Chickpea has 34 accessions of perennial *Cicer* and 8 accessions of annual *Cicer* species. In spite of this immense wealth of wild species germplasm, only two wild *Cicer* species are crossable with the cultigen and are amenable to gene transfer through wide crosses. The remaining 6 annual and 34 perennial *Cicer* species are not available for chickpea improvement.

The proven method of bringing in large variation into chickpea is through wide crosses, as the other methods of somaclonal variation and mutation breeding are cumbersome, time consuming, and unpredictable. Compatible wild species of chickpea have played an important role in broadening the genetic base through the introduction of desirable traits such as *Ascochyta* blight resistance (6), *Botrytis* gray mold resistance (7), *Helicoverpa armigera* resistance (8), nematode resistance (9), and high yield (10).

The genus *Cicer* is classified into three gene pools based on its crossability with cultigens. Based on their crossability with cultivated species, wild species, both annual and perennial, have been grouped. Using the classification proposed by Harlan and de Wet (11), a modification of the classification is proposed. Although the modification does not deviate much from the previously proposed gene pools for chickpea, the secondary gene pool is strengthened by the placement of *Cicer reticulatum*. The proposed classification is similar to the recent classification proposed by van der Maesen et al. (12). The primary gene pool consists of cultivated species and land races. The secondary gene pool consists of the progenitor species, *C. reticulatum* and *C. echinospermum*, a species that is crossable to *C. arietinum* but with reduced fertility of the resulting hybrids and progenies; nevertheless, both are cross compatible with the cultigen and do not need in vitro interventions to produce hybrids. The tertiary gene pool consists of all the annual and perennial *Cicer* species that are not crossable to cultivated *C. arietinum*. All of the perennial *Cicer* species are considered to be in the tertiary gene pool as none of the species of this group are known to cross readily with the cultivated species and produce mature seeds (Mallikarjuna and Muehlbauer, unpublished).

Many of the *Cicer* species in the tertiary gene pool harbor important traits/genes necessary for the improvement of chickpea such as *H. armigera* resistance in *C. judaicum*, *C. pinnatifidum* and *C. bijugum* (13), *Ascochyta* blight resistance in *C. judaicum*, *C. bijugum* and *C. pinnatifidum* (12), *Botrytis* gray mold resistance (14), and drought tolerance (15). There are 34 perennial wild *Cicer* species that require very specific soil and environmental conditions for growth and reproduction. Traits of interest such as resistance to *Ascochyta* blight (16), *H. armigera* (17), *Fusarium* wilt (18), and drought tolerance (19) are present in this gene pool. Perennial *Cicer* species survive the severe frost conditions and resume their vegetative growth with the onset of summer in the USDA-ARS nursery located at Washington State University, Pullman, USA. All the perennial *Cicer* species have larger plant morphology compared to the annual *Cicer* species with robust vegetative growth. The flowers are larger with multiseeded fruits/pods. Desirable traits that chickpea would benefit from perennial *Cicer* are large and robust vegetative growth, large pods with multiple seeds, drought and cold tolerance, and disease and pest resistance.

It is now known that the barriers to hybridization between cultivated chickpea and *Cicer* species in the tertiary gene pool occur after zygote formation (20, 21). Fertilizations take place, but the zygote begins to abort by 3–5 days after fertilization. Badami et al. (22) were able to postpone the abscission of pollinated pistils to 15–18 days by the application of growth regulators. This facilitated the growth of the hybrid embryo to early cotyledonary stage of development and being 0.5–1.0 mm in size (19).

Embryos of the size 0.5 mm or less did not grow directly on culture medium, while 0.3–0.4 mm size embryos responded well to specific growth hormones when cultured as *in-ovulo* embryo culture. Embryo response was maximum when zeatin (Zn) was used in combination with indole-3-acetic acid (IAA) in *in-ovulo* embryo culture medium and was evident by the emergence of embryos from the ovule after 3–4 weeks (21). Similar response was not obtained when zeatin was replaced with other cytokinins, which reduced the number of responding embryos (Mallikarjuna, unpublished).

The best time to save the aborting seeds/ovules was when the hybrid embryo had reached its maximum growth and development, being at the cotyledonary stage of development, which was 15–18 days after pollination. If left longer on the plant, the pods turned yellow, indicating abortion of the hybrid seed.

Techniques such as in vitro culture to save aborting embryos from incompatible pollinations, multiplication of the hybrid shoots and their further growth, development of plants/shoots from somatic embryos, rooting in vitro grown shoots, grafting hybrid shoots on chickpea stocks, and induction of androgenesis from the hybrid plant's microspores will be discussed in the following paragraphs.

1.1. Technical Comments

Hybrid shoots from the cross *C. arietinum* × *C. pinnatifidum* were pale yellow in color, and the scanning electron microscopic (SEM) studies showed that the chloroplasts were abnormal. Use of a cytokinin in culture medium in combination with light helped the conversion of leucoplastids to chloroplasts (22). Whereas hybrids between *C. arietinum* × *C. bijugum* and between *C. arietinum* × *C. judaicum* were green in color and albino shoots have not been observed. Hybrids between *C ari-etinum* and *C. pinnatifidum* were fragile with the leaves resembling those of *C. pinnatifidum*. The color of the flower was pale violet, resembling the violet color of the male parent, and the pollen was 100% nonviable (21).

Genotype of the cultivated chickpea, the female parent in the crossing program, was important for the success of the cross with respect to the number of hybrid ovules, in the size range of 2.5 mm or more, obtained and hybrid embryos responded. When chickpea cultivar ICCV 2 was crossed with *C. pinnatifidum* accession ICCW 37, although a larger number of pods were obtained than when crossed with cultivar ICCV 92318, the difference was larger with respect to the number of hybrid plants obtained. Many of the hybrid embryos from the crosses with ICCV 2 were small and abnormal even after ovule culture. A similar situation was observed in crosses with desi cultivars. Crosses with ICCV 6 did not set a large number of pods and ovules suitable for culture, and seedlings were not obtained. ICCV 10 set a large number of pods, and nine seedlings were obtained (21).

2. Materials

1. The hormone solution application to prevent flower abscission contains Gibberellic acid A_3, naphthaleneacetic acid, and kinetin (GA_3 50 mg/L + NAA 10 mg/L + Kn 10 mg/L).
2. Sterilizing solution: a 30% commercial bleach, Clorox solution.
3. Chickpea culture media: Murashige and Skoog's (MS) medium is generally used for chickpea tissue culture (Table 1). Modification of the MS medium known as the ML-6 medium (23) also works well for chickpea tissue culture. Major and minor salts (tissue culture grade) are prepared as 10× stocks and used to prepare medium.
4. Growth regulators: indole butyric acid (IBA), indole acetic acid (IAA), Gibberellin A_3, benzyladenine, kinetin, and zeatin. All growth regulators are filter sterilized.

Table 1
Media composition

Nutrients[a]	MS	ML-6	Ovule culture	Shoot growth	Root induction[b]
Macronutrients					
NH_4NO_3	1,650	1,000	+	+	+
KNO_3	1,900	1,000	+	+	+
$MgSO_4\cdot 7H_2O$	370	170	+	+	+
$CaCl_2\cdot 2H_2O$	440	440	+	+	+
KH_2PO_4	170	170	+	+	+
Na_2EDTA			+	+	+
$FeSO_4$			+	+	+
Micronutrients					
$MnSO_4\cdot 4H_2O$	22.3	0.0	+	+	+
$ZnSO_4\cdot H_2O$	8.6	8.6	+	+	+
H_3BO_3	6.2	6.2	+	+	+
KI	0.83	0.83	+	+	+
$Na_2MoO_4\cdot 2H_2O$	0.25	0.25	+	+	+
$CoCl_2\cdot 6H_2O$	0.025	0.025	+	+	+
Vitamins					
myo-Inositol	100	100	+	+	+
Nicotinic acid	0.5	0.5	+	+	+
Thiamine	0.5	0.5	+	+	+
Glycine	1.0	0.0	+	+	+
Growth regulators					
IBA					0.5
IAA			0.25	0.5	–
KN			–	2.0	–
Zeatin			1.0	–	–
Agar (%)			–	0.68	0.7
Sucrose (%)			3	3	0.3
pH			5.8 ± 0.2	5.8 ± 0.2	5.9 ± 0.2

[a]All ingredients are mg/L
[b]Half strength MS medium

5. Ovule culture medium (see Table 1) is a liquid medium devoid of agar.
6. Filter paper bridges for ovules (immature seeds) are prepared using Whatman No. 1 filter paper.
7. PVC tubing of 0.6 cm length and 3 mm in width was selected to hold the grafted shoot in place.

8. Laboratory supplies: forceps, cotton swab, vials, polyethylene bags, scalpels, pots, and sand.

3. Methods

3.1. Cross Pollinations with Incompatible Cicer Species

Plants used in the crossing program are cultivated under natural field conditions at 26–30°C maximum and 12–15°C minimum. These conditions give better pod set when used in the crossing program than grown in a glasshouse under similar temperature regimes. Wild species of *Cicer* take longer time to flower compared to cultivated chickpea. It is recommended to have staggered plantings of cultivated species to have continuous supply of flowers. Under ICRISAT, India conditions, wild species grown under additional light regime, after initial vegetative growth of about a month, induces profuse flowering.

Cross pollinations were carried out before 10.00 AM in the chickpea growing season at ICRISAT, India (maximum 26–28°C and minimum 12–14°C), although chickpea stigmas are receptive till late in the afternoon. It has been observed that the stigmas of cultivated chickpea remain receptive even at 35°C and till late in the afternoon.

1. Anthers were mechanically removed the previous evening from the buds chosen for pollination. All others buds at the node were removed to facilitate the growth of pollinated pistil.
2. Pollinations were carried out in the morning using fresh pollen from the male parent.
3. Soon after pollinations were carried out, cotton swab soaked with hormone solution containing Gibberellic acid A_3, Naphthaleneacetic acid, and kinetin (GA_3 50 mg/L + NAA 10 mg/L + Kn 10 mg/L) was wrapped around the base of the individual pistils to prevent flower abscission.
4. The process of hormone application was repeated for 1–3 days depending upon the retention of flowers on the plant.
5. After 15–25 days of pollination, pods from cross pollinations, which began to turn yellow from green, were harvested and prepared for ovule culture (Fig. 1).

3.2. Embryos Rescued from Interspecific Incompatible Crosses (Ovule Culture)

1. Immature pods from cross pollinations were collected from plants. Care was taken to allow immature seed to grow to its maximum, when the green pod wall began to turn pale shades of yellow (see Note 1).
2. Pods were surface-sterilized in a 30% commercial bleach (Clorox) for 15 min, and the bleach was washed off by giving three to four washes with sterilized distilled water (see Note 2).

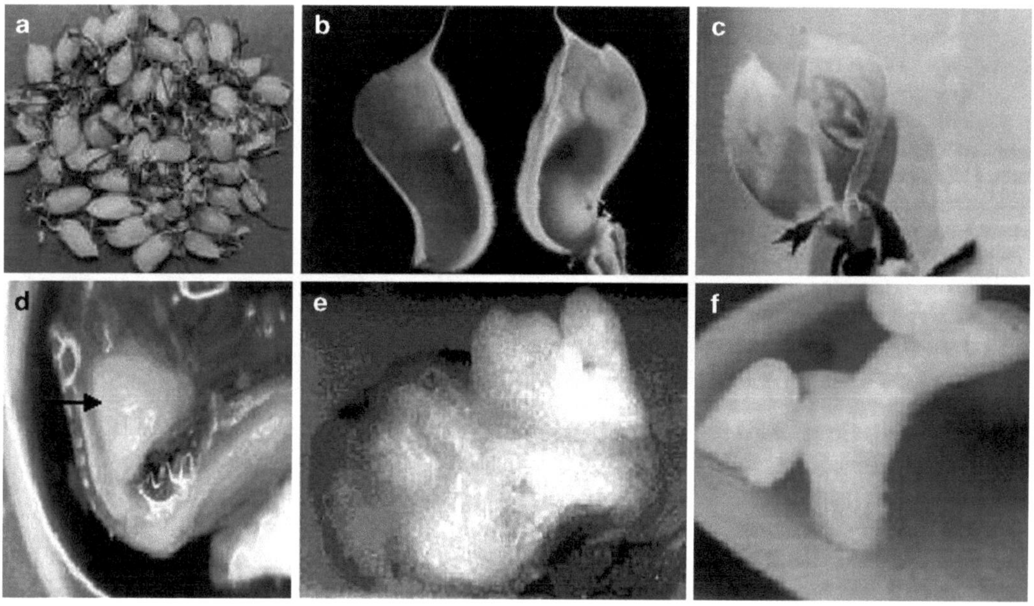

Fig. 1. Pod set, embryo abortion, and embryo rescue in chickpea wide crosses. (**a**) Pods from cross pollination *C. arietinum* × *C. pinnatifidum*. (**b**) Pod from cross pollination (*C. arietinum* × *C. pinnatifidum*), without the application of growth regulators, showing aborted seeds. (**c**) Pod from cross pollination (*C. arietinum* × *C. pinnatifidum*) showing a 4.0 mm seed/ovule. (**d**) Cross section of a pod, obtained as a result of hormone aided pollination, from the cross *C. arietinum* × *C. pinnatifidum*, showing the growth of the embryo (*arrow*). (**e**) Hybrid embryo emerging out of the cultured ovule. (**f**) Embryo dissected out of the ovule and cultured.

3. Liquid ovule culture medium was taken in small vials with filter paper bridges (see Notes 3 and 4; Table 1). It is cut as a 14 cm long and 3 mm width strip. The strip is longitudinally folded into half (1.5 mm strip) and made in the shape of an "M" with the outer borders longer than the central "V." Care is taken to see that the liquid medium does not cross the lower end of the shape "V."
4. Ovules (immature seeds) are carefully dissected out of the pods with the placental region still attached to the ovule and cultured on filter paper bridges. Care should be taken not to submerge the ovules in the liquid medium, which consisted of MS or ML-6 basal salts + 3% sucrose + IAA (0.25 mg/L) + Zn (1.0 mg/L) (Table 1 and also see Notes 5 and 6). Zeatin and IAA used is always filter-sterilized.
5. After 3 weeks of culture, ovules were transferred to fresh ovule culture medium till the embryos emerged out of the ovules (see Note 7).
6. Embryos that emerged out of the ovule were transferred to shoot growth medium, where the source of cytokinin was kinetin instead of zeatin.
7. Well-grown shoots were transferred to root induction medium, which consists of 1/2 strength MS basal salts, 1.5%

Fig. 2. Hybrid plants obtained through ovule culture in vitro. (**a**) *C. arietinum* × *C. judaicum* seedling. (**b**) *C. arietinum* × *C. bijugum* seedling. (**c**) *C. reticulatum* × *C. pinnatifidum* seedling. (**d**) *C. arietinum* × *C. pinnatifidum* seedling.

sucrose, 0.7% agar, with IBA 0.5 mg/L. Alternatively, shoots are grafted to cultivated chickpea stocks (see Note 8).

8. After a pulse treatment of 15 days on the rooting medium, shoots were transferred to the basal MS medium to induce roots on shoots.
9. Rooted shoots were transferred to pots with sieved sand and watered with sterilized tap water (see Note 9; Fig. 2).
10. It was important to maintain high humidity by covering the transferred plants with a polythene bag.

3.3. Shoot Multiplication In Vitro

1. Shoot buds were produced in clumps when embryos are transferred to embryo/shoot growth medium.
2. Individual shoots that were more than 2–3 cm long are separated from the clumps and transferred to the shoot growth medium, which was same as embryo growth medium (Table 1).
3. Shoots that grew further were transferred to root induction medium or grafted to cultivated chickpea stocks.

3.4. Somatic Embryos from Immature Cotyledons

1. Immature pods of either kabuli or desi chickpea are surface sterilized with 30% Clorox (commercial bleach).
2. Pods are washed with sterilized distilled water for four to five times to remove traces of Clorox (see Note 10).
3. Immature cotyledons (4–5 mm in size) devoid of embryo axis are isolated in a sterilized Petri dish.
4. Medium for culture [MS + 3.0% sucrose + Zn (14.0 μM/L) + IAA (5.0 μM/L) at pH 5.8] is poured into sterile Petri dishes.

5. Cotyledons are cultured on the medium with their adaxial surface up, against the medium.
6. For further growth, somatic embryos were transferred to MS medium with 2 mg/L BAP and 0.5 mg/L IAA (see Note 11).
7. Those embryos, which have grown into seedlings and without good root systems, were transferred to the rooting medium mentioned above (see Note 12).
8. Well-grown shoots were transferred to fine sand initially and later to soil: sand: farm yard manure (1:1:1) (see Note 13).

3.5. Grafting Hybrid Shoots to Chickpea Stocks

Transfer of hybrid shoots to soil is a critical step for the success of crosses with incompatible annual *Cicer* species. Although percent response with respect to the number of pollinations made and the number of hybrid plants obtained are low (21), techniques to produce hybrid shoots are in place when chickpea is crossed with *C. pinnatifidum*, *C. bijugum*, and *C. judaicum*. It is possible to multiply hybrid shoots in vitro by transferring the hybrid shoot buds to 0.7% agar solidified MS medium with 3% sucrose, 0.5 mg/L IAA, and 2.0 mg/L KN at pH 5.8 (Table 1). Shoots can be rooted on full strength or ½ MS basal medium with 1.5% sucrose, 0.5 mg/L IBA (Table 1). In most of the cases, the roots were stunted, and hence very few shoots survived the transfer to soil. To overcome the problem of rooting hybrid shoots, an alternative method of grafting hybrid shoots to chickpea stocks was standardized (see Fig. 3; (24)).

1. 15 days old chickpea seedlings were used as stocks.
2. The hybrid shoots to be grafted (scion), which were 3 cm or more, are cut into a "V" shape.
3. The Root stock plants are cut just above the base of the stem, about 2–3 cm from the soil. All axillary buds are removed to prevent the growth of axillary shoots.
4. A 0.5 cm PVC tubing whose diameter is slightly more than that of the stem is slid on the blunt end of the stem. It is essential to see that the tubing selected is flexible and can expand marginally.
5. The slit is made on the blunt send of the stem, dividing it into two equal parts.
6. The scion is inserted into the slit and secured in place with the help of the rubber tubing.
7. When the grafts had established, the tubing was slit longitudinally to free the grafted region.
8. The pot is covered with a plastic bag to maintain high humidity (see Note 14).

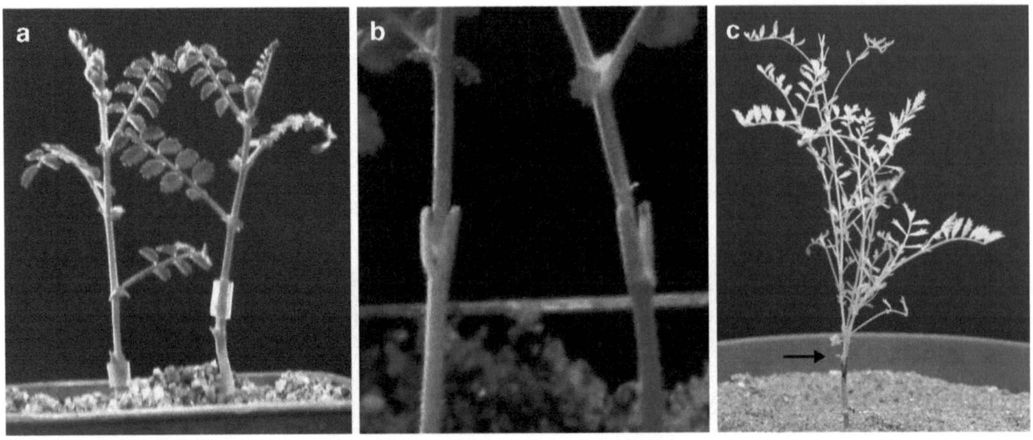

Fig. 3. Grafting technique in chickpea. (**a**) *C. arietinum* plant grafted with *C. arietinum* × *C. bijugum* hybrid shoot. Note the rubber tubing holding the grafts. (**b**) After 3–4 weeks, the rubber tubes have been removed to facilitate further growth. (**c**) The graft (*C. arietinum* × *C. pinnatifidum*) growing well. *Arrow* points to the region of the graft.

3.6. Development of Multicellular Microspores

Development of haploid plants from anther/microspore culture is now possible (25). It is also possible to get multicellular microspores from wide crosses. Mallikarjuna and Jadhav (24) and Mallikarjuna et al. (26) demonstrated that the hybrid between chickpea and *C. pinnatifidum* gave rise to flower buds with multicellular microspores in large numbers with divisions in all the microspores in some plants. As the critical step of induction of androgenesis in chickpea microspores had taken place, culturing such microspores may give rise to high frequency of haploid plants.

1. Hybrid plants from the cross *C. arietinum* × *C. pinnatifidum* need to be maintained in a growth chamber with 70% relative humidity (see Note 14).
2. Plants are watered with either de-ionized or drinking quality water with 1 mg/L zeatin. Inclusion of zeatin induces floral buds on the plants (see Note 15).
3. The buds/flower are fragile, and many of the buds have anthers with multicellular microspores with the number of cells ranging from 4 to more than 10 (see Note 16).
4. The authors have not cultured such microspores to obtain haploid plants.

4. Notes

1. Application of growth regulator combination, specified for chickpea wide crosses with any of the incompatible species is mandatory. Without its application, ovules begin to abort from third day after pollination (21).

2. Surface sterilization in alcohol or even brief wash in alcohol is not advisable as it reduces the number of responding ovules.
3. Adding filter sterilized IAA and zeatin into the ovule culture medium gives better response.
4. Embryos do not respond when cultured on semisolid medium.
5. Percent response is better when cultured with the placental region attached to the ovule.
6. Ovules respond only when cultured on the liquid medium with zeatin as the source of cytokinin. Ovules do not respond when zeatin in the culture medium is replaced with either kinetin or benzyl amino purine.
7. A second transfer of the ovules to the ovule culture medium increases the number of responding ovules.
8. Although a method to root shoots in vitro has been developed, it is not efficient, which is reflected in the percent successful transfer of rooted shoots to soil. Alternatively, grafting the shoots to chickpea stocks is very successful (>90%).
9. In vitro rooting is not very efficient for chickpea plants regenerated through multiple subcultures in vitro.
10. Traces of Clorox on the pod wall can reduce percentage of responding embryos.
11. Embryos cluster all around the cotyledons. Culturing clumps of embryos on the shoot growth medium enhances the growth of normal embryos. It is to be noted that many of the somatic embryos are abnormal without the shoot or root axis.
12. It is a straight forward process to root chickpea shoots in vitro, which have not been subcultured in vitro. Every subculture reduces rooting efficiency and by third subculture very few shoots root. Roots on the shoots from three or more subcultures are not only scanty but are unhealthy and do not withstand the transfer to soil. This may be one of the reasons why hybrid shoots, which have undergone subculture/s, do not root well on the root induction medium.
13. Hybrid shoots regenerated from immature aborting embryos are fragile and do not withstand the transfer to soil directly. They are transferred to pots with sand initially. Later, they are transferred to soil.
14. Even though the plants are maintained in an incubator at 25°C and high relative humidity, covering them with a polythene cover helps the grafts grow faster.
15. Interspecific hybrid between *C. arietinum* × *C. pinnatifidum* starts flowering when zeatin 1.0 mg/L is added to the nutrient solution or the water, which is used to water the hybrid plants (24).

16. Multicellular microspores are starting material to obtain haploid plants. Although dihaploid plants are reported for chickpea, another mode of obtaining the induction of androgenesis is by wide crosses.

References

1. Williams PC, Singh U (1987) Nutritional quality and the evaluation of quality in breeding programme. In: Saxena MC, Singh KB (eds) The chickpea. C.A.B. International, Wallingford, UK, pp 125–130
2. Abbo S, Molina C, Jungmann R, Grusak M, Berkovitch Z, Reifen R, Kahl G, Winter P, Reifen R (2005) Quantitative trait loci governing carotenoid concentration and weight in seeds of chickpea (*Cicer arietinum* L.). Theor Appl Genet 111:185–195
3. Kerem Z, Gopher A, Lev-Yadun S, Weinberg P, Abbo S (2007) Chickpea domestication in the Neolithic Levant through the nutritional perspective. J Archaeol Sci 34:1289–1293
4. Nene YL, Haware MP (1980) Screening chickpea for resistance to wilt. Plant Dis 64:379–380
5. Reed W, Cardona C, Sithanantham S, Lateef SS (1987) Chickpea insect pests and their control. In: Saxena MC, Singh KB (eds) The chickpea. C.A.B. International, Wallingford, UK, pp 283–318
6. Collard BCY, Pang ECK, Ades PK, Taylor PWJ (2003) Preliminary investigation of QTLs associated with seedling resistance to *Ascochyta* blight from *Cicer echinopsermum*, a wild relative of chickpea. Theor Appl Genet 107:719–729
7. Ramgopal D (2006) Characterization and evaluation of annual wild *Cicer* species and study the transfer of *Botrytis* gray mold and *Ascochyta* blight resistance from *Cicer echinopsermum* into cultivated species. Ph.D. Thesis, Acharya N.G. Ranga Agricultural University, Rajendra Nagar, Hyderabad, India
8. Mallikarjuna N, Sharma HC, Upadhyaya HD (2007) Exploitation of wild relatives of pigeonpea and chickpea for resistance to *Helicoverpa armigera*. SAT e-Journal 3:4–7
9. Malhotra RS, Singh KB, Di Vito M, Greco N, Saxena MC (2002) Registration of ILC 10765 and ILC 10766 chickpea germplasm lines resistant to cyst nematode. Crop Sci 42:1756
10. Singh KB, Ocampo B (1997) Exploitation of wild *Cicer* species for the yield improvement of chickpea. Theor Appl Genet 95:418–423
11. Harlan J, de Wet J (1971) Towards a rational classification of cultivated plants. Taxon 20:509–517
12. van der Maesen LJG, Maxted N, Javadi F, Coles S, Davies AMR (2007) Taxonomy of the genus *Cicer* revisited. In: Yadav SS, Redden B, Chen W, Sharma B (eds) Chickpea breeding and management. C.A.B. International, Wallingford, UK, pp 14–46
13. Sharma HC, Pampapathy G, Lanka SK, Ridsdill-Smith TJ (2005) Antibiosis mechanism of resistance to pod borer, *Helicoverpa armigera* in wild relatives of chickpea. Euphytica 142:107–117
14. Pande S, Ramgopal D, Kishore GK, Mallikarjuna N, Sharma M, Pathak M, Narayana Rao J (2006) Evaluation of wild *Cicer* species for resistance to *Ascochyta* blight and *Botrytis* gray mold in controlled environment at ICRISAT, Patancheru, India. ICPN 13:25–26
15. Bhattarai T, Fettig S (2005) Isolation and characterization of a dehydrin gene from *Cicer pinnatifidum*, a drought-resistant wild relative of chickpea. Physiol Plant 123:452–458
16. Muehlbauer FJ, Kaiser WJ, Simon CJ (1994) Potential for wild species in cool season food legume breeding. Euphytica 73:109–114
17. Sharma HC, Bhagwat MP, Pampapathy G, Sharma JP, Ridsdill-Smith TJ (2006) Perennial wild *Cicer* relatives of chickpea as potential sources of resistance to *Helicoverpa armigera*. Genet Resour Crop Evol 53:131–138
18. Kaiser WJ, Alcala-Jimenez AR, Hervas-Vargas A, Trapero-cacas JL, Jimenez-Diaz RM (1994) Screening of wild *Cicer* species for resistance to races 0 and 5 of *Fusarium oxysporum* f. sp. *ciceris*. Plant Dis 78:962–967
19. Toker C, Canci H, Yildirim T (2007) Evaluation of perennial wild *Cicer* species for drought resistance. Genet Resour Crop Evol 54:1781–1786
20. Ahmad F, Slinkard AE, Scoles GJ (1988) Investigations into the barrier(s) to interspecific hybridization between *Cicer arietinum* L. and eight other annual *Cicer* species. Plant Breed 100:193–198
21. Mallikarjuna N (1999) Ovule and embryo culture to obtain hybrids from interspecific

incompatible pollinations in chickpea. Euphytica 110:1–6
22. Badami PS, Mallikarjuna N, Moss JP (1997) Interspecific hybridization between *Cicer arietinum* and *C. pinnatifidum*. Plant Breed 116:393–395
23. Kumar AS, Gamborg OL, Nabors MW (1988) Plant regeneration from cell suspension cultures of *Vigna aconitifolia*. Plant Cell Rep 7:138–141
24. Mallikarjuna N, Jadhav DR (2008) Techniques to produce hybrid between *Cicer arietinum* L. × *C. pinnatifidum* Jaub. Indian J Genet 68:1–8
25. Grewal RK, Lulsdorf M, Croser J, Ochatt S, Vandenberg A, Warkentin TD (2009) Doubled-haploid production in chickpea (*Cicer arietinum* L.): role of stress treatments. Plant Cell Rep 28:1289–1299
26. Mallikarjuna N, Jadhav DJ, Clarke H, Coyne C, Muehlbauer F (2005) Induction of androgenesis as a consequence of wide crossing in chickpea. Int Chickpea Pigeonpea Newsl 12:12–15

Chapter 9

Muskmelon Embryo Rescue Techniques Using In Vitro Embryo Culture

Hector Gordon Nuñez-Palenius, Rafael Ramírez-Malagón, and Neftalí Ochoa-Alejo

Abstract

Among the major cucurbit vegetables, melon (***Cucumis melo***) has one of the greatest polymorphic fruit types and botanical varieties. Some melon fruits have excellent aroma, variety of flesh colors, deeper flavor, and more juice compared to other cucurbits. Despite numerous available melon cultivars, some of them are exceedingly susceptible to several diseases. The genetic background carrying the genes for tolerance and/or resistance for those diseases is found in wild melon landraces. Unfortunately, the commercial melon varieties are not able to produce viable hybrids when crossed with their wild melon counterparts. Plant tissue culture techniques are needed to surpass those genetic barriers. In vitro melon embryo rescue has played a main role to obtain viable hybrids originated from commercial versus wild melon crosses. In this chapter, an efficient and simple embryo rescue melon protocol is thoroughly described.

Key words: Anther, *Cucumis melo*, Cucurbitaceae, Melon, Ovary, Plant tissue culture, Zygotic embryos

1. Introduction

Plants belonging to the Cucurbitaceae family are commonly well known as cucurbits. The most important cultivated cucurbits – based on harvestable area and total production – around the globe are watermelon (***Citrullus lanatus*** Thunb.), cucumber (***Cucumis sativus*** L.), melon (Cantaloupe and other melons) (***Cucumis melo*** L.), pumpkin, and squash (***Cucurbita*** spp.) (1). ***C. melo*** L. is an important worldwide vegetable crop; for example, in 2007, the international melon fruit production (honeydews, cantaloupes, muskmelons, and others) was more than 26 million tons (1), representing a significant income for seedmen

Trevor A. Thorpe and Edward C. Yeung (eds.), *Plant Embryo Culture: Methods and Protocols*, Methods in Molecular Biology, vol. 710, DOI 10.1007/978-1-61737-988-8_9,

and growers. Despite the elevated production, substantial economical losses took place since some *C. melo* varieties, highly demanded by consumers, are exceedingly susceptible to pests and diseases, caused by insects, viruses, bacteria, and fungi, among others (2). According to plant breeding experts (3), the shortest, safest, and easiest way to avoid or reduce melon production losses, even better than application of pesticides and disease-preventive activities, is the use of pest- and disease-resistant melon varieties. Unfortunately, not all commercial melon varieties are tolerant or resistant to all pests and diseases, and the genetic background that could provide the tolerance or resistance for most *C. melo* diseases is not present in this species. Fortunately, the genetic pool with tolerance and/or resistance for *C. melo* diseases is found in *Cucumis*-related ones, such as *C. metuliferus* Mayer, *C. pustulatus* Naudin, *C. myriocarpus* Naudin, and *C. melo* ssp. *agrestis* (4, 5), among others. However, the ability to cross *C. melo* with its pest- and disease-resistant *Cucumis* relatives, in order to pass those desirable agronomic characteristics into commercial melon varieties, is extremely low or unfeasible by traditional breeding systems. Therefore, plant tissue culture techniques are needed to surpass those genetic barriers.

In vitro embryo culture methods have successfully been used to rescue valuable embryos in diverse plants (6–13). Nevertheless, the embryo rescue techniques may not be totally efficient, since in some cases, the embryo failed to undergo a complete differentiation and full morphogenetic process through in vitro culture, producing nonviable embryos (14, 15). Several factors, biological and physicochemical, play a main role during the in vitro embryo culture; among them genotype (13), embryo-developmental stage (16, 17), type of culture media (15), plant growth regulator and carbohydrate source (18), season when the embryos are collected (19), temperature (20), and light (21) have been reported. Numerous reports have been published on in vitro gamete (ovary and anther) and embryo culture of *Cucumis* spp. (16, 17, 20–44), and most of them have been applied to cucumber. It is generally accepted that both species, *C. sativus* and *C. melo*, have similar culture media requirements. Concerning the specific case of *C. melo*-embryo culture, this technique has been used to rescue valuable hybrids acquired from interspecies crosses (45), as well as to obtain haploid plants to select resistance to diseases such as powdery mildew (46–48), *Fusarium* wilt (49), and several viruses (13).

To apply the techniques employed for melon embryo rescue culture, a simple and efficient protocol developed for a "Galia" muskmelon (*C. melo* L. var. *reticulatus* Ser.) male parental line is outlined in this chapter. One advantage of this system is that, unlike other rescue embryo protocols, it does not require a double-layer system culture or several culture media.

2. Materials

1. Plant material: it is well recognized that healthy ***C. melo*** plants will provide the best zygotic embryos for a successful embryo culture (45). Hence, the embryo-donor melon plants must be grown under the utmost optimal conditions; for instance, mature and healthy (not showing visible damage) melon seeds are germinated on a mixture of 70% Terra Lite Plug Mix (Terra Asgrow, Apopka, FL) and 30% coarse vermiculite in polystyrene trays (cell size 2.25 cm^2 and 164 cells per tray, Speedling, Bushnell, FL). Seedlings are grown under drip irrigation in plastic pots (11.3 L) filled with soilless media (coarse grade perlite) in an evaporative-cooled fan and pad glasshouse, with temperatures maintained at 28°C day and 20°C night. Hermaphrodite and/or female flowers should be pollinated in the morning (7:00–10:00 AM), since the pollen grains show the greatest survival rate in this period (45), using at least three newly opened male flowers. A higher in vitro embryo culture success is obtained when older melon fruits [17–30 days after pollination (DAP)] are utilized; nonetheless melon seeds from fruits as young as 4 DAP may be used for embryo rescue.
2. Culture medium: salt formulations, organic additives, and plant hormones for E-21 (34) culture medium are summarized in Table 1. The culture-medium pH is adjusted to 5.9 using a few drops of KOH 0.1 N, before adding 10 g/L agar. The E-21 medium (200 mL) is placed in 500-mL Erlenmeyer flasks and sterilized at 121°C (15 lb/in.^{2}) for 20 min. Putrescine, glutamine, and coconut water are added to the sterile E-21 medium (50°C). These compounds are previously filter-sterilized using sterile Millipore® membranes (GSWP09050, 0.22 μm).
3. Sterilizing solutions: (a) 70% ethanol and (b) 1.2% sodium hypochlorite solution (15 or 20% commercial bleach containing three drops of Tween 20™ per 100 mL).
4. Phytatray™ (Sigma Chemical Co., St. Louis, MO, USA).
5. Polystyrene trays (cell size 2.25 cm^2 and 164 cells per tray, Speedling, Bushnell, FL).
6. Terra Lite Plug Mix (Terra Asgrow, Apopka, FL) and coarse vermiculite.
7. Other equipment and laboratory supplies: a steromicroscope, glass dry-seal desiccator, vacuum pump, a 2,000 mL beaker, scalpels, forceps, Petri dishes, Phytatray™ vessels, and aluminum foil.

Table 1
Components of E-21 nutrient medium

Macroelements (mg/L)		*Microelements (mg/L)*	
KNO_3	1,075.0	$MnSO_4$	11.065
NH_4NO_3	619.0	$ZnSO_4 \cdot 7H_20$	1.812
$MgSO_4 \cdot 7H_2O$	206.0	H_3BO_3	1.575
$CaCl_2 \cdot 2H_2O$	156.5	KI	0.345
KH_2PO_4	71.0	$Na_2Mo_4 \cdot 2H_2O$	0.094
$Ca(NO_3)_2 \cdot 4H_2O$	25.0	$CuSO_4 \cdot 5H_2O$	0.008
$NaH_2PO_4 \cdot 4H_2O$	19.0	$CoCl_2 \cdot 6H_2O$	0.008
$(NH_4)_2SO_4$	17.0	Na_2EDTA	37.3
KCl	3.5	$FeSO_4 \cdot 7H_2O$	27.8
Organics (mg/L)		*Plant growth regulators (mg/L) and other supplements*	
myo-Inositol	50.300	Indole-3-acetic acid	0.01
Pyridoxine–HCl	5.500	Indole-3-butyric acid	0.01
Nicotinic acid	0.700	6-Benzylaminopurine	0.01
Thiamine	0.600	Coconut water	5%
Ca-D-pantothenate	0.500	Xylose	0.02
D-Biotine	0.005	Glutamine	1
Glycine	0.100	Agar (g/L)	10
Sucrose (g/L)	20	pH	5.9
Putrescine	0.25 mM		

3. Methods

3.1. Fruit Sterilization and Seed Preparation

1. Melon fruits are harvested (using a sterile scalpel) in the morning (as early as possible), and their surface is meticulously washed (at least three times) with liquid detergent and tap water using a sponge.
2. Washed fruits are placed into a sterile laminar-air-flow cabinet. Afterward, they are surface-sterilized by treating them with 70% ethanol for 10 min into a 2,000-mL beaker. Fruits are then immersed into a 1.2% sodium hypochlorite solution for 40 min in a glass dry-seal desiccator (see Note 1). Vacuum force is applied (10 min) to the glass dry-seal desiccator to facilitate the air removal from the fruit surface (see Note 2). Wash, at least six times, with sterile distilled water until no bubbles are observed when the container is shaken.
3. Remove the surface (exo- and mesocarp) of sterile fruits until seeds are visible (see Note 3) and dissect them carefully using a scalpel, forceps and, if needed a stereomicroscope (see Note 4). Discard damaged seeds.

4. Once a seed is excised, leave it immediately in a Petri dish containing a 9% sterile (by filtration through a Millipore® membrane) sucrose solution. From this solution, seeds (up to 60) will be placed on a Petri dish containing the E-21 medium.

3.2. Embryo Culture and Transfer of Plants to Soil

1. Transfer the melon seed directly to a Petri dish containing E-21 medium (initially, the embryo is not removed from the seed for in vitro culture). It is absolutely important that the hilum must face the surface medium, as well as the seed must be partially buried into the medium (see Fig. 1).
2. Wrap the Petri dish completely with aluminum foil to avoid any incoming light, and keep the embryo cultures in a culture room at 25 ± 1.5°C for 35–40 days. During this period, the embryo developmental process will take place (see Fig. 2).
3. After the incubation period, the germinated embryos are transferred, under axenic conditions, to Phytatray™ vessels having ½ strength E-21 medium with 0.7% phytagar. These transferred embryos are cultured for 2–5 more weeks (depending on their developmental stage) under 100 μmol/m^2/s light (cool white lamps) and a 16 h photoperiod at 25 ± 1.5°C (see Fig. 2d, e).
4. Healthy well-developed seedlings (first true-leaf stage) are removed from in vitro culture vessels and planted on 70% Terra Lite Plug Mix (Terra Asgrow, Apopka, FL) and 30% coarse vermiculite in polystyrene trays (see Fig. 3). A transparent

Fig. 1. Dissection of seeds from "Galia" male parental line fruit (17 days after-pollination (DAP) stage) under axenic conditions. Notice how seeds are semiburied with the hilum facing the culture medium (*arrows*).

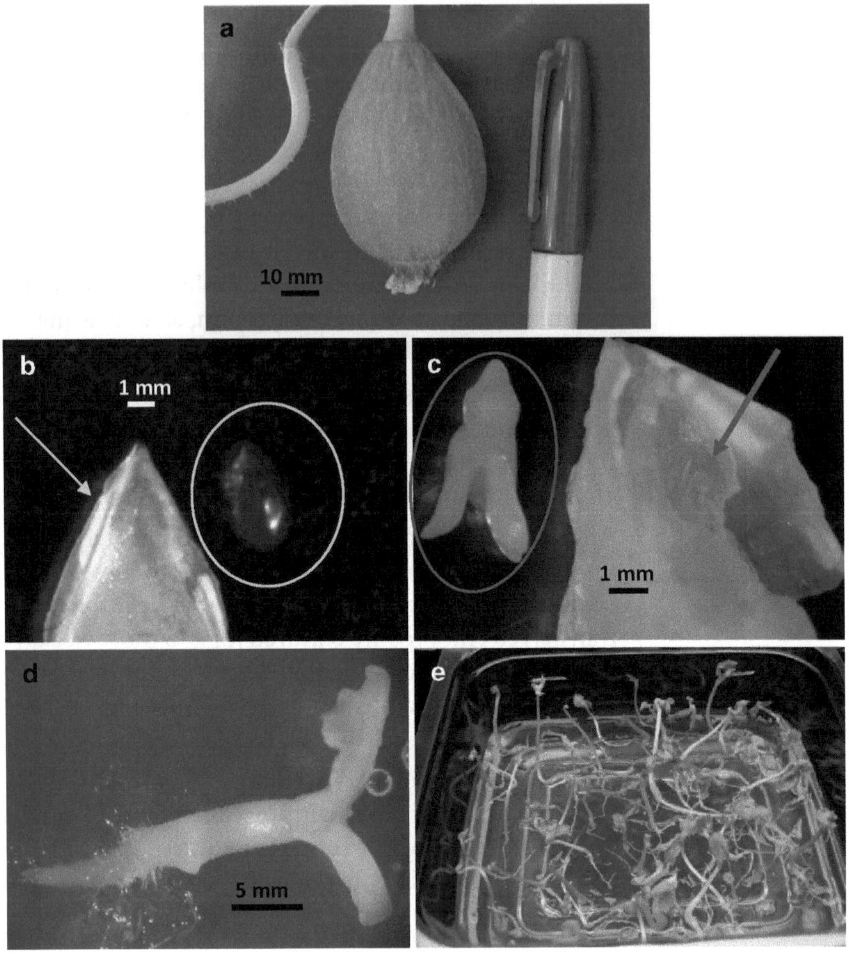

Fig. 2. "Galia" male parental line fruit at 10 DAP stage (**a**). In vitro embryo development from 10 DAP stage, (**b**), (**c**), (**d**), and (**e**) are stages after 0, 15, 21, and 35 days, respectively, of in vitro culture. (**b**) Embryo (*circle*) and seed coat (*arrow*). (**c**) Embryo (*circle*) and seed coat (*arrow*). (**e**) Seedlings growing on elongation media.

plastic cover is placed over the recently transferred seedlings and the polystyrene trays are kept in a plant growth walking-chamber (Controlled Env. Ltd, Winnipeg, Manitoba, Canada) with temperatures maintained at 28°C day and 20°C night and 16-h artificial lighting. Melon seedlings are watered as needed and plastic cover is removed after 1 week.

4. Notes

1. 400 mL of distilled water, plus 12 drops of Tween 20™ are prepared in a glass dry-seal desiccator, from which the desiccant material was previously removed. This container is sterilized at 121°C (15 lb/in.^{2}) for 20 min. Once the glass dry-seal desiccator

Fig. 3. Normal "Galia" male parental line seedlings obtained from embryo rescue, having well-developed cotyledonary (*arrows*) and true leaves (*arrows*).

has cooled down, commercial bleach (15 or 20%) is added to the water + Tween solution under a sterile laminar-air-flow cabinet.

2. It is of utmost importance to apply a vacuum force to the melon fruit when it is inside the glass dry-seal desiccator to remove all surface air and allow a more efficient penetration of the disinfectant solution, since pathogens are able to thrive when air bubbles are not removed from melon fruit surface.

3. When melon fruits below 12 days after-pollination (DAP) are used, special care must be taken, since seeds have not developed their hard testa; consequently, it is very easy to damage the immature seeds. If fruit above 15 DAP are used, it is easier to cut the melon fruit longitudinally and scoop out the seeds (see Fig. 1).

4. Sizes of melon seeds and physical state of placental tissues will depend on fruit age. Consequently, if melon fruits are above 15 DAP, seeds are easily scooped from a semiliquid placenta, and 300–500 seeds may be collected in less than 2 h. On the other hand, if melon fruits are below 15 DAP, a scalpel and smooth-tip microforceps should be used to carefully remove seeds from placenta tissues and the procedure to obtain 100 seeds will take more than 4 h.

References

1. Food and Agriculture Organization (FAO) (2007) FAOSTAT data. http://www.fao.org/corp/statistics/en/. Retrieved 3 March 2009
2. Zitter TA, Hopkins DL, Thomas CE (1998) Compendium of cucurbit diseases. APS, St. Paul, MN
3. Fehr WR, Fehr EL, Jessen HJ (1987) Principles of cultivar development, vol 1, Theory and technique. Collier Macmillan, New York
4. Guillaume R, Boissot N (2001) Resistance to ***Diaphania hyalinata*** (Lepidoptera: Crambidae) in ***Cucumis*** species. J Econ Entomol 94:719–723
5. De Dias CS, Pico B, Espinos A, Nuez F (2004) Resistance to melon vine decline derived from ***Cucumis melo*** ssp. ***agrestis***: genetic analysis of root structure and root response. Plant Breed 123:66–72
6. Martinez L, Agüero CB, Galmarini CR (1997) Obtention of haploid plants by ovaries and ovules culture in onion (***Allium cepa*** L.). Acta Hortic 433:447–454
7. Xynias IN, Zamani IA, Gouli-Vavdinoudi E, Roupakias DG (2001) Effect of cold pretreatment and incubation temperature on bread wheat (***Triticum aestivum*** L.) anther culture. Cereal Res Commun 29:331–338
8. Weymann K, Urban K, Ellis DM, Novitzky R, Dunder E, Jayne S, Pace G (1993) Isolation of transgenic progeny of maize by embryo rescue under selective conditions. In Vitro Cell Dev Biol 29P:33–37
9. Eijlander R, Stiekema WJ (1994) Biological containment of potato (***Solanum tuberosum***) – outcrossing to the related wild-species black nightshade (***Solanum nigrum***) and bittersweet (***Solanum dulcamara***). Sex Plant Reprod 7:29–40
10. Faure N, Serieys H, Kaan F, Berville A (2002) Partial hybridization in crosses between cultivated sunflower and the perennial ***Helianthus mollis***: effect of in vitro culture compared to natural crosses. Plant Cell Rep 20:943–947
11. Alemanno L, Guiderdoni E (1994) Increased doubled haploid plant-regeneration from rice (***Oryza sativa*** L.) anthers cultured on colchicine-supplemented media. Plant Cell Rep 13:432–436
12. Hoekstra S, Vanzijderveld MH, Louwerse JD, Heidekamp F, Vandermark F (1992) Anther and microspore culture of ***Hordeum vulgare*** L. cv Igri. Plant Sci 86:89–96
13. Lotfi M, Alan AR, Henning MJ, Jahn MM, Earle ED (2003) Production of haploid and doubled haploid plants of melon (***Cucumis melo*** L.) for use in breeding for multiple virus resistance. Plant Cell Rep 21:1121–1128
14. Marcellan ON, Camadro EL (2000) Preliminary results on embryo rescue for circumventing hybridization barriers in ***Asparagus***. Biocell 24:247–251
15. de Oliveira ACB, Maluf WR, Pinto JEBP, Azevedo SM (2003) Resistance to papaya ringspot virus in summer squash ***Cucurbita pepo*** L. introgressed from an interspecific ***C. pepo*** x ***C. moschata*** cross. Euphytica 132:211–215
16. Ezura H, Kikuta I, Oosawa K (1994) Production of aneuploid melon plants following in vitro culture of seeds from a triploid x diploid cross. Plant Cell Tissue Organ Cult 38:61–63
17. Beharav A, Cohen Y (1995) Effect of kinetin and GA_3 on ***in vitro*** ovule embryo culture of ***Cucumis melo*** L. Plant Growth Regul 16:267–269
18. Metwally EI, Moustafa SA, El-Sawy BI, Shalaby TA (1998) Haploid plantlets derived by anther culture of ***Cucurbita pepo***. Plant Cell Tissue Organ Cult 52:171–176
19. Sauton A (1988) Effect of season and genotype on gynogenetic haploid production in muskmelon, ***Cucumis melo*** L. Sci Hortic (Amsterdam) 35:71–75
20. Metwally EI, Moustafa SA, El-Sawy BI, Haroun SA, Shalaby TA (1998) Production of haploid plants from ***in vitro*** culture of unpollinated ovules of ***Cucurbita pepo***. Plant Cell Tissue Organ Cult 52:117–121
21. Gémes-Juhász A, Balogh P, Ferenczy A, Kristóf Z (2002) Effect of optimal stage of female gametophyte and heat treatment on ***in vitro*** gynogenesis induction in cucumber (***Cucumis sativus*** L.). Plant Cell Rep 21:105–111
22. Norton JD (1981) Embryo culture of ***Cucumis*** species. HortScience 16:69
23. Dryanovska OA, Ilieva IN (1983) ***In vitro*** anther and ovule cultures in muskmelon (***Cucumis melo*** L.). C R Acad Bulg Sci 36:1107–1110
24. Lazarte JE, Sasser CC (1982) Asexual embryogenesis and plantlet development in anther culture of ***Cucumis sativus*** L. HortScience 17:88
25. Przyborowski J, Niemirowicz-Szczytt K (1994) Main factors affecting cucumber (***Cucumis sativus*** L.) haploid embryo development and haploid plant characteristics. Plant Breed 112:70–75

26. Chen JF, Staub JE, Tashiro Y, Isshiki S, Miyazaki S (1997) Successful interspecific hybridization between ***Cucumis sativus*** L. and ***C. hystrix*** Chakr. Euphytica 96:413–419
27. Malepszy S, Sarreb DA, Mackiewicz HO, Narkiewicz M (1998) Triploids in cucumber: I. Factors influencing embryo rescue efficiency. Gartenbauwissenschaft 63:34–37
28. Chen JF, Staub J, Adelberg J, Lewis S, Kunkle B (2002) Synthesis and preliminary characterization of a new species (amphidiploid) in ***Cucumis***. Euphytica 123:315–322
29. Ondrej V, Nvrátilová B, Lebeda A (2002) ***In vitro*** cultivation of ***Cucumis sativus*** ovules after fertilization. Acta Hortic 588:339–343
30. Chen J, Staub J, Qian C, Jiang J, Luo X, Zhuang F (2003) Reproduction and cytogenetic characterization of interspecific hybrids derived from ***Cucumis hystrix*** Chakr. x ***Cucumis sativus*** L. Theor Appl Genet 106:688–695
31. Ashok-Kumar HG, Murthy HN, Paek KY (2003) Embryogenesis and plant regeneration from anther cultures of ***Cucumis sativus*** L. Sci Hortic (Amsterdam) 98:213–222
32. Gémes-Juhász A, Venczel G, Sági Z, Gajdos L, Zatykó L, Kristóf Z, Vági P (2006) Production of doubled haploid breeding lines in case of paprika, spice paprika, eggplant, cucumber, zucchini and onion. Acta Hortic 725:845–853
33. Dolcet-Sanjuan R, Claveria E, Garcia-Mas J (2006) Cucumber (***Cucumis sativus*** L.) dihaploid line production using ***in vitro*** rescue of ***in vivo*** induced parthenogenic embryos. Acta Hortic 725:837–844
34. Nuñez-Palenius HG, Klee HJ, Cantliffe D (2006) Embryo rescue culture of the 'Galia' muskmelon (***Cucumis melo*** L. var. ***reticulatus*** Ser.) male parental line. Plant Cell Tissue Organ Cult 85:345–352
35. Skálová D, Dziechciarková M, Lebeda A, Navrátilová B, Křístková E (2007) Interspecific hybridization of ***C. anguria*** x ***C. zeyheri***, ***C. sativus*** x ***C. melo***, and ***C. sativus*** x ***C. metuliferus*** with the use of embryo cultures. Acta Hortic 731:77–82
36. Skálová D, Dziechciarková M, Lebeda A, Křístková E, Navrátilová B (2008) Interspecific hybridization of ***Cucumis anguria*** and ***C. zeyheri via*** embryo-rescue. Biol Plant 52:775–778
37. Smiech M, Sztangret-Wisniewska J, Galecka T, Korzeniewska A, Marzec L, Kolakowska G, Piskurewicz U, Niemirowicz-Szczytt K (2008) Potential use of RAPD markers in characteristics of cucumber (***Cucumis sativus*** L.) haploids and double-haploids. Acta Soc Bot Pol 77:29–34
38. Skálová D, Navratilova B, Lebeda A (2008) Embryo rescue of cucumber (***Cucumis sativus***), muskmelon (***C. melo***) and some wild ***Cucumis*** species (***C. anguria***, ***C. zeyheri***, and ***C. metuliferus***). J Appl Bot Food Qual 82:83–89
39. Mackiewicz HO, Malepszy S, Sarreb DA, Narkiewicz M (1998) Triploids in cucumber: II. Characterization of embryo rescue plants. Gartenbauwissenschaft 63:125–129
40. Kumar HGA, Murthy HN, Paek KY (2003) Embryogenesis and plant regeneration from anther cultures of ***Cucumis sativus*** L. Sci Hortic (Amsterdam) 98:213–222
41. Ondrej V, Navratilova B, Lebeda A (2002) Influence of GA_3 on the zygotic embryogenesis of ***Cucumis*** species in vitro. Biologia 57:523–525
42. Ezura H, Amagai H, Oosawa K (1993) Efficient production of triploid melon plants by in-vitro culture of abnormal embryos excised from dried seeds of diploid x tetraploid crosses and their characteristics. Jpn J Breed 43:193–199
43. Cuny F, Grotte M, Devaulx RD, Rieu A (1993) Effects of gamma-irradiation of pollen on parthenogenetic haploid production in muskmelon (***Cucumis melo*** L.). Environ Exp Bot 33:301–312
44. Diao WP, Jia YY, Song H, Zhang XQ, Lou QF, Chen JF (2009) Efficient embryo induction in cucumber ovary culture and homozygous identification of the regenerants using SSR markers. Sci Hortic (Amsterdam) 119:246–251
45. Robinson RW, Decker-Walters DS (1999) Cucurbits. CAB International, Wallingford, NY
46. Kuzuya M, Hosoya K, Hayato-Masuya Y, Ezura H (2000) Histological observations of powdery mildew resistance in diploid and haploid melons. Acta Hortic 510: 71–75
47. Kuzuya M, Hosoya K, Tomita K, Ezura H (2002) Selection of powdery mildew resistance among haploid plants generated from hybrids of resistant and susceptible melon genotypes. Acta Hortic 588:331–338
48. Kuzuya M, Hosoya K, Yashiro K, Tomita K, Ezura H (2003) Powdery mildew (***Sphaerotheca fuliginia***) resistance in melon is selectable at the haploid level. J Exp Bot 54:1069–1074
49. Ficcadenti N, Sestili S, Annibali S, Campanelli G, Belisario A, Maccaroni M, Corazza L (2002) Resistance to ***Fusarium oxysporum*** f. sp. ***melonis*** race 1, 2 in muskmelon lines Nad-1 and Nad-2. Plant Dis 86:897–900

Chapter 10

Phaseolus Immature Embryo Rescue Technology

Pascal Geerts, André Toussaint, Guy Mergeai, and Jean-Pierre Baudoin

Abstract

Predominant among the production constraints of the common bean *Phaseolus vulgaris* are infestation of *Ascochyta* blight, Bean Golden Mosaic virus (BGMV), and Bean Fly. Interbreeding with *Phaseolus coccineus* L. and/or *Phaseolus polyanthus* Greenm has been shown to provide *P. vulgaris* with greater resistance to these diseases. For interspecific crosses to be successful, it is important to use *P. coccineus* and *P. polyanthus* as female parents; this prevents rapid reversal to the recurrent parent *P. vulgaris*. Although incompatibility barriers are post-zygotic, early hybrid embryo abortion limits the success of F1 crosses. While rescue techniques for globular and early heart-shaped embryos have improved in recent years, success in hybridization remains very low. In this study, we describe six steps that allowed us to rescue 2-day-old *P. vulgaris* embryos using a pod culture technique. Our methods consisted of (i) pod culture, (ii) extraction and culture of immature embryos, (iii) dehydration of embryos, (iv) germination of embryos, (v) rooting of developed shoots, and (vi) hardening of plantlets.

Key words: Dehydration conditions, Embryo rescue, Micropropagation, *Phaseolus*, Pod culture, Secondary gene pool, Zygotic embryos

1. Introduction

In many parts of Latin America, common bean (*Phaseolus vulgaris* L.) is considered to be one of the most valuable sources of plant protein. Common bean also contributes substantially to the sustainability of traditional cropping systems. However, dry production of *P. vulgaris* remains very low, averaging less than 600 kg/ha/year (1, 2). Several factors contribute to this low productivity. For one, high-energy investment is required for both the production of large amounts of protein in the beans and rhizobial nitrogen fixation in the root nodules. This competes directly with sequestration of photosynthate in the pods. Severe yield losses are also caused by high

Trevor A. Thorpe and Edward C. Yeung (eds.), *Plant Embryo Culture: Methods and Protocols*, Methods in Molecular Biology, vol. 710, DOI 10.1007/978-1-61737-988-8_10,

incidences of pests and diseases, and by unfavourable climatic and edaphic conditions prevalent in the region.

One of the greatest factors reducing the dry seed yield of common bean is the lack of improved cultivars adapted for traditional multiple cropping systems. On small-scale subsistence farms typical of the tropical and subtropical regions of Latin America, *P. vulgaris* and other *Phaseolus* cultivars are often cultivated in association with one or several other companion crops. These may include cereals, root and tuber plants, vegetables, fruit trees, and other commodities. Multiple cropping systems constitute risk insurance for small farmers and offer biological and socioeconomic benefits (3–5).

Generally, breeders have concentrated their efforts on the characterization and utilization of *P. vulgaris* landraces crossed to wild forms for genetic improvement programmes. However, there is insufficient genetic variation within the common bean primary gene pool to overcome several major production constraints (6). Better sources of resistance to these have been identified in alien germ plasm, namely, in the secondary gene pool. Common bean's secondary gene pool consists of the species *P. coccineus* and *P. polyanthus*. Both species are well adapted to highlands (above 2,000 m) and express useful agronomic traits (e.g. plant architecture, rusticity, disease resistance, and cold and acid soil tolerance) lacking in *P. vulgaris*. To improve beans for multiple cropping systems genetically, these two legumes could be bred as distinct crops or hybridized with common bean (7).

We undertook an in-depth evaluation of worldwide germ plasm collections of *P. coccineus* and *P. polyanthus* in representative stations and identified lines showing strong resistance to *Phoma exigua* var. *diversispora*, *Colletotrichum lindemuthianum*, and *Phaeisariopsis griseola* (6, 8). The most striking finding of our screening was the strong and stable field resistance of the whole *P. polyanthus* collection to *Ascochyta* leaf blight. However, in breeding the hybrids, we experienced great difficulty in rupturing the genetic linkage in *P. polyanthus* between high levels of disease resistance and unfavourable traits such as lateness in flowering, profuse branching, and low harvest index. In these crosses, the use of *P. vulgaris* as female parent increased the abundance of successful hybrids; however, the presence of *P. vulgaris* cytoplasm caused a quick reversal to the recurrent species at the expense of the donor species (9).

Given these circumstances, a critical case study was undertaken to examine methods of integrating *Ascochyta* blight resistance from *P. polyanthus* into *P. vulgaris*. We describe two alternative methods to develop new hybrids using *P. polyanthus* as the mother parent.

The first option is to increase the number of cross pollinations, which are affected by parental combination and environmental conditions. In crosses between *P. polyanthus* and *P. vulgaris*, the use of *P. polyanthus* cytoplasm avoids a quick reversal to the recurrent parent *P. vulgaris* explained above, but up to 60% of globular embryos failed to develop due to as yet undefined incompatibility barriers between the embryo and the mother plant. Among more than 3,000 crosses, only one hybrid could be obtained using embryo culture, and its seed production was relatively low (10). This hybrid is the only one obtained so far using *P. polyanthus* as the female parent in cross-pollination programmes.

The second alternative requires that we further our knowledge of the cause and developmental stage of the hybrid embryo's abortion. Such knowledge could allow for the development of customized in vitro embryo rescue culture techniques. Geerts et al. (11) determined that differences between early embryo abortions in reciprocal crosses are generally related to endosperm development. Rapid division of the primary endosperm nucleus (PEN) is observed in *P. vulgaris* × *P. polyanthus* (PvPp) seeds. This allows greater development of the embryo, which is initiated 2–3 days after pollination (DAP). However, PEN remains uninucleated in PvPp seeds during the first four DAP, limiting nutrient exchange between maternal tissues and the zygote for the second through fourth DAP (11). Moreover, Geerts et al. (11) showed that zygotes of PvPp seeds were still able to divide five DAP when PEN had divided at least once. This suggests that embryo abortion in PvPp seeds could be related to low nutrient exchange during early development. Therefore, manipulating the time at which cell divisions occur may be a means to overcome incompatibilities between hybrid embryo and endosperm.

On the basis of these results, rescue of PvPp embryos could be facilitated by the development of an in vitro technique of culturing early globular embryos. We will present the steps that led Geerts et al. (12) to rescue 2-day-old *P. vulgaris* embryos using a pod culture technique. Production of plantlets via immature embryos' rescue requires at least six stages: (i) pod culture, (ii) extraction and culture of immature embryos, (iii) dehydration of embryos, (iv) germination of embryos, (v) rooting of developed shoots, and (vi) hardening of plantlets.

2. Materials

1. Plant materials: genotypes of *P. polyanthus* were chosen based on their ability to cross with *P. vulgaris* and on their resistance to one of the following pathogens: *Ascochyta* blight, BGMV or Bean fly. *Phaseolus vulgaris* genotypes were selected for high productivity and for their origin (13).
2. Media: stock salt formulations, organic additives, and phytohormones for in vitro culture of pods and embryos are detailed in Table 1. All stock solutions are stored at 4°C for up to

Table 1
Composition of the media: $P_0$0 and $P_0$1 used for young *Phaseolus* pod culture and $P_0$1, G6, and G7 used for Phaseolus embryo culture (15)

Elements	$P_0$0[a]	$P_0$1[a]	G6[a]	G7[a]
Major elements (mg/L)				
$CaCl_2 \cdot 2H_2O$	300	600	150	150
$MgSO_4 \cdot 7H_2O$	220	435	250	250
KNO_3	1,050	2,100	2,500	2,500
$NaH_2PO_4 \cdot H_2O$	42	85	150	150
$(NH_4)_2SO_4$	–	–	134	134
KH_2PO_4	162	325	–	–
NH_4NO_3	500	1,000	400	–
Minor elements (mg/L)				
$MnSO_4 \cdot 7H_2O$	7.50	15	10	10
$ZnSO_4 \cdot 7H_2O$	2.50	5	2	2
$CuSO_4 \cdot 5H_2O$	0.050	0.1	0.025	0.025
$CoCl_2 \cdot 6H_2O$	0.050	0.1	0.025	0.025
KI	0.50	1	0.75	0.75
H_3BO_3	3.70	5	3	3
$NiCl_2 \cdot 6H_2O$	0.018	0.04	–	–
$Na_2MoO_4 \cdot 2H_2O$	0.2	0.4	0.25	0.25
$FeSO_4 \cdot 7H_2O$	12.50	25	27.85	27.85
Na_2EDTA	18.62	37.25	37.25	37.25
Vitamins (mg/L)				
Thiamin	0.25	1	1	1
Nicotinic acid	1.25	5	0.1	5
Pyridoxine	0.125	0.5	0.1	0.5
myo-Inositol	25	100	100	100
Sugars (g/L)				
Sucrose : 580 mosm	143	–	–	–
Sucrose : 450 mosm	–	102	–	–
Sucrose : 350 mosm	–	80	–	–
Sucrose	–	–	100	30
Amino acids (g/L)				
L-Glutamine	0.25	1	–	0.1
Casein hydrolysate	0.25	1	–	0.1
Hormones (μM)				
N6-benzylaminopurine (BAP)	–	1	–	1

(continued)

Table 1 (continued)

Elements	$P_0 0$[a]	$P_0 1$[a]	G6[a]	G7[a]
Gibberellic acid (GA3)	–	–	–	0.18
Abscisic acid (ABA)	0.095	–	–	–
Tryptophan	5.5	–	–	–
1-Naphthalene acetic acid (NAA)	0.1	0.1	–	–
Adenin	–	10.0	–	–
Gel and other (g/L)				
DIFCO agar	5	5	8	5
Activated charcoal	–	–	5	–

[a] $P_0 0$, $P_0 1$, G6, and G7 represent, respectively, two new modified Phillips et al. (32) media used for pod and embryo culture, the dehydration medium as described by Hu and Zanettini (19) modified, and the Mergeai et al. (15) rooting medium modified

4–6 weeks. All reagents were obtained from Duchefa except for Tween-20, which was purchased from Sigma.

3. Solutions for sterilization: 70% ethanol; 12% calcium hypochlorite solution containing Tween-20 (two drops per 100 mL solution) as a surfactant.
4. Plant preservative mixture (PPM, Plant Cell Technology, Inc, Washington, DC 20036 USA).
5. Other laboratory equipment and supplies: sterile glass beads (3 mm), Petri dishes (15 × 4 cm), Reynolon film, 20 × 160 mm, borosilicate glass culture tubes, forceps, scalpels, dissecting needles, peristaltic pump, and stereomicroscope.

3. Methods

The potential of various in vitro culture techniques to rescue immature embryos (14) was investigated. Our objectives were twofold: (i) to rescue 2- to 4-day-old embryos and (ii) to develop in vitro culture techniques adapted to *P. vulgaris*/*P. polyanthus* hybrids. There are two common, major challenges that arise during the culture of immature embryos: (i) selecting a medium that meets the complex nutritional requirements of very small embryos and (ii) extracting very small embryos without damaging their suspensors. To reduce suspensor damage, we applied in ovulo pod culture, which protects the embryos, for 5–10 days before extracting them. Embryos are thus extracted at a stage when the role of the suspensor is much less important, i.e. the late heart-shaped or cotyledonary stage. Such evolution is particularly well observed in pods cultivated 4 DAP. Some embryos at the globular developmental stage proceed to develop to the cotyledonary stage. For these 2-day-old embryos, further rescue techniques are used. For our study, we also examined the impact of manipulating osmolality of the media.

3.1. Plant Maintenance

Several bean genotypes (see Note 1) were grown in a controlled growth chamber set to 24 and 20°C day and night temperatures, respectively. The chamber was set to 75% relative humidity, 580 μmol/m^2/s light intensity (measurements at 60 cm from 400 W Grolux lamps), and 11.5 h day length.

3.2. Media Preparation

The mineral salt composition of germination media is one of the most important factors that influence the success of in vitro immature embryo rescue. Our research (12, 15–18) has allowed us to compile a series of media for the in vitro rescue of immature embryos. We found that applying high and variable osmotic conditions similar to those observed in vivo produced the best results in terms of ovule and embryo development during pod culture and before embryo extraction (see Table 2).

Most of the major salts, the amino acids, myo-inositol, and sucrose are added directly to the medium at concentrations varying between 0.1 and 143 g/L. Minor salts, vitamins, and growth regulators are taken from stock solutions concentrated 50×. All of these ingredients are dissolved in 600-mL high-purity demineralised water in a 1-L beaker on a magnetic stirrer. The pH of the medium is adjusted to 5.7 by drop-wise addition of 0.1 N KOH. For solidification, plant agar is pre-dissolved in 400 mL of

Table 2
In vitro culture steps of pod and embryo culture

Stages	Media
Pod culture (1 week)	$P_0$0: Philips et al. (32) modified at 580 mosm $P_0$1: Philips et al. (32) modified at 450 and 350 mosm
Embryo maturation (2 weeks)	$P_0$1: Philips et al. (32) modified at 350 mosm
Embryo dehydration (2 weeks)	G6: Hu and Zanettini (19) modified
Germination and rooting (1 week)	G7: Mergeai et al. (15) modified + IAA (1 μM)
Growth (±2 weeks)	G7: Mergeai et al. (15) modified

high-purity demineralised water and warmed until clarification, and then added to the medium. The latter is then autoclaved (120°C for 20 min) to obtain 1 L of solid medium.

After homogenisation and cooling, but before solidification, 30 mL media is aliquoted into sterile Petri dishes (55 mm × 1.5 cm) in a sterile laminar flow cabinet. Petri dishes are sealed with a Reynolon film when gel is solidified (about 4 h later). Prepared Petri dishes are stored at room temperature in the dark (see Note 2).

3.3. Pod Culture

1. Harvest young pods 2–5 DAP during the early morning when plants are not subject to water or temperature stress. Carry the pods from the growth chamber to the laboratory in a sealed plastic box maintained at 21°C to avoid desiccation.
2. Dip the pods in 70% ethanol and mix gently for 1 min. Drop the pods in 12% calcium hypochlorite solution containing Tween-20 (two drops per 100 mL solution) as a surfactant, and mix gently for 2–3 min. Wash three times with sterile distilled water (see Note 3).
3. Liquid medium is used to provide an environment in which constant changes of osmolality can be obtained during the first week of culture (Fig. 1). The pods are supported on sterile glass beads (3 mm) in Petri dishes (15 × 4 cm), each containing 100 mL liquid $P_0$0 medium (osmolality adapted to pre-globular embryos: 580 mosm; see Note 4). Petri dishes are connected via a peristaltic pump to 1 L bottles containing $P_0$1 medium (osmolality adapted to cotyledonary

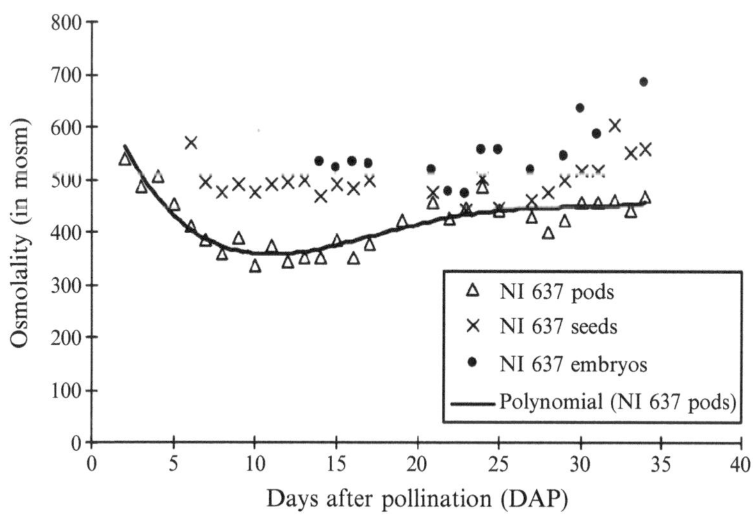

Fig. 1. Evolution of osmolality in pods, seeds, and embryos of *P. vulgaris* (NI637) by Geerts (26). There is a gradient between liquid endosperm, embryo, seed, and pod, as reported in the literature (26). Modifications of osmolality values occur at two different periods: immediately after pollination, up to 11 DAP, when embryos reached cotyledonary stage, and 22 DAP, corresponding to dehydration of seeds (reproduced by permission of P. Geerts).

Fig. 2. Experimental design used for *in vitro* culture of young *Phaseolus* pods and permitting a constant evolution of the culture medium. On the *right side*, culture of young pods on solid medium (see Note 7) (Picture: A. Toussaint).

embryos: 350 mosm) (Fig. 2). A Millipore filter attached to the pump prevents contamination of media flowing from the bottle to the plates (Fig. 2; see Note 5).

4. Place the Petri dishes under light (60 μmol/m^2/s light intensity; 11.5 h day length) at 26°C and 100% relative humidity.
5. During the first 5 days of culture, 100 mL of medium per day is dripped from the bottle decanter into each Petri dish, enabling a constant evolution of the culture medium osmolality. A lateral aperture permits the discharge of excess liquid, maintaining a constant volume within the Petri dish (see Notes 6 and 7). After 5 days, the osmolality of the liquid medium should reach 350 mosm.

3.4. Extraction and Culture of Immature Embryos

1. Dissect the pods and extract fertilized ovules under a stereomicroscope after 5–7 days of culture. The microscope should be fitted with a 12× magnification ocular micrometer.
2. Transfer fertilized ovules to a sterile water solution with 120 g/L sucrose and 1.75 g/L agar. In this environment, extract the embryos from ovules, reaching at least 2 mm using two dissecting needles under a binocular with 40× magnification (see Note 8).
3. After extraction, aspirate the embryos from the sterile dissection solution using a Pasteur pipette. Transfer each embryo with two droplets of sterile solution to the Petri dishes containing maturation medium ($P_0 1$), where they will continue the maturation processes. The above three operations are carried out in a sterile laminar flow cabinet.
4. Divide the Petri dishes into eight sections by a pencil mark line on the bottom of the Petri dish after medium preparation, and mark each with a reference indicating the culture medium

characteristics and the number of each of the embryos isolated at the same early heart-shaped stage. Seal the dishes with Reynolon.

5. Incubate the immature embryos in darkness at 26°C for 2–3 weeks (16) while maturation proceeds (see Note 9).

3.5. Dehydration of Embryos

Transfer the mature embryos to dehydration medium G6 under the laminar flow hood (19) (see Note 10). Maintain the Petri dishes in darkness at 26°C for 1–2 weeks.

3.6. Germination of the Embryos

1. Transfer the embryos to germination medium G7 with salts, as per the procedure described by Gamborg et al. (20), and 1 μM indole acetic acid (IAA). Incubate for 1 week, or until germination takes place.
2. Place the Petri dishes under light (60 μmol/m^2/s light intensity; 11.5 h day length) at 26°C (see Note 11).

3.7. Rooting of Developed Shoots

1. Transfer the developed shoots to 20×160 mm borosilicate glass culture tubes containing 20 mL sterile rooting medium G7 (without IAA). At this stage, the embryos are 3–5 mm in length.
2. Cover the culture tubes with plastic caps (not sealed with Reynolon, as this would suffocate the plantlets).
3. Place the culture tubes in an incubator (Lumınincube II, Analis) under controlled conditions: 60 μmol/m^2/s light intensity, 11.5 h day length, and 26°C. Roots should appear after 2–3 weeks of culture (Fig. 3) (see Note 12).

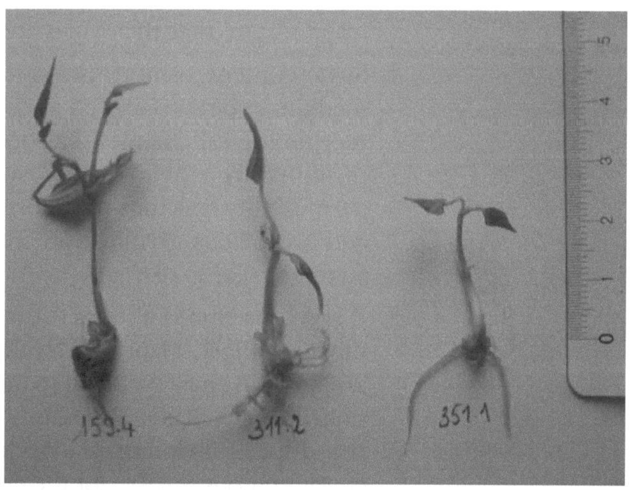

Fig. 3. After a dehydration period, germinated embryos are transferred in tubes for rooting. After rooting and a period of in vitro growth, plantlets are transplanted in Jiffy pot for acclimatisation (Picture: A. Toussaint).

3.8. Hardening of Plantlets

1. Plantlets sufficiently developed, as indicated by a good rooting ramification and at least one trifoliate leaf, are transferred to Jiffy pots for acclimatisation. These are placed in closed glass boxes with 100% relative humidity (see Note 13).
2. Carry out progressive acclimatisation of plantlets by increasing the atmospheric water vapour deficit.
3. Water plantlets once a week with an aqueous solution containing one half the concentration of Murashige and Skoog (21) minerals.
4. Transfer plantlets after 15 days of growth to a controlled chamber with 24°C day temperature, 20°C night temperature, and 12.5 h day length (60 μE/m^2/s; see Note 14).

4. Notes

1. Seven cultivated (NI 637, PVA 773, G 6620, G 9545, G 22090, G 17723, and Altense) and one wild (G 21245) *P. vulgaris* genotypes, as well as eight *P. polyanthus* genotypes (NI 429, ×1059, NI 1015, NI 1029, G 35526, G 35345, G 35547, and G 35348) were used for our experiments (1). *Phaseolus* embryos of different origins reacted similarly to modifications of the in vitro conditions (16).
2. For preparation of stock solutions, various reagents as well as laboratory equipments and supplies are required (12, 16–18).
3. After the treatment with calcium hypochlorite, ensure that the pods remain green. If the pods appear discolored, this may indicate that the ovules are burnt.
4. Sugar content, osmotic pressure, and ABA level of the growth medium appear to exert a strong effect on physiological and morphological changes that occur during early development of immature embryos and ovules. In particular, membrane permeability, calcium transport, endogenous ABA synthesis, and precocious germination are sensitive to changes in these factors (16, 22–24).
5. A low concentration (from 0.1 to 1 mL/L) of plant preservative mixture (PPM, Plant Cell Technology, Inc, Washington, DC 20036 USA) may be added to media to control pod contamination. PPM contains a mixture of two isothiazolones and is a broad-spectrum industrial biocide reported as non-phytotoxic at concentrations suitable for the prophylactic control of microbial contaminants in plant tissue culture. Using this protocol, contamination was almost completely eliminated.
6. Pod culture method was adapted from Lazaridou et al. (25) and Mergeai et al. (15).

7. Osmotic pressure of culture media should change depending on pod age (26, 27). When using solid media, this transition is carried out by applying a series of solid media: $P_0$0 medium at 580 mosm during the first day, $P_0$1 medium at 450 for 2 days more, and lastly, basal medium 350 mosm for 4 days (Fig. 2).
8. During our experiments, no germination occurred when the suspensor was damaged during transfer to culture medium. Yeung (28) and Brady and Comb (29) have demonstrated the active role of the suspensor in *P. coccineus* and *P. vulgaris* embryonic development.
9. In general, pod growth was greater in liquid medium than in solid medium. However, pod growth was not correlated with the development of ovules or embryos; the number of pods containing at least one developed ovule was not different between solid and liquid media. Moreover, the number of ovules longer than 2 mm per pod was greater in solid medium (mean of 3.7 developed ovules per pod) compared to that in liquid medium (mean of 2 developed ovules per pod). The number of extracted embryos was also higher in solid medium (3.8 per pod) than that in liquid medium (1.9 per pod). The developmental stage of extracted embryos was quite variable, ranging from globular to cotyledonary. Mean embryo length was higher when extracted from pods of *P. vulgaris* cultured in liquid medium than in solid medium (17). *P. polyanthus* yielded a higher percentage of germinated embryos and plantlets in acclimatisation when pods were cultivated in solid medium (30).
10. The incorporation of a dehydration medium, which reflects the natural process of seed dehydration, considerably increases the number of plantlets regenerated. The importance of this medium is well detailed in Gamborg and Philips (31).
11. Germination of extracted embryos is higher in *P. vulgaris* (68.7%) than in *P. polyanthus* (28.4% in NI 1015 and 20.7% in G 35348). Pod age at the time of in vitro culture does not appear to influence the germination capacity of the embryos (30).
12. One factor responsible of the loss of embryos during in vitro culture is the development of a callus on the root apex, accompanied by vitrification. The callus appears between the 25th and 30th day of culture at the end of the dehydration phase in more than 85% of developed embryos.
13. A better acclimatisation rate is observed when damaged embryos are cultivated in germination medium containing GA_3. A 71.29% germination rate was observed with GA_3, compared to 47.36% in control medium.
14. The ratio between the number of plantlets undergoing acclimatisation and the initial number of germinated embryos is higher in *P. polyanthus* (76.2% for NI 1015 and 73.7% for G 35348) than in *P. vulgaris* (51.1%). Six weeks after the

onset of acclimatisation, the percent of plantlets growing out of the number of extracted embryos is higher in *P. vulgaris* (>30%) than that in *P. polyanthus* (≤5%) (30).

References

1. Woolley J, Davis JHC (1991) The agronomy of intercropping with beans. In: van Schoonhoven A, Voysest O (eds) Common beans: research for crop improvement. CAB International/CIAT, Wallingford, pp 707–735
2. Lobo M (1994) *Phaseolus* production in Colombia. Grain Legumes AEP 5:26–27
3. Francis CA (1986) Multiple cropping systems. Macmillan, New York, p 383
4. Baudoin JP, Maréchal R (1991) Wide crosses and taxonomy of pulse crops, with special emphasis on *Phaseolus* and *Vigna*. In: Ng NQ, Perrino P, Attere A, Zedan H (eds) Crop genetic resources of Africa. IITA/IBPGR/UNEP/CNR, Nigeria, pp 287–302
5. Baudoin JP, Camarena F (1994) Food legumes adapted for multiple cropping systems. Grain Legumes AEP 5:23–25
6. Baudoin JP, Camarena F, Lobo M (1995) Amélioration de quatre espèces de légumineuses alimentaires tropicales *Phaseolus vulgaris*, *P. coccineus*, *P. polyanthus* et *P. lunatus*. Sélection intra et interspécifique. In: Dubois J, Demarly Y (eds) Actes des IVe Journées Scientifiques du Réseau Biotechnologies Végétales de l'AUPELF-UREF. John Libbey Eurotext, Paris, pp 31–49
7. Baudoin JP, Camarena F, Schmit V (1992) Contribution à une meilleure connaissance de la position phylétique de la légumineuse alimentaire *Phaseolus polyanthus* Greenm. Bull Rech Agron Gembloux 27:167–198
8. Schmit V, Baudoin JP (1992) Study of several intraspecific crosses in the food legume *Phaseolus polyanthus* Greenm. Bull Rech Agron Gembloux 27:473–494
9. Baudoin JP (2001) Contribution des ressources phytogénétiques à la sélection variétale de légumineuses alimentaires tropicales. Biotechnol Agron Soc Environ 5:221–230
10. Camarena F, Baudoin JP (1987) Obtention des premiers hybrides interspécifiques entre *Phaseolus vulgaris* et *Phaseolus polyanthus* avec le cytoplasme de cette dernière forme. Bull Rech Agron Gembloux 22:43–55
11. Geerts P, Toussaint A, Mergeai G, Baudoin JP (2002) Study of the early abortion in reciprocal crosses between *Phaseolus vulgaris* L. and *Phaseolus polyanthus* Greenm. Biotechnol Agron Soc Environ 6:109–119
12. Geerts P, Khaled S, Mergeai G, Baudoin JP (2000) Development of an in vitro pod culture technique for young pods of *Phaseolus vulgaris* L. In Vitro Cell Dev Biol- Plant 36:481–487
13. Lecomte B (1997) Etude du développement embryonnaire in vivo et in vitro dans le genre *Phaseolus* L. Thèse de doctorat. Faculté universitaire des Sciences agronomiques, Gembloux (Belgium), p 186
14. Sharma DR, Kaur R, Kumar K (1996) Embryo rescue in plants – a review. Euphytica 89:325–337
15. Mergeai G, Schmit V, Lecomte B, Baudoin JP (1997) Mise au point d'une technique de culture *in vitro* d'embryons immatures de *Phaseolus*. Biotechnol Agron Soc Environ 1:49–58
16. Geerts P, Mergeai G, Baudoin JP (1999) Rescue of early heart-shaped embryos and plant regeneration of *Phaseolus polyanthus* Greenm. and *P. vulgaris* L. Biotechnol Agron Soc Environ 3:141–148
17. Geerts P, Toussaint A, Mergeai G, Baudoin JP (2001) Culture of very young *Phaseolus vulgaris* L. pods and plantlet regeneration. Acta Hort 560:411–416
18. Toussaint A, Geerts P, Clément F, Mergeai G, Baudoin JP (2004) Early abortion in reciprocal crosses between *Phaseolus vulgaris* and *Phaseolus polyanthus*, and in vitro culture of immature embryos from these species. Belg J Bot 137:47–54
19. Hu CY, Zanettini MHB (1995) Embryo culture and embryo rescue for wide cross hybrids. In: Gamborg OL, Phillips GC (eds) Plant cell tissue organ culture: fundamental methods. Springer, Berlin, pp 129–141
20. Gamborg OL, Miller RA, Ojima K (1968) Nutrient requirements of suspension cultures of soybean root cells. Exp Cell Res 50:151–158
21. Murashige T, Skoog F (1962) A revised medium for rapid growth and bio-assays with tobacco tissue culture. Physiol Plant 15:473–497
22. Salisbury FB, Ross CW (1992) Plant physiology, 4th edn. Wadsworth, Belmont, CA, pp 357–403
23. Liu S, Kriz A (1996) Tissue-specific and ABA-regulated maize *Glb1* gene expression in transgenic tobacco. Plant Cell Rep 16:158–162
24. Pacheco-Moisès F, Valencia-Turcotte L, Rodriguez-Sotres R (1997) Regulation of

acyltransferase activity in immature maize embryos by abscisic acid and the osmotic environment. Plant Physiol 114:1095–1101

25. Lazaridou TB, Roupakias DG, Economou AS (1993) Embryo rescue in *Vicia faba* and *Vicia narbonensis*. Plant Cell Tiss Organ Cult 33:297–301
26. Geerts P (2001) Study of embryo development in *Phaseolus* in order to obtain interspecific hybrids. Thèse de doctorat, Faculté universitaire des Sciences agronomiques de Gembloux, Belgium, pp 183
27. Yeung EC, Brown DCW (1982) The osmotic environment of developing embryos of *Phaseolus vulgaris*. Z Pflanzenphysiol 106:149–156
28. Yeung EC (1980) Embryogeny of *Phaseolus*: the role of the suspensor. Z Pflanzenphysiol 96:17–28
29. Braddy T, Comb SH (1980) The suspensor is a major route of nutrients into proembryo, globular and heart stage *Phaseolus vulgaris* embryos. In: Cresti M, Gori P, Pacini E (eds) Sexual reproduction in higher plants. Springer, Heidelberg, pp 419–424
30. Toussaint A, Clément F, Mergeai G, Baudoin JP (2002) In vitro culture of immature embryos of *Phaseolus polyanthus* Greenm. and *P. vulgaris* L. *Annu.* Rep Bean Improv Coop 45:244–245
31. Gamborg OL, Phillips GC (1995) Plant cell, tissue and organ culture – fundamental methods. Springer, Berlin, p 358
32. Phillips GC, Collins GB, Taylor NL (1982) Interspecific hybridization of red clover (*Trifolium pratense* L.) with *T. sarosiense* Hazsl. using in vitro embryo rescue. Theor Appl Genet 62:17–24

Chapter 11

Wide Crossing in Lentil through Embryo Rescue

Richard Fratini and Maria L. Ruiz

Abstract

Lentil seeds have provided an appreciated source of protein, carbohydrates and fibre to the diet of humans since the dawn of agriculture. Low amounts of variation have been detected in the cultivated lentil germplasm collections. Interspecific crosses allow for the introgression of important alleles of agricultural interest from wild species, such as the resistance or tolerance to abiotic and biotic stresses. Interspecific crosses within the genus *Lens* generally abort and embryo rescue techniques are necessary to recover hybrids. The in vitro culture procedure to rescue interspecific hybrids of *Lens* consists of at least four different stages: (1) in ovulo embryo culture (2), embryo culture, (3) plantlet development and finally, (4) the gradual habituation to ex vitro conditions of the recovered interspecific hybrid plantlets. In this chapter, the approach to rescue interspecific hybrids in the genus *Lens* is outlined.

Key words: Interspecific hybridization, Legume, *Lens*, Ovule-embryo culture, Pulse crop

1. Introduction

Pollination followed by fertilization normally leads to the production of an embryo, which in the intact plant is linked with normal seed development. Crossability is defined (1) as the potential for intercrossing individuals belonging to the same or different taxa and for producing embryos or seeds that can give rise to an F_1 plant. Crossability is either limited by incompatibility or by incongruity; the sexual barriers belonging to the second aspect have been divided into pre- and post-fertilization barriers (2). Part of post-fertilization barriers may be overcome by using in vitro embryo rescue methods (3), although depending on plant species, the process can entail the culture of ovaries immediately after pollination and/or in ovulo embryo culture and/or embryo culture. Ladizinsky (1) explained that success in lentil crosses depends on the interaction between the parental genomes in the hybrid zygote, embryo or

Trevor A. Thorpe and Edward C. Yeung (eds.), *Plant Embryo Culture: Methods and Protocols*,
Methods in Molecular Biology, vol. 710, DOI 10.1007/978-1-61737-988-8_11,

endosperm and between the hybrid tissue and the surrounding maternal tissue. The crossability between lentil and its wild relatives is hampered by pre- and post-fertilization barriers (4–8).

Lentil (*Lens culinaris* Medik.) belongs to the group of crops first domesticated in the Near East Fertile Crescent, concretely in the foothills of the mountains of southern Turkey and northern Syria (9, 10). It has been cultivated for at least 10,000 years in the most difficult agricultural environments, growing in poor soils and withstanding a high degree of drought and cold resistance. Throughout the ages lentil seeds have contributed an appreciated source of protein, carbohydrates and fibre to human food consumption habits. As a pulse crop belonging to the cool season food legumes, lentil is also valued in crop rotations with cereals to replenish soil nitrogen levels. In addition, the whole plant may also be used as animal fodder. Lentil flowers are complete with a typical structure of the sub-family *Papilionaceae* of the *Leguminosae* family, the cultivated species with cleistogamous flowers (11) is self-pollinated with a degree of outcrossing which ranges between 0.06 and 5.12% depending on cultivar, location and year (12, 13).

An essential aspect for breeding genetically improved lentil cultivars is to possess abundant amounts of genetic variability. Unfortunately, low levels of variation have been identified in the cultivated lentil germplasm collections (14–19). Artificial cross-pollination in a highly self-pollinated crop species, such as lentil, is important to generate genetic variability. With regard to wide crosses, interspecific hybridization allows for the introgression of important alleles of agricultural interest from wild species to cultivars, as for instance, the resistance or tolerance to abiotic and biotic stresses (20–22).

The genus *Lens* comprises of the cultivated lentil (*L. culinaris* subsp. *culinaris*), which includes small seeded (microsperma) or large seeded (macrosperma) varieties, its wild ancestor *L. culinaris* subsp. *orientalis*, as well as the species *L. odemensis*, *L. nigricans*, *L. ervoides*, *L. tormentosus* and *L. lamottei* (10, 22). Intraspecific crosses between cultivated lentils produce viable descendants (23–27). With regard to intersubspecific hybrids of lentil, it has been reported that the domesticated lentil is readily crossable with subspecies *orientalis* (10, 21, 22, 27, 28), although the fertility of the hybrids depends on the chromosome arrangement of the wild parent (28, 29). Interspecific crosses within the genus *Lens* abort (4, 7) and embryo rescue techniques are necessary to recover hybrids (30–32). Nonetheless, by applying GA_3 to developing pods viable interspecific *Lens* hybrids have been obtained (33). Hybridization by hand-pollination followed by embryo rescue is in some cases the only practical method to recover interspecific hybrids of *Lens*. However, a major bottleneck of the method is to obtain enough embryos to rescue in vitro. Emasculation and artificial crossing are difficult due to different characteristics of the reproductive system of lentil (small and fragile cleistogamous flowers and low seed per pod set), manual crossing is tedious and often unsuccessful or entails a low success rate (27, 32).

So far, only two embryo rescue protocols have been published in lentil to recover hybrids of wide crosses (30, 32). In the first case (30), ovule-embryos were collected 2 weeks after pollination and cultured on MS medium (34) supplemented with 10% sucrose and 0.2 mg/L IAA + 0.5 mg/L ZEA + 0.5 mg/L GA_3 and 0.9% agar. After 1 week in culture, embryos were excised from the ovular integuments and placed on MS medium supplemented with 3% sucrose and 0.2 mg/L IAA + 0.2 mg/L ZEA and 0.9% agar, the method has allowed for the recovery of interspecific hybrids between the cultivated lentil and *L. ervoides* and *L. nigricans*. Afterwards, using the same embryo rescue procedure, interspecific hybrids between the cultivated lentil and *L. ervoides* were again obtained (31). With regard to the second embryo rescue technique (32), interspecific hybrids between the cultivated lentil and *L. odemensis*, *L. ervoides* and *L. nigricans* have been recovered in our laboratory. The main differences between both methods are the number of media used and the amount of carbohydrate added to the media. Whereas in the first study (30), media were supplemented with 10 and 3% of sucrose, the medium used in our procedure contained only 1% sucrose (32). With regard to the phytohormones, both procedures have approximately equivalent concentrations of auxins and cytokinins, even though ZEA (30) and KN (32) were respectively used.

In order to illustrate the approaches used to obtain hybrids of wide crosses in the genus *Lens*, the details of our work carried out by performing hybridizations and embryo rescue between Spanish landraces of lentil (*L. culinaris* M.) and *L. odemensis*, *L. ervoides* and *L. nigricans* will be shown. Taking into account the rescue of interspecific *Lens* hybrids, the in vitro culture procedure consists of at least four different stages: (1) in ovulo embryo culture, (2) embryo culture, (3) plantlet development and finally, (4) the gradual habituation to ex vitro conditions of the recovered interspecific hybrid plantlets.

2. Materials

1. Lentil rescue medium (LRM). LRM medium is based on the Murashige and Skoog (MS) (34) basal medium (Sigma) supplemented with 1 μM Indole-3-acetic acid sodium salt (IAA) (Sigma) + 0.8 μM 6-Furfurylaminopurine (KN) (Sigma) and 1% (w/v) sucrose (10 g/L) plus 0.8% (w/v) American bacteriological agar (Pronadisa) (8 g/L). Adjust the pH of the culture medium to 5.7 before adding the agar, autoclave for 20 min at 121 °C. After autoclaving add the phytohormones

previously sterilized using a 0.22 μm nitrocellulose filter (Millipore).

2. Phytohormone stock solutions can either be prepared at a 1,000× (1 μM/mL) or 10,000× (10 μM/mL) strength relative to the final concentration and are stored cold (5 °C). Culture medium is placed in Petri dishes (20 mL) and in test tubes (20 mL) (see Note 1).
3. 70% (v/v) ethanol and 70% commercial bleach (5% NaOCl) for surface sterilization.
4. Pots of approximately 1 and 10 L, enriched peat (COMPO SANA Universal; www.compo.es) and vermiculite are need at various stages.
5. General laboratory equipment and supplies: forceps, microscope slides, pots and plastic bags.

3. Methods

3.1. Obtaining Material, Explant Preparation and Sterilization

1. In order to harvest pollen, remove anthers from flowers with a relation of petals to sepals of ¾, 1 to 1 and open flowers; place anthers on a microscopic slide and carefully squash with the forceps to release pollen grains (see Note 2).
2. Carry out emasculation and hybridization when the petals of flowers have reached three-quarters of the length of the sepals. Hold the flower bud between the thumb and the forefinger with the standard facing the operator, take particular care not to bend or twist the peduncle; use sharp-pointed forceps to remove the sepals and to twist back the standard, then incise the upper end of the keel to open the flower and to remove all of the ten anthers, pay attention not to touch the stigma with anthers or forceps so as to avoid selfing or damage of the stigma, pollinate manually immediately after emasculation; push out the stigma of the flower previously emasculated and rub three consecutive times on the microscope slide carrying the pollen; after pollination, return the standard, the keel and the wings to their original position around the pistil and tag flower buds (see Note 3).
3. Harvest pods of wide crosses 18 days after pollination (DAP) (see Note 4).
4. Surface sterilize pods by immersion for about 5 s in 70% ethanol followed by immersion in the bleach solution for 5–10 min.
5. After surface sterilization, rinse pods three consecutive times with sterile water in a laminar flow cabinet.

6. In order to obtain ovule-embryos, cut and split open lentil pods of wide crosses with a scalpel at the seam opposite to that where the ovules via the funiculus are attached to the pericarp (see Note 5).

3.2. Ovule-Embryo Culture

1. Place only one ovule-embryo in a Petri dish containing the LRM medium so as to avoid loss of other explants due to contamination, only the micropylar end of the ovule should be in direct contact with the medium (see Note 6).
2. Culture ovule-embryos for 2 weeks in a growth chamber at 25 ± 1 °C exposed under a 12 h photoperiod to a photosynthetic photon flux density (PPFD) of 30 μmol/m^2/s (see Note 7).

3.3. Embryo Excision and Culture

1. Cut open ovules at the end opposite to that of the micropyle and carefully excise embryos from the ovular integuments (see Note 5).
2. Place only one embryo per Petri dish containing LRM medium to avoid loss of other explants due to contamination, culture embryos in an upright position with the medium reaching up to the hypocotyl (see Note 8).
3. Culture embryos for about 2 weeks (see Note 9) in a growth chamber at 25 ± 1 °C and exposed under a photoperiod of 12 h (see Note 10) to a light intensity of 60 μmol/m^2/s.
4. Transfer the rescued embryos to test tubes holding LRM medium, place in upright position with radicle entirely submerged and medium reaching up to the hypocotyl (see Note 8).
5. Maintain for approximately a month the test tubes containing the interspecific hybrid embryos in a growth chamber at 25 ± 1 °C exposed to a 12 h photoperiod of 75 μmol/m^2/s, allow embryos to develop into plantlets (see Notes 10 and 11).

3.4. Plantlet Hardening

1. Transfer the hybrid lentil plantlets recovered from test tubes to a greenhouse programmed to maintain 25 ± 5 °C and place in 1 L pots containing a 1:1 mixture of enriched peat and vermiculite and covered with plastic bags. Leave covered for at least a week (see Note 12).
2. Perforate progressively the plastic bags for about another week to slowly acclimatize plantlets to ex vitro conditions until pots are finally left uncovered.
3. Transplant the recuperated interspecific hybrids to large 10 L pots containing a 1:1 mixture of clay soil and enriched peat (see Note 13).

4. Notes

1. For stock solution preparation, usage of implements, equipment, and so on, as well as requirements for a tissue culture facility (see (35–37)).
2. Use preferentially vigorous plants as pollen donors; fertilization may be aided by collecting pollen in different development stages ranging from flowers in the ¾ stage of petals to sepals up to the blossom stage.
3. Only vigorous plants should be used for emasculation and hybridization, careful attention must be made so as not to damage the tiny and delicate lentil flowers. Hybridization success in lentil is generally favoured by high relative humidity (RH) conditions, shading and temperatures ranging between 20 and 25°C (24, 25, 27). After wide crosses are made, plants should be pruned of open flowers and selfed pods (27). For more information regarding hybridization technique in lentil (see (11, 23–27, 38)).
4. Abortion of wide crosses in lentil usually takes place around 18 DAP (32), depending on environmental conditions and cross combination. Nevertheless, crosses with certain *L. odemensis* accessions do not need rescue (29), while crosses with *L. nigricans*, *L. ervoides*, *L. tormentosus* and *L. lamottei* might abort earlier and pods should be harvested around 14 DAP (4, 7, 30, 31).
5. Proceed immediately to the next step to avoid explant dehydration.
6. Placing ovules flat on medium might result in swelling of the embryo's radicle and/or hypocotyl or overall callus formation depending on cross combination and on time elapsed until the moment of rescue (with younger tissues more sensitive to medium components) (32).
7. After a 2-week culture period, ovular integuments become pale green and holding the ovule against light allows us to clearly spot the embryo which has increased in size. These ovules are ready to proceed to the next step. In contrast, immature embryos which have not increased in size may continue culture for another 1–2 weeks.
8. Maintaining the natural geotropism of embryos aids rescue, horizontal embryo culture on medium might result in radicle and/or hypocotyl swelling or abnormal development (32).
9. Culture period depends on rate of embryo development, the apical meristem should not be allowed to touch the lid of the Petri dish in order to avoid the vitrification of tissues, proceed to the following step of hybrid plantlet culture in a test tube according to epicotyl elongation.

10. As a cool season pulse legume, lentil is a long-day plant characterized by a flowering induction at circa 14 h of daylight; keeping the in vitro cultures exposed to a 12-h photoperiod maintains the juvenile phase and allows for a prolonged vegetative development of the interspecific hybrids.
11. Culture period depends on rate of plantlet development, plantlets should be allowed to develop a functional radicular system with secondary roots and the epicotyl should possess at least four nodes before proceeding to the next stage of gradual habituation to ex vitro conditions (see Fig. 1 of ref. (32)).
12. Ex vitro acclimatization conditions are extremely important; survival and vigour of interspecific hybrid lentil plantlets are maximized by bagging each pot individually, on the contrary, acclimatization tunnels with a continuously operating fog system are not well tolerated.
13. Lentil plants grow well on sandy loamy to heavy soils provided they are well drained. However, vitality is maximized by transferring interspecific hybrid lentil plantlets into the large pots containing a non-sterilized equal mixture of clay soil and enriched peat to assure symbiotic nodulation. Establishing plants early in the season allows for an extended vegetative development phase which is suitable to maximize the production of F_2 seeds. Nevertheless, due to the limited number of interspecific lentil hybrids obtained in certain desirable cross-combinations, together with the fact that in determined interspecific hybrids the F_2 seed production might be scarce, the in vitro propagation of interspecific hybrids is an appropriate method to multiply and construct large F_1 populations which produce abundant numbers of segregating F_2 descendants useful for further genetic study and breeding (39).

References

1. Ladizinsky G (1992) Crossability relations. In: Kalloo G, Chowdhury JB (eds) Distant hybridization of crop plants, monographs on theoretical and applied genetics 16. Springer, Dordrecht, pp 15–31
2. Stebbins GL (1958) The inviability, weakness, and sterility of interspecific hybrids. Adv Genet 9:14–215
3. Yeung EC, Thorpe TA, Jensen CJ (1981) In vitro fertilization and embryo culture. In: Thorpe TA (ed) Plant tissue culture, methods and applications in agriculture. Academic, New York, pp 253–271
4. Abbo S, Ladizinsky G (1991) Anatomical aspects of hybrid embryo abortion in the genus *Lens* L. Bot Gaz 152:316–320
5. Ladizinsky G (1993) Wild lentils. Crit Rev Plant Sci 12:169–184
6. Ladizinsky G, Abbo S (1993) Cryptic speciation in *Lens culinaris*. Genet Resourc Crop Evol 40:1–5
7. Abbo S, Ladizinsky G (1994) Genetical aspects of hybrid embryo abortion in the genus *Lens* L. Heredity 72:193–200
8. van Oss H, Aron Y, Ladizinsky G (1997) Chloroplast DNA variation and evolution in the genus *Lens* Mill. Theor Appl Genet 94:452–457
9. Sandhu JS, Singh S (2007) History and origin. In: Yadav SS, McNeil D, Stevenson PC (eds) Lentil: an ancient crop for modern times. Springer, Dordrecht, pp 1–10

10. Cubero JI, Pérez de la Vega M, Fratini R (2009) Origin, phylogeny, domestication and spread. In: Erskine W, Muehlbauer F, Sharma B (eds) The lentil: botany, production and uses. CABI, Oxfordshire, pp 13–33
11. Wilson VE (1972) Morphology and technique for crossing *Lens esculenta* Moench. Crop Sci 12:231–232
12. Wilson VE, Law AG (1972) Natural crossing in *Lens esculenta* Moench. J Am Soc Hortic Sci 97:142–143
13. Horneburg B (2006) Outcrossing in lentil (*Lens culinaris*) depends on cultivar, location and year, and varies within cultivars. Plant Breed 125:638–640
14. Alvarez MT, García P, Pérez de la Vega M (1997) RAPD polymorphism in Spanish lentil landraces and cultivars. J Genet Breed 51:91–96
15. Ford R, Pang ECK, Taylor PWJ (1997) Diversity analysis and species identification in *Lens* using PCR generated markers. Euphytica 96:247–255
16. Ferguson ME, Newbury HJ, Maxted N, Ford-Lloyd BV, Robertson LD (1998) Population genetic structure in *Lens* taxa revealed by isozyme and RAPD analysis. Genet Resour Crop Evol 45:549–559
17. Ferguson M (2000) *Lens* spp: conserved resources, priorities for collection and future prospects. In: Knight R (ed) Linking research and marketing opportunities for pulses in the 21st century, vol. 34, Current plant science and biotechnology in agriculture. Kluwer, Dordrecht, pp 613–620
18. Sonante G, Pignone D (2001) Assessment of genetic variation in a collection of lentil using molecular tools. Euphytica 120:301–307
19. Durán Y, Pérez de la Vega M (2004) Assessment of genetic variation and species relationships in a collection of *Lens* using RAPD and ISSR. Span J Agric Res 2:538–544
20. Erskine W, Tufail M, Russell A, Tyagi MC, Rahman MM, Saxena MC (1994) Current and future strategies in breeding lentil for resistance to biotic and abiotic stresses. Euphytica 73:127–135
21. Ocampo B, Conicella C, Moss JP (2000) Wide crossing: opportunities and progress. In: Knight R (ed) Linking research and marketing opportunities for pulses in the 21st century, Vol. 34, Current plant science and biotechnology in agriculture. Kluwer, Dordrecht, pp 411–419
22. Davis PA, Lülsdorf MM, Ahmad M (2007) Wild relatives and biotechnological approaches. In: Yadav SS, McNeil D, Stevenson PC (eds) Lentil: an ancient crop for modern times. Springer, Dordrecht, pp 225–240
23. Malhorta RS, Balyan HS, Gupta PK (1978) Crossing technique in lentils. Lens Newslet 5:7–8
24. Solh MB, Paredes OM, Tiwari AS (1980) Crossing technique in lentil under field conditions. Lens Newslet 7:9–14
25. Mera M, Erskine W (1982) Crossing techniques for lentil under field conditions. Lens Newslet 9:11–15
26. Kumar A, Singh DP (1998) Hybridization technique in lentil under field conditions. Lens Newslet 25:1–3
27. Fratini R, Ruiz ML, Pérez de la Vega M (2004) Intra-specific and inter-sub-specific crossing in lentil (*Lens culinaris* Medik.). Can J Plant Sci 84:981–986
28. Ladizinky G (1979) The origin of lentil and its wild gene pool. Euphytica 28:179–187
29. Ladizinsky G, Braun D, Goshen D, Muehlbauer FJ (1984) The biological species of the genus *Lens* L. (*Lens nigricans*). Bot Gaz 145:253–261
30. Cohen D, Ladizinsky G, Ziv M, Muehlbauer FJ (1984) Rescue of interspecific *Lens* hybrids by means of embryo culture. Plant Cell Tiss Organ Cult 3:343–347
31. Ladizinsky G, Cohen D, Muehlbauer FJ (1985) Hybridization in the genus *Lens* by means of embryo culture. Theor Appl Genet 70:97–101
32. Fratini R, Ruiz ML (2006) Interspecific hybridization in the genus *Lens* applying in vitro embryo rescue. Euphytica 150:271–280
33. Ahmad M, Fautrier AG, McNeil DL, Burritt DJ, Hill GD (1995) Attempts to overcome postfertilization barrier in interspecific crosses of the genus *Lens*. Plant Breed 114:558–560
34. Murashige T, Skoog F (1962) A revised medium for rapid growth and bioassays with tobacco tissue cultures. Physiol Plant 15:473–497
35. Biondi S, Thorpe TA (1981) Requirements for a tissue culture facility. In: Thorpe TA (ed) Plant tissue culture – methods and application in agriculture. Academic, NY, pp 1–20
36. Brown DCW, Thorpe TA (1984) Organization of a plant tissue culture laboratory. In: Vasil IK (ed) Cell culture and somatic cell genetics of

plants, Vol. 1, Laboratory procedures and their applications. Academic, NY, pp 1–12
37. Pierik RLM (1998) In vitro culture of higher plants, 4th edn. Springer, Dordrecht
38. Muehlbauer FJ, Slinkard AE, Wilson VE (1980) Lentil. In: Fehr WR, Hadley HH (eds) Hybridization of crop plants. American Society of Agronomy and Crop Science Society of America, Madison, WI, pp 417–426
39. Fratini R, Ruiz ML (2008) Micropropagation of intra and interspecific *Lens* hybrids. Euphytica 159:199–206

Chapter 12

Generation of Interspecific Hybrids of *Trifolium* Using Embryo Rescue Techniques

Ajoy Kumar Roy, Devendra Ram Malaviya, and Pankaj Kaushal

Abstract

The genus *Trifolium* Leguminosae (Fabaceae), commonly called clovers, includes 237–290 annual and perennial species, of which about 20 are important as cultivated and pasture crops. Taxonomic distribution supported by molecular analysis indicates that Mediterranean region is one of the main centers of distribution of the genus and also a center of domestication and breeding. Self-incompatibility is prevalent in the genus, controlled by a single, multiallelic gene expressed gametophytically in the pollen. It was suggested that hybridity did not play a major role in the evolution of the genus due to the poor cross-ability of the species under natural conditions. Interspecific hybridization in the genus *Trifolium* by conventional crossing techniques has been largely unsuccessful. Post-zygotic barriers appear to be a primary cause of the reproductive isolation, associated with endosperm disintegration and consequent abnormal differentiation and starvation of the hybrid embryo. As hybridization using conventional techniques has almost failed in *Trifolium*, embryo culture technique was used by breeders to obtain new combinations of interspecific hybrids. Embryo culture has been effectively used in developing interspecific hybrids in *Trifolium ambiguum*, *T. pratense*, *T. montanum*, *T. occidentale*, *T. isthomocarpum*, *T. repens*, *T. nigrescens*, *T. uniflorum*, *T. sarosiense*, *T. alexandrinum*, *T. apertum*, *T. resupinatum*, *T. constantinopolitanum*, *T. rubens*, and *T. alpestre* in various combinations. The successful embryo rescue and development of hybrid plantlets requires skilled techniques of tissue culture and field practices. It includes hybridization in field; excision of hybrid embryos at appropriate stage; disinfection and culture in suitable culture media to allow maturation of embryo, multiplication of shoots, and rooting; hardening of the plantlets; inoculation with suitable *Rhizobium* culture; and transfer to field.

Key words: Clovers, Embryo rescue, Forage, Interspecific hybridization, Pasture, Tissue culture, *Trifolium*, Zygote

1. Introduction

The genus *Trifolium* of tribe Trifolieae of the Leguminosae (Fabaceae) is very important for its agricultural value. The genus *Trifolium*, commonly called clover, includes 237–290 annual and

Trevor A. Thorpe and Edward C. Yeung (eds.), *Plant Embryo Culture: Methods and Protocols*, Methods in Molecular Biology, vol. 710, DOI 10.1007/978-1-61737-988-8_12,

perennial species, of which about 20 are agriculturally important as cultivated and pasture crops.

The important perennial pasture clovers *T. repens* (white clover), *T. hybridum* (alsike clover), *T. pratense* (red clover), and *T. ambiguum* (Caucasian clover) are widely distributed in the temperate and subtemperate regions of the world. The annual types *T. alexandrinum* (Egyptian clover or Berseem), *T. resupinatum* (Persian clover or Shaftal), and *T. subterraneum* (subterranean clover) are commonly cultivated as winter annuals in the subtropical regions such as Egypt, India, Pakistan, Turkey, and the Mediterranean countries.

1.1. Center of Origin

Mediterranean region with its 110 species belonging to 7 sections is accepted as one of the main centers of distribution of the genus and also as a center of domestication and breeding (1). Molecular analysis also supported the Mediterranean origin of the genus with new world clade embedded with old world species (2). Based on nrDNA and cpDNA markers, it was suggested that the genus has a monophyletic origin in the Mediterranean region (3). Another center of distribution is the Californian region which is considered as a primary center of speciation of the genus, although the number of the species in this region is lower. It was suggested that some of the species native to the western part of America migrated to Asia and then spread to the Mediterranean area, where they created a highly diversified speciation center (4, 5).

1.2. Self-Incompatibility

Self-incompatibility is prevalent in the genus and is reported to be controlled by a single, multiallelic gene expressed gametophytically in the pollen. Populations of self-incompatible species of clover contain a large number of S-alleles (6, 7). However, true self-compatibility is conferred by the very rare "Sf" allele (8).

Populations of *T. repens* have been reported to contain about 100 alleles and those of *T. pratense* contain up to twice this number, while *T. hybridum* populations possesses only 17 S-alleles (9). In *T. alexandrinum* (Egyptian clover), using different pollination methods ranging from selfing, tripping, and controlled bee visit to open pollination, it was proved that genotypes can be grouped into four classes ranging from total self-fertile to total self-incompatible class. Thus in this species, several populations with different breeding mechanisms exist (10).

1.3. Interspecific Compatibility

Interspecific hybridization in the genus *Trifolium* has been largely unsuccessful. Near absolute failure of interspecific hybridization in *Trifolium* is of great evolutionary interest. Great variation in chromosome complexes in *Trifolium* has been suggested to be a result of mutational changes in species, which have become isolated by intersterility rather than the result of hybridization. Thus, while early in the evolution of *Trifolium*, natural hybridizations have taken place, later mutational and chromosomal changes were dominant factors in the speciation of this genus. This resulted in the cytological incompatibility between the species which is responsible for the failure of most interspecific crosses in *Trifolium* (11). Isozymes study in 134 accessions belonging to 25 species indicated the existence of a strong incompatibility barrier among the species (12). It was suggested that hybridity did not play a major role in the evolution of the genus *Trifolium* due to the poor crossability of the species under natural conditions (4).

Reports on the development of interspecific hybrids in *Trifolium* under natural conditions are meager. Successful hybridization between *T. repens* and *T. nigrescens* without embryo culture has been reported (13). *T. occidentale*, *T. uniflorum*, and *T. ambiguum* have been reported to produce occasional hybrids with *T. repens* (14, 15).

Interspecific hybridization in *Trifolium* is very difficult to obtain by conventional crossing techniques, since strong pre- and postfertilization barriers exist in the genus. Post-zygotic barriers appear to be a primary cause of the reproductive isolation, which is commonly associated with endosperm disintegration and consequent abnormal differentiation and starvation of the hybrid embryo (16–18).

In many interspecific crosses, the embryo starts growing normally only for 2–4 days. The hybrid embryo grows up to heart-shaped stage in *T. ambiguum* × *T. repens* (18) and up to globular stage in *T. semiplosum* × *T. repens* (17). Slower mitotic rate was recorded after 4–5 days in *T. repens* × *T. medium* hybrid embryo (19). These studies suggest that the action of deleterious genes is initiated at the time of fertilization or shortly thereafter. Endosperm disintegration is another phenomenon associated with failure of embryo development. In *T. ambiguum* × *T. repens* crosses, the endosperm develops only up to 128 nucleate stage (18).

1.4. Embryo Rescue Technique for Interspecific Hybridization in Trifolium

As hybridization using conventional techniques has almost failed in *Trifolium*, embryo culture technique was used by breeders to obtain new combinations of interspecific hybrids.

The first successful *Trifolium* embryo culture was reported by Keim (20), by obtaining normal mature plants by culturing immature embryos (8 days post-pollination) of red clover. Fifteen days post-pollination, the embryos were cultured from *T. repens* × *T. nigrescens* cross and mature hybrid embryos were obtained. Hybrids of *T. ambiguum* × *T. hybridum* were obtained utilizing embryo culture technique; however, these hybrids did not flower (20).

Embryo culture has been effectively used in developing interspecific hybrids in *T. ambiguum*, *T. pratense*, *T. montanum*, *T. occidentale*, *T. isthomocarpum*, *T. repens*, *T. nigrescens*, *T. uniflorum*, *T. sarosiense*, *T. alexandrinum*, *T. apertum*, *T. resupinatum*, and *T. constantinopolitanum* in different combinations (15, 21–34).

Embryo rescue technique was used in interspecific crosses involving red clover with zigzag clover, *T. rubens*, *T. alpestre*,

T. incarnatum, *T. lupinaster*, and *T. hybridum*. Embryos were available 8–15 days after pollination in crosses with zigzag clover, *T. rubens*, *T. alpestre*, and *T. hybridum*. However, embryos developed abnormally and no hybrid plants were rescued (34).

Generation of new plants from *T. repens* crossed with Kura clover, *T. nigrescens*, *T. uniflorum*, and *T. isthomocarpum* has been reported (35). Successful rescue of immature hybrid embryos of diploid red clover crosses with *T. sarosiense* ($2n = 48$) by in vitro culture (36) and of crosses of red clover with zigzag clover and *T. alpestre* was reported (23). Successful hybridization and recovery of plants following embryo rescue have been reported in case of *T. alexandrinum* × *T. constantinopolitanum* (29), *T. alexandrinum* × *T. apertum* (30), and *T. alexandrinum* × *T. resupinatum* (31).

Transplanted nurse endosperm was utilized to culture heart-shaped embryos excised 14–16 days after pollination from white clover by tetraploid kura clover crosses (15). Success was achieved in 10% of the embryos and plantlets were produced after 4–6 months. Embryo culture with the aid of nurse endosperm was used in *Trifolium* (37).

Ovule culture has been used for the production of interspecific hybrid between *T. ambiguum* and *T. repens* (38). It was reported that ovules of *T. repens* could be cultured as early as 1 day after pollination when the proembryo was at the two-celled stage, but only on a medium supplemented with the juice of water melon or young cucumber (39).

White clover has been successfully hybridized with six *Trifolium* species such as *T. nigrescens* (34, 40), *T. uniflorum* (14, 24, 34, 41), *T. argutum* (syn *T. xerocephalum* (42)), *T. occidentale* (43), *T. isthomocarpum* (44), and *T. ambiguum* (25, 45). In crosses with *T. repens* × *T. nigrescens* and *T. repens* × *T. uniflorum*, immature embryos excised 12–25 days after pollination were cultured and hybrid plants were raised. In crosses of white clover with *T. alexandrinum*, *T. subterraneum*, and *T. arvense*, embryos were cultured, but showed either abnormal or no growth (34). Triploid male and female sterile hybrids were obtained from crosses of *T. pratense* ($2n = 28$) and *T. pallidum* ($2n = 16$) (46).

Embryo and ovule culture were successfully used to develop interspecific hybrids in various combinations of *Trifolium* species (22). A sterile hybrid was obtained using immature embryo culture in *T. alpestre* × *T. pratense* cross (21). Development of hybrids of *T. repens* with *T. hybridum* using *in ovulo* embryo and embryo culture was reported. Ovules containing hybrid embryos were excised 12–14 days after pollination and cultured for 5–6 days on Nitsch medium (47) supplemented with 15% young cucumber juice. The embryos were subsequently excised and transferred to a hormone-free EG medium. A total of 118 hybrid seedlings were obtained, all of which showed chlorophyll deficiency (26).

Methods for the rescue of heart-staged hybrid embryos of red clover prior to in situ abortion were developed and standardized. A series of defined culture media were adjusted for the osmotic sensitivity of immature embryos and for maturation of embryos, promotion of shoot germination and development, multiplication of shoots, and rooting. Plant regeneration was also achieved from some embryos which produced only callus (36).

2. Materials

1. Hybridization: Glass house or field conditions for raising crops. Needle, forceps, brush, alcohol, hand lens, etc.
2. Embryo rescue and culture: Tissue culture facility including laminar air flow, inoculation chamber, tissue culture racks, standard tissue culture media, and chemicals as per Table 1.
3. Hardening of plants: A glasshouse facility will be better for hardening and establishment.
4. Culture media: The compositions of various media used are detailed in Table 1. The basal media used are EG, MS, L2, and RL. The EC3 medium is composed of the Murashige and Skoog (MS) basal medium supplemented with 2.3 μM kinetin and 3% sucrose. The LSP3 medium is based on the L2 basal medium supplemented with 4.3 μM α-naphthalene acetic acid (NAA), 0.66 μM 6-benzyladenine(BA; see Note 1), and 2.5% sucrose. The RL1 medium is based on the RL basal medium supplemented with 1.2 μM indole-3-acetic acid (IAA) and 2.5% sucrose. The LIH medium is composed of the L2 basal medium supplemented with 0.365 M sucrose, 25 nM picloram, and 15 μM adenine. The LSP2 medium is composed of L2 basal medium supplemented with 4 nM picloram and 0.66 μM BA.
5. Cucumber juice preparation: Juice from the fresh green tender cucumber is made in distilled water, sterilized by membrane filtration, and added post-autoclaving to the medium as per the procedure detailed by Nakajima et al. (39); 15% cucumber juice is recommended.
6. *Rhizobium* solution: The easiest way is to take out young green plants of same species carefully from field. From the roots of these plants, young pinkish nodules are taken out, washed, and crushed in distilled water. This solution contains enough strains of *Rhizobium* to inoculate the young plants at the time of hardening.

Table 1
Composition of L2, MS, RL, and EG basal media (26, 48, 49)

Components	EG medium (mg/L)	MS basal	L2 basal	RL basal
KNO_3	950	18.8 mM	20.8 mM	10.4 mM
NH_4NO_3	600	20.6 mM	12.5 mM	6.25 mM
KH_2PO_4	170	1.25 mM	2.34 mM	2.34 mM
$MgSO_4 \cdot 7H_2O$	185	1.5 mM	1.8 mM	0.9 mM
$CaCl_2 \cdot 2H_2O$	166	3.0 mM	4.1 mM	2.0 mM
NaH_2PO_4	–	–	0.6 mM	0.3 mM
$FeSO_4 \cdot EDTA \cdot 7H_2O$	27.85	100 μM	90 μM	90 μM
$Na_2 \cdot EDTA \cdot 2H_2O$	37.25	100 μM	–	–
$MnSO_4 \cdot 4H_2O$	2.23	100.0 μM	90 μM	45 μM
H_3BO_3	0.62	100.0 μM	82 μM	41 μM
$ZnSO_4 \cdot 7H_2O$	0.86	30.0 μM	18 μM	9 μM
KI	0.083	5.0 μM	6 μM	3 μM
$Na_2MoO_4 \cdot 2H_2O$	0.025	1.03 μM	1.7 μM	0.85 μM
$CoCl_2 \cdot 6H_2O$	0.0025	0.105 μM	0.42 μM	0.21 μM
$CuSO_4 \cdot 5H_2O$	0.0025	0.1 μM	0.4 μM	0.2 μM
Myo-inositol	–	100 mg/L	1.4 mM	0.7 mM
Thiamine HCl	–	0.1 mg/L	6 μM	3.0 μM
Pyridoxine HCl	–	0.5 mg/L	2.4 μM	1.2 μM
Nicotinic acid	–	0.5 mg/L	–	8.5 μM
3-Aminopyridine	–	–	–	24 μM
Sucrose	–	87.6 mM	73 mM	44 μM
Glucose	30 g/L	–	–	–
Agar	7 g/L	0.7%	0.8%	0.65%
pH	5.8	5.8	5.8	5.8

7. *Rhizobium trifolii*: Any native strain will serve the purpose. It will vary from location to location and from species to species.
8. Sterilizing solutions: 0.1% mercuric chloride; 40% commercial bleach.
9. Supplies: Waxed paper bags, needles, brushes, dissection tools, cellophane paper, and 0.2-μm filter units.

3. Methods

Protocols for developing interspecific hybrids are detailed below. This section has been divided into three parts depending upon the female parent used: (a) *T. alexandrinum*, (b) *T. pratense*, and (c) *T. repens*.

3.1. *T. alexandrinum* as Female Parent

3.1.1. Hybridization in Field

1. The crop (both male and female parents (for female parent, see (29–31) and personal experiences)) should be raised as per standard agronomic practice recommended for the crop in that particular zone. Staggered sowing should be done if the flowering time differs in the parents. It will ensure availability of pollen and receptive stigma for crossing.
2. Emasculate flower buds prior to anther dehiscence in early morning hours. Remove the anthers and bag the emasculated flowers using waxed paper bags to prevent drying up (see Note 2).
3. Pollinate emasculated flower buds by applying pollen gently to the stigma using a needle or a brush. For pollination, collect pollen from freshly opened flowers of the male parent.
4. After pollination, cover the flowers with waxed paper bags and label properly.
5. Watch for indicators of fertilization such as petal weathering, swollen ovary, etc.

3.1.2. Embryo Rescue and Culture

1. The time of excision of flowers depends on different cross combinations. Usually, in interspecific hybrids, growth of embryo is slow. Embryo excised at the heart-shaped stage (10–12 DAP) was found to respond best.
2. Take the pollinated flowers to the laboratory, dissect the swollen ovaries, and surface sterilize for 2 min using 0.1% mercuric chloride.
3. Dissect the embryos from ovaries and culture on EC3 medium containing MS (48) basal medium supplemented with 2.3 μM kinetin and 3% sucrose.
4. The embryos are initially kept in the dark for 2 days; after germination, they are cultured at 25 ± 2°C under a 16/8-h (light/dark) photoperiod.
5. Subculture the germinated embryos in the shoot-inducing LSP3 medium supplemented with 4.3 μM NAA, 0.66 μM BA (see Note 1), and 2.5% sucrose. It helps in accelerated growth of the plantlets and multiple shoot formation.
6. For root induction, split multiple shoots and transfer them separately in RL1 medium (49) supplemented with 1.2 μM IAA (see Note 3) and 2.5% sucrose.

3.1.3. Hardening of Hybrid Plantlets, Transfer to Field, and Their Growth

1. Keep the embryo-rescued plants in the culture tubes out of the culture room for 2–3 days at room temperature (30°C).
2. Remove the plants from the culture tubes, free them of media, and keep for an additional day with the roots submerged in sterilized distilled water. High humidity should be maintained by covering the plant with cellophane paper.
3. Inoculate with *Rhizobium* culture suitable for the female parent (see Note 4).
4. Transfer the regenerated plants to sterile soil in pots (see Note 5).
5. Protect the plants from direct sunlight for the first 3–5 days in the field.

3.2. *T. pratense* as Female Parent (49)

3.2.1. Embryo Rescue and Culture

1. Hand-pollinate target female flowers with desired genotypes/species.
2. Collect the female pollinated florets 14–19 days after pollination. The time varies for different species combinations.
3. Disinfect the florets by rinsing with water, immerse in 70% ethanol for 1–2 min, transfer to 40% commercial bleach (2% sodium hypochlorite) for 5–8 min, and rinse in sterile, deionized distilled water for 5 min.
4. Excise immature embryos from each floret and place them individually onto LIH medium for 8–14 days at 25°C under low-intensity light. It helps in embryo maturation.
5. Transfer the embryos to LSP2 medium for shoot emergence and development (see Note 6).

3.2.2. Rooting of Shoots

1. Place individual shoots on RL medium at 25°C under low-intensity light. Normally, roots appear within 2–4 weeks.
2. An additional 1 month culture on fresh RL medium encourages further root and plant development.

3.2.3. Hardening

1. Free the plants of agar by using forceps and rinse gently in lukewarm water.
2. Pot the plants in a mixture of soil, peat, and prewashed vermiculite (1:1:1, by volume). Root tips must be planted in a downward orientation (see Note 5). Inoculation with *R. trifolii* is performed at this time.
3. Conditions of high humidity must be provided for about 2 weeks. After about 2 weeks, most plants adjust to normal greenhouse conditions and can then be transplanted in the field.

3.3. *T. repens* as Female Parent (26)

1. Make crosses by hand pollination. In case of self-compatible plants, emasculate the flowers, or where the plants were known to be self-incompatible or male sterile, no emasculation is required.

2. Remove pods 12–14 days after pollination and surface sterilize for 90 s in 70% alcohol and then for 10 min in 20% "Janol" (Commercial bleach, 5% sodium hypochlorite), followed by two rinses in sterile distilled water.
3. Dissect stimulated ovules aseptically and culture for 5–6 days on Nitsch medium (47) supplemented with 15% young cucumber juice (39). Sterilize the juice by membrane filtration and add to the medium aseptically post-autoclaving.
4. Check all the ovules for the presence of embryos after 5–6 days. Gently remove the ovule containing an embryo to nutrient medium in a drop of sterile water.
5. Culture immature embryos in EG medium developed principally for *T. repens.*
6. Maintain in vitro cultures in a temperature-controlled growth room and at a light intensity of 600 Lux for 16 h with 18°C at night.
7. Transfer developing embryos to fresh medium every 3 weeks. Transfer plants with four to five trifoliate leaves to a half-strength MS medium (48) for hardening.
8. After development of four to five normal foliage leaves, remove the plants from sterile culture and transplant to soil. This needs a hardening step similar to that used for *T. pratense* as described under Subheading 3.2.3. Keep in a humidity chamber for 7–10 days and subsequently transfer to the glass house.

4. Notes

1. BA is dissolved in 1 N NaOH and final volume made up with H_2O.
2. In case of Egyptian clover, emasculation in the morning hours when temperature is 20–25°C was found to be the best. Pollination 48 h after emasculation was found to be most appropriate.
3. IAA solutions should be stored in amber-colored bottles in dark.
4. For inoculation with *Rhizobium*, green plants in the field were carefully uprooted. Young pinkish live nodules were taken out from root surface, washed with distilled water, and crushed in the water solution. Roots from tissue culture-derived plants are dipped in this solution for 24 h before transplanting to the field. It was found to be quite effective for nodulation of new plantlets.
5. During transplanting in pots, extra care should be taken that root tips do not break and roots are pointed downwards.

6. As the crossing and culture protocols are difficult and hybrids are generally obtained in very small numbers, it is desirable to multiply the hybrids. It is more important as the hybrids are often obtained when the crop growing season is over. So the hybrids are multiplied and maintained in vitro till the next growing season when they are transplanted in the field.

References

1. Gillett JM (1952) The genus *Trifolium* in southern Arabia and in Africa south of Sahara. Kew Bull 7:367–404
2. Watson LE, Sayed-Ahmed H, Badr A (2000) Molecular phylogeny of old world *Trifolium* (Fabaceae), based on plastid and nuclear markers. Plant Syst Evol 224:153–171
3. Ellison NW, Liston A, Steiner JJ, Williams WM, Taylor NL (2006) Molecular phylogenetics of the clover genus (*Trifolium* – Leguminosae). Mol Phylogenet Evol 39:688–705
4. Zohary M (1972) Origin and evolution in the genus *Trifolium*. Bot Not 125:501–511
5. Zohary M, Heller D (1984) The genus *Trifolium*. The Israel Academy of Sciences and Humanities, Jerusalem
6. Atwood SS (1940) Genetics of cross-incompatibility among self-incompatible plants of *Trifolium repens*. J Am Soc Agron 32:955–968
7. Williams W (1951) Genetics of incompatibility in alsike clover, *Trifolium hybridum*. Heredity 5:51–73
8. Atwood SS (1944) Oppositional alleles in natural populations of *Trifolium repens*. Genetics 29:428–435
9. Lawrence MJ (1996) Number of incompatibility alleles in clover and other species. Heredity 76:610–615
10. Roy AK, Malaviya DR, Kaushal P (2005) Pollination behaviour in different breeding populations in Egyptian clover. Plant Breed 124:171–175
11. Wexelsen H (1928) Chromosome numbers and morphology in *Trifolium*. Calif Univ Agric Sci 2:355–376
12. Malaviya DR, Roy AK, Kaushal P, Kumar B, Tewari A (2008) Genetic similarity among *Trifolium* species based on isozyme banding pattern. Plant Syst Evol 276:125–136
13. Marshall AH, Michaelson-Yeats TPT, Aluka P, Meredith M (1995) Reproductive characters of interspecific hybrids between *Trifolium repens* L. and *T. nigrescens* Viv. Heredity 74:136–145
14. Gibson PB, Chen CC, Gillingham JT, Barnett OW (1971) Interspecific hybridization of *Trifolium uniflorum* L. Crop Sci 11:895–899
15. Williams E (1978) A hybrid between *Trifolium repens* and *T. ambiguum* obtained with the aid of embryo culture. N Z J Bot 16:499–506
16. Chen CC, Gibson PR (1971) Seed development following the mating of *Trifolium repens*× *T. uniflorum*. Crop Sci 11:667–672
17. White DWR, Williams E (1976) Early seed development after crossing of *Trifolium semiplosum* and *T. repens*. N Z J Bot 14:161–168
18. Williams E, White DWR (1976) Early seed development after crossing of *Trifolium ambiguum* and *T. repens*. In: Proceedings of the 3rd international congress of SABRAO, vol 2, pp 26–30
19. Kazimierska EM (1978) Embryological studies of cross compatibility in the genus *Trifolium* L. II Fertilization, development of embryo and endosperm in crossing *T. repens* L. with *T. medium* L. Genet Pol 19:15–24
20. Keim WF (1953) Interspecific hybridization in *Trifolium* utilizing embryo culture techniques. Agron J 45:601–606
21. Phillips GC, Grosser JW, Berger S, Taylor NL, Collins GB (1992) Interspecific hybridization between red clover and *Trifolium alpestre* using *in vitro* embryo rescue. Crop Sci 32:1113–1115
22. Ferguson NH, Rupert EA, Evans PT (1990) Interspecific *Trifolium* hybrids produced by embryo and ovule culture. Crop Sci 30: 1145–1149
23. Collins GB, Taylor NL, Phillips GC (1981) Successful hybridization of red clover with perennial *Trifolium* species *via* embryo rescue. In: Smith TA, Hays VW (eds) International grassland congress, Lexington, vol 14. Westview, Boulder, CO, pp 168–170
24. Pandey KK, Grant JE, Williams EG (1987) Interspecific hybridization between *Trifolium*. Aust J Bot 35:171–182
25. Yamada T, Fukuoka H (1986) Production of interspecific hybrids between *Trifolium*

ambiguum M. Bieb. and T. repens L. by ovule culture. Jpn J Breed 36:233–239

26. Przywara L, White DWR, Sanders PM, Maher D (1989) Interspecific hybridization of *Trifolium repens* with *T. hybridum* using *in ovulo* embryo and embryo culture. Ann Bot 64:613–624
27. Sawai A, Ueda S, Gau M, Uchiyama K (1990) Interspecific hybrids of *Trifolium medium* L. ×4× *T. pratense* L. obtained through embryo culture. J Jpn Soc Grassl Sci 35:267–272
28. Sawai A, Yamaguchi H, Uchiyama K (1995) Fertility and morphology of the chromosome-doubled hybrid *Trifolium medium* × *T. pratense* (red clover) and backcross progeny. Grassl Sci 41:122–127
29. Roy AK, Malaviya DR, Kaushal P, Kumar B, Tiwari A (2004) Interspecific hybridization of *T. alexandrinum* with *T. constantinopolitanum* using embryo rescue. Plant Cell Rep 22:605–610
30. Malaviya DR, Roy AK, Kaushal P, Kumar B, Tiwari A (2004) Development and characterization of interspecific hybrids of *Trifolium alexandrinum* × *T. apertum* using embryo rescue. Plant Breed 123:536–542
31. Kaushal P, Malaviya DR, Roy AK, Kumar B, Tiwari A (2005) *Trifolium alexandrinum* × *T. resupinatum* – interspecific hybrids developed through embryo rescue. Plant Cell Tissue Organ Cult 83:137–144
32. Selim AK, Abdel-Tawab FM, Fahmy EM (1977) Phylogenetic relationship in genus *Trifolium* L. I. Interspecific crossability and serological affinities. Egypt J Genet Cytol 6:274–283
33. Trimble JP, Hovin AW (1960) Interspecific hybridization of certain *Trifolium* species. Agron J 52:405
34. Evans AM (1962) Species hybridization in *Trifolium*. II. Investigating the pre-fertilization barriers to compatibility. Euphytica 11:256–262
35. Rupert EA, Evans PT (1980) Embryo development after interspecific cross-pollinations among species of *Trifolium*, section Lotoidea, Agronomy abstracts. American Society of Agronomy, Madison, WI, p 68
36. Phillips GC, Collins GB, Taylor NL (1982) Interspecific hybridization of red clover (*Trifolium pratense*) with *T. sarosiense* using *in vitro* embryo rescue. Theor Appl Genet 62:17–24
37. Williams EG, De Lautour G (1980) The use of embryo culture with transplanted nurse endosperm for the production of interspecific hybrids in pasture legumes. Bot Gaz 141:252–257
38. Yamada T, Fukoka H (1985) Application of ovule culture to interspecific hybridization between *Trifolium repens* and *T. ambiguum*. In: Proceedings of the XVth International Grassland Congress, pp 241–243. held at Kyoto August 24–31, 1985 Published by The Science Council of Japan. The Japanese Society of Grassland Science.
39. Nakajima T, Doyama Y, Matsumoto H (1969) *In vitro* culture of excised ovules of white clover *Trifolium repens* L. Jpn J Breed 19:373–378
40. Brewbaker JL, Keim WF (1953) A fertile interspecific hybrid in *Trifolium*. Am Nat 8:323–326
41. Pandey K (1957) A self compatible species in *Trifolium*. J Hered 48:278–281
42. Kazimierski T, Kazimierska EM (1968) Investigations of hybrids of the genus *Trifolium* L., I sterile hybrid *T. repens* × *T. xerocephalum* (in Polish). Acta Soc Bot Pol 37:549–560
43. Gibson PB, Beinhart G (1969) Hybridization of *Trifolium occidantale* with two other species of clover. J Hered 60:93–96
44. Kazimierski T, Kazimierska EM (1973) Investigations of hybrids of the genus *Trifolium* L., IV. Cytogenetics of the cross *T. repens* × *T. isthmocarpum* Brot. (in Polish). Acta Soc Bot Pol 41:127–147
45. Williams E, Verry IM (1981) A partially fertile hybrid between *Trifolium repens* and *T. ambiguum*. N Z J Bot 19:1–7
46. Armstrong KC, Cleveland RW (1970) Hybrids of *Trifolium pratense* × *Trifolium pallidum*. Crop Sci 10:354–357
47. Nitsch JP (1951) Growth and development *in vitro* of excised ovaries. Am J Bot 38:566–577
48. Murashige T, Skoog F (1962) A revised medium for rapid growth and bioassays with tobacco tissue cultures. Physiol Plant 15:473–497
49. Phillips GC, Collins GB (1984) Red clover and other forage legumes. In: Sharp WR, Evans DA, Ammirato PV, Yamada Y (eds) Handbook of plant cell culture, vol 2. Macmillan, New York, pp 169–210

Part III

Cryopreservation of Embryos

Chapter 13

Cryopreservation of Embryos: An Overview

Florent Engelmann

Abstract

Cryopreservation (liquid nitrogen, −196°C) is the only safe and cost-effective option for long-term conservation of genetic resources of non-orthodox seed species. Cryopreservation protocols have been developed for various materials including seeds, dormant buds, cell suspensions, calli, apices, zygotic, and somatic embryos of numerous plant species. Zygotic embryos or embryonic axes of almost 100 different species and somatic embryos of almost 40 different species from both temperate and tropical climates, comprising crops, fruit, and forest trees as well as wild species, whose seeds displayed orthodox, intermediate, and recalcitrant storage characteristics, have been successfully cryopreserved. With zygotic embryos and embryonic axes, the desiccation technique has been used with the majority of the species tested, leading to highly variable survival and recovery after freezing, especially during earlier experiments. More recently, new cryopreservation techniques viz. encapsulation-dehydration and vitrification have been employed, leading to generally improved results. With somatic embryos, different cryopreservation methods have been used viz. desiccation, pre-growth-desiccation, encapsulation-dehydration, vitrification, encapsulation-vitrification, and droplet-vitrification. There are also a few examples of the utilisation of slow controlled freezing, which correspond to the earlier experiments performed with somatic embryos. The development and application of cryopreservation is significantly more advanced for somatic embryos, in comparison with zygotic embryos, mainly because of the different origin and characteristics of the species treated. In most cases, zygotic embryos originate from tropical, wild species, for which knowledge and techniques relevant to the development of cryopreservation protocols are limited, or even non-existent. By contrast, somatic embryos are generally produced from cultivated species, which have already been studied extensively and for which propagation techniques are already operational. A number of technical possibilities to explore exist in order to improve the development of cryopreservation protocols for zygotic embryos and embryonic axes. For both categories of materials, the utilisation of analytical techniques has proved to be extremely useful to assist in the development of cryopreservation protocols.

Key words: Conservation, Cryopreservation, Genetic resources, Somatic embryos, Zygotic embryos

1. Introduction

Many of the world's major food plants produce seeds that undergo maturation drying and are thus tolerant to extensive desiccation and can be stored dry at low temperature. Seeds of this type are

Trevor A. Thorpe and Edward C. Yeung (eds.), *Plant Embryo Culture: Methods and Protocols*,
Methods in Molecular Biology, vol. 710, DOI 10.1007/978-1-61737-988-8_13,

termed "orthodox" (1). Storage of such orthodox seeds is the most widely practised method of ex situ conservation of plant genetic resources since 90% of the 6.1 million accessions stored in genebanks are maintained as seed (2). Techniques have been devised which allow seeds of many species to be conserved in this way for several decades. These techniques involve drying seeds to low moisture content (3–7% fresh weight basis, depending on the species) and storing them, in hermetically sealed containers, at low temperature, preferably at −18°C or cooler (3). All relevant techniques are well established and a series of practical documents have been published which cover the main aspects of seed conservation (4–7). A recent significant application of the orthodox seed storage technology is the establishment of the Svalbard Global Seed Vault (http://www.croptrust.org/main/). Another technical achievement in the area of orthodox seed conservation concerns the development of the so called "ultra-dry" seed storage technology (8), which is based on the principle that desiccating seeds to much lower moisture contents (MC) than those generally used in standard procedures will allow their storage for extended periods at room temperature, thereby avoiding the requirement for refrigeration facilities.

In contrast to orthodox seeds, a considerable number of species, predominantly tropical or sub-tropical in origin, such as coconut, cacao and many forest and fruit tree species, produce seeds which do not undergo maturation drying and are shed at relatively high moisture content (9). Such seeds are unable to withstand desiccation and are often sensitive to chilling. They therefore cannot be maintained under the conventional seed storage conditions described above, that is storage at low moisture content and low temperature. Seeds of this type are called "recalcitrant" and have to be kept in moist, relatively warm conditions to maintain viability (1, 10). Even when recalcitrant seeds are stored in an optimal manner, their lifespan is limited to weeks, occasionally months. Of more than 7,000 species for which information on seed storage behaviour has been published (11), approximately 3% are recorded as recalcitrant and an additional 4% as possibly recalcitrant.

More recent investigations have identified species exhibiting "intermediate" storage behaviour. While such seeds can tolerate desiccation to fairly low MC, once dried, they become particularly susceptible to injury caused by low temperature (12, 13). The storage life of intermediate seeds can be prolonged by further drying, but it remains impossible to achieve the long-term conservation of orthodox seeds. About 1% of the aforementioned 7,000+ species studied and included in the Compendium on Seed Storage Behaviour are reported as producing intermediate seeds and another 1% have been characterised as possibly intermediate (11). Included in this category are some economically important species such as coffee, citrus, rubber, oil palm, and many tropical forest tree species. It should be noted that the percentages of intermediate and recalcitrant seed-producing species cited above are likely to be largely underestimated. These figures are based on scientific and technical publications, which, by default, concern mainly temperate species. In addition, it can be expected that a large proportion of the species for which no information is available, which are predominantly from tropical or sub-tropical origin, exhibit recalcitrant, or to a lesser extent intermediate seed storage behaviour.

There are other species for which conservation as seed is problematic. First, there are those that do not produce seeds at all and, consequently are propagated vegetatively, for example banana and plantain (*Musa* spp.). Second, there are crops such as potato (*Solanum tuberosum*), other root and tuber crops such as yams *(Dioscorea* spp.), cassava *(Manihot esculenta)* and sweet potato *(Ipomoea batatas)*, and sugarcane (*Saccharum* spp.) that have either some sterile genotypes and/or some that produce orthodox seeds. However, these seeds are highly heterozygous and, therefore, of limited utility for the conservation of particular genotypes. These crops are usually propagated vegetatively to maintain genotypes as clones.

Traditionally, the field genebank has been the ex situ storage method of choice for the aforementioned "problem materials". Around 527,000 accessions are maintained in field genebanks (2). In some ways, this method offers a satisfactory approach to conservation. The genetic resources under conservation can be readily accessed and observed, thus permitting detailed evaluation. However, there are certain drawbacks that limit its efficiency and threaten its security (14). The genetic resources are exposed to pests, diseases, and other natural hazards such as drought, weather damage, human error, and vandalism. In addition, they are not in a condition that is readily conducive to germplasm exchange because of the great risks of disease transfer through the exchange of vegetative material. Field genebanks are costly to maintain and, as a consequence, are prone to economic decisions that may limit the level of replication of accessions, the quality of maintenance and even their very survival in times of economic stringency. Even under the best circumstances, field genebanks require considerable inputs in the form of land (often needing multiple sites to allow for rotation), labour, management and materials, and, in addition, their capacity to ensure the maintenance of much diversity is limited.

Tissue culture techniques are of great interest for the collecting, multiplication and storage of plant germplasm (14). Tissue culture systems allow propagation of plant material with high multiplication rates in an aseptic environment. During the last 40 years, in vitro propagation techniques, mainly based on micropropagation and somatic embryogenesis, have been extensively developed and applied to well over 1,000 different species. Virus-free plants can be obtained through meristem culture in combination with thermotherapy, thus ensuring the production of disease-free stocks and simplifying quarantine procedures for the international exchange of germplasm. The miniaturisation of explants allows reducing space requirements and consequently labouring costs for the maintenance of germplasm collections. Different in vitro conservation methods are employed, depending on the storage duration required (15). For short- and medium-term storage, various techniques have been devised, which allow reduction of growth and which increase the intervals between subcultures. In vitro conservation techniques using slow growth storage have been developed for a wide range of species, including temperate woody plants, fruit trees, horticultural species, as well as numerous tropical species. However, despite the availability of such techniques, only around 38,000 accessions are conserved in vitro worldwide (2), because many conservation programmes are unable to meet requirements for relatively sophisticated equipment, reliable electricity supply, and trained staff. In addition, only a limited amount of genetic diversity can be maintained in vitro. Slow growth storage is used routinely in national, regional, and international germplasm conservation centres with species including, notably, banana, root and tuber crops, and temperate fruits.

2. Cryopreservation

For long-term storage, cryopreservation, that is storage at ultra low temperature (liquid nitrogen, −196°C), is employed. At this temperature, all cellular divisions and metabolic processes are stopped. The plant material can thus be stored without alteration or modification for a theoretically unlimited period of time. Moreover, cultures are stored in a small volume, protected from contamination, and requiring very limited maintenance. Cryopreservation currently offers the only safe and cost-effective option for the long-term conservation of genetic resources of problem species (16).

Some materials, such as orthodox seeds or dormant buds, display natural dehydration processes and can be cryopreserved without any pre-treatment. However, most of the experimental systems employed in cryopreservation (cell suspensions, calluses, shoot tips, embryos, etc.) contain high amounts of cellular free water and are thus extremely sensitive to freezing injury since most of them are not inherently freezing-tolerant. Cells have thus to be dehydrated artificially to protect them from damage caused by crystallisation of intracellular water into ice (17). The techniques employed and the physical mechanisms upon which they are based are different in classical and new cryopreservation techniques (15). Classical techniques involve freeze-induced dehydration, whereas new techniques are based on vitrification. Vitrification can be defined as the transition of water directly from the liquid phase into an amorphous phase or glass, whilst avoiding the formation of crystalline ice (18).

Cryopreservation protocols are now available for various materials including seeds, dormant buds, cell suspensions, calluses, apices, and zygotic and somatic embryos of several hundreds of species of temperate and tropical origin. Thanks to the development of new cryopreservation procedures for apices and embryos, reports involving a larger number of genotypes/varieties are becoming more frequent (19, 20) There are an increasing number of cases where cryopreservation is currently used in a genetic resources conservation framework (21).

In this chapter, we first briefly describe the current cryopreservation techniques available for freezing plant tissues and organs. We then review their application for the long-term conservation of zygotic and somatic embryos.

3. Classical Cryopreservation Techniques

Classical cryopreservation techniques involve slow cooling down to a defined pre-freezing temperature, followed by rapid immersion in liquid nitrogen. With temperature reduction during slow cooling, cells and the external medium initially supercool, followed by ice formation in the medium (17). The cell membrane acts as a physical barrier and prevents the ice from seeding the cell interior and the cells remain unfrozen but supercooled. As the temperature is further decreased, an increasing amount of the extracellular solution is converted into ice, thus resulting in the concentration of intracellular solutes. Since cells remain supercooled and their aqueous vapour pressure exceeds that of the frozen external compartment, cells equilibrate by loss of water to external ice. Depending on the rate of cooling and the pre-freezing temperature, different amounts of water will leave the cell before the intracellular contents solidify. In optimal conditions, most or all intracellular freezable water is removed, thus reducing or avoiding detrimental intracellular ice formation upon subsequent immersion of the specimen in liquid nitrogen. However, too intense freeze-induced dehydration can incur different damaging events due to concentration of intracellular salts and changes in the cell membrane. Rewarming should be as rapid as possible to avoid the phenomenon of recrystallization in which ice melts and reforms at a thermodynamically favourable, larger and more damaging crystal size (17).

Classical freezing procedures include the following successive steps: pre-growth of samples, cryoprotection, slow cooling (0.5–2.0°C/min) to a determined pre-freezing temperature (usually around −40°C), rapid immersion of samples in liquid nitrogen, storage, rapid rewarming, and recovery. Classical techniques are generally operationally complex since they require the use of sophisticated and expensive programmable freezers. In some cases, their use can be avoided by performing the slow freezing step with a domestic or laboratory freezer.

Classical cryopreservation techniques have been successfully applied to undifferentiated culture systems such as cell suspensions and calluses (15). In the case of differentiated structures, these techniques can be employed for freezing apices of cold-tolerant species (22).

4. New Cryopreservation Techniques

In vitrification-based procedures, cell dehydration is performed prior to freezing by exposure of samples to concentrated cryoprotective media and/or air desiccation. This is followed by rapid cooling. As a result, all factors that affect intracellular ice formation are avoided. Glass transitions (changes in the structural conformation of the glass) during cooling and rewarming have been recorded using thermal analysis (23). Vitrification-based procedures offer practical advantages in comparison to classical freezing techniques. Like ultra-rapid freezing (above), they are more appropriate for complex organs (shoot tips, embryos) which contain a variety of cell types, each with unique requirements under conditions of freeze-induced dehydration. By precluding ice formation in the system, vitrification-based procedures are operationally less complex than classical ones (e.g. they do not require the use of controlled freezers) and have greater potential for broad applicability, requiring only minor modifications for different cell types (24).

A common feature to all these new protocols is that the critical step to achieve survival is the dehydration step, and not the freezing step, as in classical protocols. Therefore, if samples to be frozen are amenable to desiccation down to sufficiently low water contents (which vary depending on the procedure employed and the type and characteristics of the propagule to be frozen) with no or little decrease in survival in comparison to non-dehydrated controls, no or limited further drop in survival is generally observed after cryopreservation (24).

Eight different vitrification-based procedures can be identified: (1) encapsulation-dehydration, (2) vitrification, (3) encapsulation-vitrification, (4) dehydration, (5) pre-growth, (6) pre-growth-dehydration, (7) droplet freezing, and (8) droplet-vitrification.

The encapsulation-dehydration procedure is based on the technology developed for the production of artificial seeds. Explants are encapsulated in alginate beads, pre-grown in liquid medium enriched with sucrose for 1–7 days, partially desiccated in the air current of a laminar air flow cabinet or with silica gel to a water content around 20% (fresh weight basis), then frozen rapidly. Survival is high and growth recovery of cryopreserved samples is generally rapid and direct, without callus formation. This technique has been applied to apices of numerous species from temperate and tropical origin as well as to cell suspensions and somatic embryos of several species (21).

Vitrification includes the following steps: pre-culture of samples on medium enriched with cryoprotective substances, treatment with a loading solution (e.g. a mixture of 2 M glycerol and 0.4 M sucrose, dehydration with a highly concentrated vitrification solution such as the glycerol-based PVS2 solution (23) which has a molarity of 7.8 M, rapid freezing and thawing, and removal of cryoprotectants and recovery. This procedure has been developed for apices, cell suspensions, embryogenic tissues and somatic embryos of numerous species (25).

Encapsulation-vitrification is a combination of encapsulation-dehydration and vitrification procedures, where samples are encapsulated in alginate beads, then subjected to freezing by vitrification. It has been applied to apices of an increasing number of species (25).

Dehydration is the simplest procedure since it consists of dehydrating explants, then freezing them rapidly by direct immersion in liquid nitrogen. This technique is mainly used with zygotic embryos or embryonic axes extracted from seeds. It has been applied to embryos of a large number of recalcitrant and intermediate species (26–28). Desiccation is usually performed in the air current of a laminar airflow cabinet, but more precise and reproducible dehydration conditions are achieved by using a flow of sterile compressed air or silica gel.

The pre-growth technique consists of cultivating samples in the presence of cryoprotectants, then freezing them rapidly by direct immersion in liquid nitrogen. The pre-growth technique has been developed for *Musa* meristematic cultures (29).

In a pre-growth-dehydration procedure, explants are pre-grown in the presence of cryoprotectants, dehydrated under the laminar airflow cabinet or with silica gel, and then frozen rapidly. This method has been applied notably to asparagus stem segments, oil palm polyembryonic cultures, and coconut zygotic embryos (16).

The droplet-freezing technique has presently been applied to apices of potato, asparagus, and apple (16). Apices are pre-treated with liquid cryoprotective medium, then placed on an aluminium foil in minute droplets of cryoprotectant and frozen slowly (apple) or rapidly (potato) in liquid nitrogen.

Finally the newly developed droplet-vitrification technique consists in treating explants with loading and vitrification solutions like in a vitrification protocol and in freezing them ultra-rapidly in a droplet of vitrification solution placed on an aluminium foil, as described for the droplet-freezing technique. Droplet-vitrification is being successfully applied to an increasing number of species (25).

5. Cryopreservation of Zygotic Embryos

Table 1 presents a list of species whose embryos or embryonic axes have been reported to withstand cryopreservation. It includes almost 100 different species from both tropical and temperate climates, comprising crops, fruit, and forest trees as well as wild species. Various observations can be made concerning this list.

5.1. Characteristics of Plant Material

Cryopreservation experiments have been performed with embryos sampled from seeds displaying all three categories of storage behaviour, viz. orthodox (e.g. maize, wheat), intermediate (e.g. citrus, coffee, oil palm), and recalcitrant (e.g. coconut, cocoa, rubber tree). As regards the latter category, it should be noted, however, that recalcitrance is a dynamic concept which evolves with research on the biology of species and improvement in classical storage procedures. As a result, some species previously classified as recalcitrant have thus been moved to the intermediate or even sub-orthodox categories and stored using classical or new storage techniques (27).

Depending on the species, whole embryos or embryonic axes are employed for cryopreservation. Important parameters for freezing are the size, histological composition, and developmental stage of explants. If explants are too large, the hydration and thermal gradients created within the samples during pre-treatment with cryoprotectants or physical desiccation and during freezing, respectively, will result in structural damage (16). In order to avoid such problems, cotyledons are often removed from embryos and only the embryonic axis is used for cryopreservation. However, if excision of cotyledons results in the production of smaller and more homogenous explants, of a size more favourable for freezing, it has been observed with a variety of tropical/sub-tropical woody species that shoot development fails to occur from such explants, with the apical meristem becoming necrotic (30). This lack of regrowth, which is a consequence of excision injury, can be overcome by leaving blocks of cotyledon tissue attached to the axis, even though this increases explant thermal mass unfavourably. A similar observation had been made by Kim et al. (31) with tea zygotic embryos, for which cotyledonary embryonic axes (excised cotyledons containing embryonic axis) showed higher regrowth rate than excised embryonic axes.

Zygotic embryos are histologically highly heterogeneous structures. However, embryos are often of very complex tissue composition which display differential sensitivity to desiccation

Table 1
List of species for which cryopreservation protocols have been developed using embryos and embryonic axes

Species	Reference
Acer platanoides	(91)
A. pseudoplatanus	(91)
Aesculus hippocastanum	(92)
A. glabra	(93)
Amaryllid	(94)
Anadenantha colubrina	(95)
Arachis hypogaea	(96)
Araucaria hunstenii	(97)
Artocarpus heterophyllus	(35, 40, 98)
Azadirachta indica	(99, 100)
Baccaurea motleyana	(37, 101)
B. polyneura	(37, 101)
Bactris gasipaes	(44)
Bletilla striata	(102)
Brassica napus	(103)
Byrsonima intermedia	(104)
Calamus mannan	(105)
Camellia japonica	(106)
C. sinensis	(31, 33, 35, 51–53, 107)
Capsella bursa-pastoris	(108)
Carva	(92)
Carya illinoiensis	(109)
Castanea sativa	(45, 84, 92, 93)
Citrus aurantifolia	(110)
C. latipes	(111)
C. madurensis	(48, 50, 112)
C. macroptera	(111)
C. sinensis	(113, 114)
C. suhuiensis	(115)
Cocos nucifera	(32, 116–118)

(continued)

Table 1 (continued)

Species	Reference
Coffea spp.	(119–121)
C. arabica	(122)
C. liberica	(123)
Corylus avelana	(124, 125)
Durio zibenthinus	(39)
Ekebergia capensis	(126)
Elaeis guineensis	(127–129)
Elateriospermum tapos	(130)
Euphoria longan	(131)
Fagus	(91, 92)
Fraxinus excelsior	(132)
Hevea brasiliensis	(46, 49, 133)
Hopea odorata	(38)
Hordeum vulgare	(103)
Howea foeteriana	(134)
Ilex brasiliensis, *I. brevicuspis*, *I. dumosa*, *I. intergerrima*, *I. paraguariensis*, *I. pseudoboxus*, *I. taubertiana*, *I. theezans*	(42)
Juglans regia	(92)
Landolphia kirkii	(41)
Lansium domesticum	(101)
Livistona chinensis	(135)
Manihot esculenta	(136)
Melia azedarach	(53)
Musa acuminata, *M. balbisiana*	(137)
Nephelium lappaceum	(39)
Olea europaea	(138)
Paeonia lactiflora	(43)
Phaseolus vulgaris	(139)
Pinus radiata	(140)

(continued)

Table 1
(continued)

Species	Reference
Pisum sativum	(58)
Poncirus trifoliata	(141, 142)
Prunus amygdalus	(143)
Prunus persica	(144)
Ptychospermum macarthurii	(37)
Quercus faginea	(145)
Q. falcata	(93)
Q. ilex	(146)
Q. macrocarpa	(93)
Q. nigra	(93)
Q. palustris	(93)
Q. robur	(147)
Q. rubra	(93)
Q. suber	(146)
Ricinus communis	(148)
Sechium edule	(149)
Shorea leprosula	(38)
Shorea odorata	(38)
S. ovalis	(37)
S. parvifolia	(37)
Sterculia cordata	(36, 47)
Swietenia macrophylla	(150)
Theobroma cacao	(151)
Triticum aestivum	(139, 152)
Veitchia merillii	(133)
Vigna	(123)
Zea mays	(153, 154)
Zizania palustris	(155)
Z. texana	(156)

and freezing, the root pole seeming more resistant than the shoot pole. Due to the characteristics of their cells (small size, low vacuolation, dense cytoplasm, high nucleocytoplasmic ratio), meristematic zones (root and shoot pole) withstand desiccation and freezing much better than more differentiated tissues such as cotyledons which are highly hydrated and are therefore severely harmed or killed by desiccation and freezing. A histological study of coconut embryos during freezing has shown that the haustorium (cotyledon) is destroyed during the cryopreservation process (32). As a result, no development of the haustorium is observed during regrowth of frozen embryos, in contrast to non-frozen controls.

As regards the importance of the developmental status of embryos, there is generally an optimal developmental stage, which leads to higher survival after cryopreservation. In the case of coconut embryos, survival after cryopreservation could be obtained with both immature (7–8 month-old) embryos and fully mature (12–13 month-old) embryos (31, 33). However, despite their smaller size, which is a favourable element for cryopreservation, only a few plantlets could be regenerated from immature embryos, whereas cryopreserved mature embryos consistently produced high percentages of fully developed plantlets. This result is related to the high difficulties faced with in vitro culture of immature embryos, for which very complex culture media and sequences are required in order to allow their development to take place (34). Comparable results have been notably obtained during cryopreservation of tea, jackfruit (35), and *Sterculia* (36) embryos, with embryos at an intermediate development stage producing the highest recovery percentages.

5.2. Techniques Employed

Various cryopreservation techniques have been employed for cryopreservation of embryos and embryonic axes. However, the desiccation technique has been used with the majority of the species tested, especially during the earlier experiments performed in the 1980s and 1990s (28). In these experiments, survival was extremely variable, regeneration frequently restricted to callusing or incomplete development of plantlets and the number of accessions tested per species was generally very low, as demonstrated by the in-depth critical analysis performed by Engelmann (28). A number of reasons have been mentioned to explain this situation. Most of the species studied were wild species, in their majority tropical forest trees. As a consequence, no or little is known on the biology, and all the more so on the seed storage behaviour of many of these species. In cases where some information on seed storage behaviour was available, tissue culture protocols, including inoculation in vitro, germination and growth of plantlets, propagation and acclimatisation which are needed for regrowth of embryos and embryonic axes after freezing, were often non-existent or not fully operational. Seeds and embryos of recalcitrant species also display very important variations in moisture content and maturity stage between provenances, between and among seed lots, as well as between successive harvests, which make their cryopreservation difficult.

The optimal embryo/embryonic axis MC for cryopreservation vary depending on the species, between, for example 4 and 8% MC (fresh weight basis) for *Shorea* (37, 38) and 30 and 33% MC for *Nephelium lappaceum* (39) and *Artocarpus heterophyllus* (40). The group of Prof. Patricia Berjak (University of Natal, Durban, South Africa) has demonstrated that very rapid desiccation (a technique termed Flash-drying) followed by ultra-rapid freezing was very effective for cryopreservation of several species including tea and *Landolphia kirkii* (41).

More recent experiments have capitalised upon the newer cryopreservation techniques developed during the 1990s and 2000s, which have been described in a previous section of this chapter. Better results have generally been obtained compared with earlier works, thanks to the higher efficiency of these techniques and to the fact that, for most of the plant species studied, in vitro culture protocols for their embryos and/or embryonic axes were already sufficiently developed when cryopreserved experiments were initiated. As a result, high survival percentages and recovery of whole plantlets from cryopreserved material have been more frequently reported.

The new cryopreservation techniques most frequently employed for freezing embryos are vitrification and encapsulation-dehydration. With both techniques, optimal conditions vary depending on the sensitivity of the species to exposure to cryoprotectant solutions and desiccation. For encapsulation-dehydration, embryos of various *Ilex* species were excised from seeds, pre-cultured for 1 week on medium with 0.3 M sucrose, encapsulated in 3% calcium alginate, pre-treated in liquid medium with daily increasing sucrose concentration (from 0.5 to 1.0 M), then desiccated to around 25% MC (fresh weight basis) before freezing (42), leading to 18–83% survival depending on species. In the case of *Poenia* embryos, optimal conditions consisted of pre-culture on standard medium for 1 day followed by encapsulation in calcium alginate and treatment for 1 h in medium with 2 M glycerol and 0.5 M sucrose (43). For peach palm cryopreservation, embryos were encapsulated in alginate with 2 M glycerol and 0.4 M sucrose, then pre-treated for 24 h in medium with 1.0 M sucrose and desiccated to 20% MC before rapid freezing (44). Under these conditions, 29% of frozen embryos withstood cryopreservation and developed plantlets.

When the vitrification technique was tested for freezing zygotic embryos and embryonic axes, the PVS2 solution (30% glycerol + 15% ethylene glycol (EG) + 15% dimethyl sulfoxide (DMSO) + 0.4 M sucrose) was generally employed and an optimal PVS2 treatment duration had to be determined. In the case of *Castanea* embryos, 68% recovery was achieved by 3-day pre-culture on high sucrose medium followed by 60 min application of PVS2 vitrification solution prior to cryogenic storage (45). With *Hevea* zygotic embryos, the PVS2 vitrification solution was more effective than the other solutions tested (PVS: 22% glycerol + 15% EG + 15% propylene glycol + 7% DMSO + 0.4 M sucrose; and L: 20% glycerol + 30% EG + 10% DMSO + 15% sucrose + 10 mM $CaCl_2$) to achieve survival after freezing in liquid nitrogen (46). The optimal PVS2 exposure duration was 80 min. By contrast, when freezing *Sterculia* embryos, no optimal PVS2 treatment could be identified among the durations tested, which varied between 1 and 3 h (47). It should be noted that with *Sterculia*, the PVS2 treatment was non-toxic to the embryos, in contrast to what is generally observed with other plant species. A very comprehensive study has been performed with *Citrus madurensis* embryonic axes (48). Among the seven different loading solutions tested, the solution containing 2 M glycerol + 0.4 M sucrose was the most efficient. Of the six vitrification solutions tested, the PVS2 vitrification solution, applied for 20 min at 25°C or for 60 min at 0°C, ensured the highest survival. A three-step vitrification protocol, involving treatment of embryonic axes at 0°C with half-strength PVS2 solution for 20 min, then with full-strength PVS2 for an additional 40 min was more efficient than a two-step protocol that involved treatment of axes directly with full-strength PVS2 solution for 60 min. After rapid immersion in liquid nitrogen, rapid rewarming, unloading in a 1.2-M sucrose solution for 20 min, culture on solid medium with 0.3 M sucrose for 1 day and growth recovery for 4 weeks on standard medium, survival of *C. madurensis* embryonic axes reached 85% following the three-step process, compared with 70% for the two-step process.

In some cases, only one cryopreservation technique proved to be effective for a given material, whereas in other cases it was possible to achieve survival of the same material using different techniques, developed in the same or in different laboratories. In the case of *Hevea*, vitrification was very effective for freezing zygotic embryos, whereas desiccation of embryos, embryos with sucrose pre-treatment, and encapsulated embryos with sucrose pre-treatment was relatively ineffective (46, 49). By contrast, *Castanea* embryos were successfully cryopreserved using vitrification and also desiccation to 24–20% MC (FWB), giving between 93 and 100% survival (45). *Citrus madurensis* embryonic axes were successfully frozen using both the vitrification (48) and the encapsulation-dehydration technique (50) with optimal survival reaching 85% for vitrification and 65% for encapsulation-dehydration. Cryopreservation of tea embryonic axes has been successfully achieved by several research teams in various countries. In Korea, embryonic axes were cryopreserved using flash-drying (31) and desiccation (33). Flash-drying was also employed in South Africa (51) and desiccation in India (52). More recently, tea axes were also cryopreserved using the encapsulation-dehydration technique by an Iranian group (53).

5.3. Prospects

There are various options to consider for improving storage of zygotic embryos and embryo axes. First of all, basic knowledge of the plant material studied, including its biology, physiology, etc., is a pre-requisite to any cryopreservation project. Indeed, it has been demonstrated that physiological parameters, for example the developmental stage of embryos, are of critical importance for the success of any cryopreservation experiment (35, 54). An operational in vitro culture protocol for the plant material studied including disinfection, inoculation in vitro, germination of embryos or embryonic axes, plantlet development, and possibly limited propagation also needs to be established.

Most importantly, different analytical techniques are now available to describe and understand the physical and biological processes which take place in explants during cryopreservation (16). These techniques are extremely useful in cryopreservation protocol development, and allow to move from the more empirical approach followed in earlier cryopreservation works to a more scientific and rational approach for the establishment of a cryopreservation protocol. The most directly relevant technique is differential scanning calorimetry (DSC), which allows, among various possibilities, to measure the thermal events occurring in samples during cooling and warming. There is indeed a narrow window of hydration levels within which survival is possible, and the optimal water content for cryopreservation usually corresponds to the unfrozen water content, as shown notably with coffee and citrus seeds (55, 56).

The new vitrification-based cryopreservation techniques described and discussed in previous sections of this chapter including pre-growth-desiccation, encapsulation-dehydration, vitrification, encapsulation-vitrification, and droplet-vitrification offer interesting possibilities for cryopreserving zygotic embryos and embryonic axes. Manipulating the plant material before cryopreservation may also have a positive impact on cryopreservation results. Notably, pre-growth on media containing cryoprotective substances may confer on the tissues increased tolerance to further desiccation and reduce the heterogeneity of the material (27, 57). Other medium manipulations may be envisaged. Mycock (58) indicated that the addition of calcium and magnesium ions in the cryoprotectant solutions employed for freezing pea embryos improved recovery by reducing callus formation. The reduction in callus formation is proposed to be a result of the stabilisation of the subcellular cytoskeletal system by the added elements. More recently, Uchendu et al. (59) showed the beneficial effect of adding the antioxidants vitamin C and E to various media employed for freezing *Rubus* shoot tips.

The very precise control of sample desiccation, which is a crucial step of cryopreservation protocols, by using saturated salt solutions, which has been successfully employed for freezing coffee and citrus seeds (56, 60) may also be instrumental in improving recovery. Preliminary experiments performed recently with coffee embryos (Dussert, 2009, personal communication) have shown promising results.

Finally, with species for which attempts to freeze whole embryos or embryonic axes have proven unsuccessful, it has been suggested to use shoot apices sampled on the embryos, adventitious buds or somatic embryos induced from the embryonic tissues (57). This might be the only solution for species which do not have well-defined embryos but this will request that more sophisticated tissue culture procedures be developed and mastered. A recent example of this strategy has been provided by Varghese et al. (61) for freezing *Trichilia emetica*, a tropical tree species producing recalcitrant seeds, whose embryos and embryonic axes have proven impossible to cryopreserve (62). Shoot tips sampled on in vitro plantlets originating from in vitro germinated seeds were successfully cryopreserved following treatment with sucrose and glycerol, cryoprotection with the PVS2 solution, and slow cooling, with 71% of cryopreserved shoot tips producing shoots.

6. Cryopreservation of Somatic Embryos

6.1. Characteristics of Plant Material

In Table 2, we have listed the plant species for which cryopreservation work has been performed using somatic embryos only, but not embryogenic cultures, which are usually cryopreserved using the classical techniques established for cell suspensions and calluses (16).

Somatic embryo cultures are generally highly heterogenous since they consist of embryos at different developmental stages. However, embryos at one particular stage are selected for freezing experiments. As a consequence, the material employed for cryopreservation becomes highly homogenous in terms of size, water content, and histological composition.

The plant species included in Table 2 originate from tropical and temperate areas; their seeds can fall in the orthodox, intermediate, or recalcitrant categories. They are generally cultivated plants for which the establishment of large-scale propagation protocols is relevant. In most cases, the primary objective for the establishment of a cryopreservation protocol is the improvement of large-scale production management; plant genetic resource conservation is only a secondary objective.

As in all cryopreservation experiments, the developmental stage of the plant material is a very important parameter, as it is correlated with size, bearing the risk that explants become too large, with the occurrence of detrimental dehydration and thermal gradients and with their histological structure, with the problems

Table 2
List of species for which cryopreservation protocols have been developed using somatic embryos

Species	Reference
Abies nordmanniana	(157)
Aesculus hippocastanum	(158)
Asparagus officinalis	(159, 160)
Brassica napus	(161)
Camellia japonica	(106)
C. sinensis	(162)
Carya illinoensis	(163)
Castanea sativa	(84)
Citrus grandis	(164)
C. junos	(164)
C. platymamma	(164)
C. sinensis	(73, 77, 165)
Clitoria ternatea	(166)
Cnidium officinale	(167)
Coffea arabica	(63, 71, 72, 88)
C. canephora	(82, 87)
Coriandrum sativum	(168)
Cucumis melo	(74, 169)
Daucus carota	(79, 170, 171)
Elaeis guineensis	(75, 76, 172)
Iris nigricans	(173)
Juglans regia	(68, 163, 174)
Macropidia fulginosa	(83)
Manihot esculenta	(88, 175)
Melia azedarach	(78, 176)
Olea europea	(80)
Paeonia lactiflora	(65)
Phoenix dactylifera	(88)
Picea mariana	(177)
P. glauca	(177, 178)

(continued)

Table 2 (continued)

Species	Reference
P. glauca x engelmannii	(178)
Pinus patula	(179)
Picea sitchensis	(180)
Pisum sativum	(88)
Quercus suber	(181)
Saccharum spp.	(86)
Theobroma cacao	(64, 81, 182)
Vitis vinifera	(183)

created by complex tissue composition, which will have different desiccation/freezing sensitivities. In the case of carrot somatic embryos, Florin et al. (63) showed that embryos cryopreserved at the heart and torpedo stages (with a size up to 0.50 mm) displayed a better development and conversion rate than larger embryos. Smaller embryos were not or little injured by freezing whereas larger embryos withstood irreversible freezing injury. The increased cellular differentiation in larger embryos leads to different cell types with different water contents, making controlled osmotic dehydration for different tissues difficult to achieve. With cocoa, early cotyledonary somatic embryos withstood freezing better than globular, heart, and torpedo embryos (64). In the case of *Paeonia*, embryos of intermediate size displayed higher survival than smaller or larger embryos (65).

6.2. Techniques Employed

It is important to mention that it is with somatic embryos that DSC was first employed to assist in the development of cryopreservation protocols. Using DSC, it was shown with oil palm somatic embryos that recovery of cryopreserved samples increased in line with decreasing freezable water content in samples (66). Recovery was optimal when all freezable water had been extracted from embryos during conditioning treatment (sucrose pre-treatment followed by desiccation), that is only glass transitions were recorded during cooling and warming, indicating that vitrification of internal solutes occurred. Similar results have been obtained notably with carrot (67) and *Juglans* (68) somatic embryos.

Depending on the material, different cryopreservation techniques (mostly vitrification-based) have been used for freezing somatic embryos, viz. desiccation, pre-growth-desiccation, encapsulation-dehydration, vitrification, encapsulation-vitrification, and droplet-vitrification. There are also a few examples of the utilisation of slow controlled freezing, which correspond to the earlier experiments performed with somatic embryos.

It is interesting to note that a number of the protocols developed for temperate and/or even some tropical plants include a treatment of the somatic embryos with abscisic acid (ABA) before cryopreservation. ABA is an important stress hormone which increases tolerance of plants to desiccation and low temperature. Its mode of action includes maintenance of water balance in cells (69) and triggering specific genes that initiate the production of antifreeze products (70). The ABA concentration employed and treatment duration vary depending on the species. In the case of *Brassica* somatic embryos, the ABA treatment consisted of a 7-day application of 10 mg/L ABA. Coffee somatic embryos were treated with 1 µM ABA for 6 weeks before freezing (71), whereas the optimal treatment for *Aesculus* was 4 days with 0.75 µM ABA.

The first experiments performed for somatic embryo cryopreservation employed classical cryopreservation protocols, including pre-treatment of material with cryoprotectant mixtures consisting of sucrose and dimethylsulfoxide (DMSO), slow pre-freezing in a programmable freezer followed by immersion of samples in liquid nitrogen. Examples include, notably, coffee somatic embryos (72) which were pre-treated with 0.5 M sucrose and 5% DMSO, then pre-frozen at 0.5°C/min to –40°C before immersion in liquid nitrogen, and *Citrus sinensis* embryos (73), which were treated with 0.1 M sucrose and 10% DMSO then cooled at 0.5°C/min to –42°C, before immersion in liquid nitrogen.

The desiccation technique has produced very good results with embryos of desiccation-tolerant species such as conifers, melon, *Brassica*, or *Paeonia*. Embryos are placed in containers with controlled relative humidity (RH) produced using saturated solutions until equilibration of their water content with the external RH, then frozen rapidly. Optimal conditions for *Picea glauca* and *P. mariana* included equilibration of embryos for 48 h in 97% RH, resulting in a moisture content decrease to 0.23 g/g H_2O and a recovery percentage of 93% after cryopreservation. In the case of melon somatic embryos, equilibration in 60% RH reduced water content to 11.8% and resulted in 65% survival after cryopreservation (74).

Pre-growth-desiccation, which includes culture of embryos on medium with high sucrose concentration followed by desiccation, was first established with oil palm somatic embryos (75, 76). Optimal conditions vary depending on the material. With oil palm, embryos were pre-grown for 7 days on medium with 0.75 M sucrose before desiccation to around 30% MC (76). Coffee embryos required a 2-week treatment with 0.44 M sucrose followed by desiccation to 25% MC (71), whereas citrus embryos were treated for only 1 day with 0.75 M sucrose, then dehydrated to 20–25% MC (77). *Melia* embryos were pre-treated with daily increasing sucrose concentrations, from 0.5 to 1 M, desiccated to 19% MC, then frozen slowly at 1°C/min to −30°C before immersion in liquid nitrogen (78).

Encapsulation-dehydration was first applied to carrot somatic embryos (79). The main parameters of the technique, viz. duration of sucrose treatment, concentration of sucrose and optimal water content vary depending on the material employed. With carrot somatic embryos, optimal conditions were pre-culture in medium with 0.3 M sucrose for 18 h followed by desiccation to 19% MC, ensuring 92% survival after cryopreservation (79). After treatment for 4 days with 0.75–1.25 M sucrose, desiccation to 21% MC and rapid freezing, survival of olive somatic embryos was 40% (80). In the case of *Theobroma cacao*, embryos were treated for 7 days with 1 M sucrose and desiccated 16% MC. After freezing, regrowth was between 25 and 72% with the four genotypes tested (81). An original protocol was established for coffee somatic embryos (82). Indeed, naked embryos were pre-treated on media with progressively increasing sucrose concentration, from 0.3 to 0.8 M, and only then encapsulated in alginate beads with 0.5 M sucrose. After desiccation to 13% MC, direct regrowth was observed on around 30% of cryopreserved embryos.

All vitrification protocols developed to date with somatic embryos use the PVS2 vitrification solution. *Macropidia* embryos were successfully cryopreserved with the following procedure: they were pre-treated for 2 days on medium with 0.8 M glycerol, then treated with PVS2 at 0°C for 30–40 min before rapid freezing (83), thus achieving 90% regrowth. In the case of *Castanea* (84), optimal conditions including pre-treatment with 0.3 M sucrose for 3 days, dehydration with PVS2 for 60 min ensured 68% regrowth of somatic embryos after cryopreservation. A recent study performed on *Theobroma cacao* showed that high survival (74%) of somatic embryos was achieved after treatment with 0.5 M sucrose, loading, PVS2 treatment for 60 min at 0°C, and rapid freezing (85).

Encapsulation-vitrification has been successfully applied to olive and sugarcane somatic embryos (80, 86). With olive, 64% survival was achieved after pre-culture of encapsulated embryos for 4 days in 0.75–1.25 M sucrose, 3 h treatment with PVS2 vitrification solution and rapid freezing. With sugarcane, somatic embryos were loaded in 2 M glycerol and 0.4 M sucrose for 20 min, treated with PVS2 solution for 80 min at 0°C and frozen rapidly, ensuring 30% recovery after cryopreservation.

Until now, droplet-vitrification has been experimented only with sugarcane embryos (86). Embryos were loaded with 1.5 M glycerol + 0.3 M sucrose, dehydrated with PVS2 vitrification solution for 20–40 min, and cooled rapidly. Under these conditions, viability reached 55%.

It has been shown in several cases that different techniques could be developed for the same material, either in the same laboratory or in different laboratories. The best example is provided by coffee for which somatic embryos have been cryopreserved in France using slow freezing (72, 87) and pre-growth-desiccation (63), in South Africa using desiccation (88) and in Japan using encapsulation-dehydration (83). In the case of sugarcane, encapsulation-dehydration and droplet-vitrification were successful, the highest results being obtained with the latter technique, but no survival was achieved with vitrification (86). Finally, *Castanea* embryos were cryopreserved using pre-growth-desiccation and vitrification, giving 33% survival with the former technique and 68% with the latter (84).

6.3. Prospects

There are very good prospects for the future development and application of cryopreservation to somatic embryos. Indeed, the protocols available are generally very effective, thanks to the positive impact of the new, vitrification-based, cryopreservation techniques. They have been developed for a broad range of species and there are already examples of their routine application, as in the case of coffee (89). The use of analytical tools, especially DSC, has been very instrumental in facilitating their optimisation in cases where they have been employed. With the current state of the art, these protocols are relatively easy to establish. The information published indicates indeed that, for any new material, there is at least one technique which produces positive results among the techniques tested. Today, the main bottleneck seems to lie more with the establishment of somatic embryo cultures from any new material than with the development of a cryopreservation protocol for this material.

7. Conclusion

In conclusion, we have reviewed in this chapter the past and current research on cryopreservation of zygotic and somatic embryos. The development and application of cryopreservation is significantly more advanced for somatic embryos, in comparison with zygotic embryos. This is mainly due to the different origin and characteristics of the species treated. In the majority of cases, zygotic embryos originate from tropical, wild species, for which knowledge and techniques relevant to the development of cryopreservation protocols are limited, or even non-existent. By contrast, somatic embryos are generally produced from cultivated species, which have already been studied extensively and for which propagation techniques are already operational. A number of technical possibilities to explore have been identified in order to improve the development of cryopreservation protocols for zygotic embryos and embryonic axes.

For both categories of materials, the utilisation of analytical techniques has proved to be extremely useful to assist in the development of cryopreservation protocols. It is obvious that a better understanding of the biological and physical processes underlying the tolerance of plant tissues to desiccation and freezing have had a highly significant impact on cryopreservation research. This stresses the importance of increasing the level of fundamental research related to cryopreservation. The first International Symposium on cryopreservation in horticultural species, held in Belgium in April 2009 (http://www.biw.kuleuven.be/dtp/tro/ISHSPlantCryo/) has shown that the number of researchers worldwide working on different fundamental areas related to cryopreservation is increasing, which is an encouraging signal.

Concomitantly to cryopreservation, a strong research focus should be directed towards the development of improved seed storage techniques. Indeed, it can be expected that progress made in this area will result in the possibility of storing genetic resources of an increasing number of plant species in seed form. The results obtained recently with coffee and citrus seeds (56), for which cryopreserved storage has now become a reality, thanks to the very precise control of seed desiccation, are very good examples of such possibilities.

Finally, attention should also focus on two important areas related to the development and use of cryopreservation. The first one is the analysis of the costs of cryopreserved storage in comparison with other storage techniques. A very thorough study has been published recently, which clearly demonstrates the cost-efficiency of cryopreserved storage for plant genetic resource conservation (90). The second one concerns the integration of cryopreservation as an additional technology towards the development of complementary plant genetic resource conservation strategies. It is indeed necessary to start envisaging now how the future availability of a cryopreservation protocol for a given plant will impact on the respective utilisation of the other existing conservation technologies.

References

1. Roberts EH (1973) Predicting the viability of seeds. Seed Sci Technol 1:499–514
2. FAO (1996) Report on the State of the World's Plant Genetic Resources for Food and Agriculture. Food and Agriculture Organization of the United Nations, Rome
3. FAO and IPGRI (1994) Genebank Standards. Food and Agriculture Organization of the United Nations, IPGRI, Rome
4. Cromarty AS, Ellis RH, Roberts EH (1982) The design of seed storage facilities for genetic conservation. Handbooks for Genebanks No 1. IPGRI, Rome
5. Ellis RH, Hong TD, Roberts EH (1985) Handbook of seed technology for genebanks Vol. I: principles and methodology. Handbooks for Genebanks No. 2. IPGRI, Rome

6. Ellis RH, Hong TD, Roberts EH (1985) Handbook of seed technology for genebanks Vol. II: compendium of specific germination information and test recommendations. Handbooks for Genebanks No. 3. IPGRI, Rome, Italy
7. Kameswara Rao N, Hanson J, Dulloo ME, Ghosh K, Nowell D, Larinde M (2006) Manual of seed handling in genebanks. Handbooks for Genebanks No. 8. IPGRI, Rome
8. Walters C (ed) (1998) Ultra-dry technology: perspective from the National Seed Storage Laboratory, USA. Seed Sci Res 8, Suppl. No 1:11–14
9. Chin HF (1988) Recalcitrant seeds: a status report. IPGRI, Rome
10. Chin HF, Roberts EH (eds) (1980) Recalcitrant crop seeds. Tropical Press Sdn. Bhd, Kuala Lumpur
11. Hong TD, Linington S, Ellis RH (1996) Seed storage behaviour: a compendium. Handbooks for Genebanks: No. 4. IPGRI, Rome
12. Ellis RH, Hong T, Roberts EH (1990) An intermediate category of seed storage behaviour? I. Coffee. J Exp Bot 41:1167–1174
13. Ellis RH, Hong T, Roberts EH, Soetisna U (1991) Seed storage behaviour in *Elaeis guineensis*. Seed Sci Res 1:99–104
14. Engelmann F (1997) In vitro conservation methods. In: Ford-Lloyd BV, Newburry JH, Callow JA (eds) Biotechnology and plant genetic resources: conservation and use. CAB International, Wallingford, pp 119–162
15. Withers LA, Engelmann F (1998) In vitro conservation of plant genetic resources. In: Altman A (ed) Biotechnology in agriculture. Marcel Dekker, New York, pp 57–88
16. Engelmann F (2003) Current research status and utilization of plant cryopreservation. In: Proceedings of the International Workshop on Cryopreservation of Bio-genetic Resources, International Technical Cooperation Center, RDA, Suwon, Korea, 3–5 June 2003, pp 19–40
17. Mazur P (1984) Freezing of living cells: mechanisms and applications. Am J Physiol 143:C125–C142
18. Fahy GM, MacFarlane DR, Angell CA, Meryman HT (1984) Vitrification as an approach to cryopreservation. Cryobiology 21:407–426
19. Engelmann F, Takagi H (eds) (2000) Cryopreservation of tropical plant germplasm – current research progress and applications. Japan International Centre for Agricultural Sciences, Tsukuba, IPGRI, Rome
20. Reed BM (2008) Plant cryopreservation – a practical guide. Springer, New York
21. Gonzalez-Arnao MT, Panta A, Roca WM, Escobar RE, Engelmann F (2008) Development and large scale application of cryopreservation techniques for shoot and somatic embryo cultures of tropical crops. Plant Cell Tiss Org Cult 92:1–13
22. Reed BM, Chang Y (1997) Medium- and long-term storage of in vitro cultures of temperate fruit and nut crops. In: Razdan MK, Cocking EC (eds) Conservation of plant genetic resources in vitro, vol 1, General aspects. Science Publishers, Enfield, pp 67–105
23. Sakai A, Kobayashi S, Oiyama I (1990) Cryopreservation of nucellar cells of navel orange (*Citrus sinensis* Osb. *var. brasiliensis* Tanaka) by vitrification. Plant Cell Rep 9:30–33
24. Engelmann F (1997) In vitro conservation methods. In: Ford-Lloyd BV, Newburry JH, Callow JA (eds) Biotechnology and plant genetic resources: conservation and use. CABI, Wellingford, pp 119–162
25. Sakai A, Engelmann F (2007) Vitrification, encapsulation-vitrification and droplet-vitrification: a review. Cryo Letters 28:151–172
26. Engelmann F (1997) Importance of desiccation for the cryopreservation of recalcitrant seed and vegetatively propagated species. Plant Genet Res Newslett 112:9–18
27. Engelmann F (2000) Importance of cryopreservation for the conservation of plant genetic resources. In: Engelmann F, Takagi H (eds) Cryopreservation of tropical plant germplasm – current research progress and applications. Japan International Centre for Agricultural Sciences, Tsukuba, IPGRI, Rome, pp 8–20
28. Engelmann F (1999) Alternative methods for the storage of recalcitrant seeds – an update. In: Marzalina M, Khoo KC, Tsan FY, Krishnapillay B (eds) IUFRO Seed Symposium 1998 "Recalcitrant Seeds". FRIM, Kuala Lumpur, pp 159–170
29. Panis B, Strosse H, Van den Henda S, Swennen R (2002) Sucrose preculture to simplify cryopreservation of banana meristem cultures. Cryo Letters 23:375–384
30. Berjak P, Pammenter NW, Goveia M, Sershen ED (2009) Cryo-conservation of genetic diversity of recalcitrant-seeded species via zygotic embryonic axes: successes and problems. In: 1st International Symposium "Cryopreservation in Horticultural Species", Leuven, Belgium, 5–8 April 2009, p 33 (abstract)

31. Kim HH, Cho KS, Cho KT, Cha YS, Baek HJ, Gwag JK, Cho HJ (1998) Cryopreservation of tea (*Camellia sinensis* L.) seed embryos by flash drying method. RDA J Indust Crop Sci 40:18–24
32. Assy-Bah B, Engelmann F (1992) Cryopreservation of mature embryos of coconut (*Cocos nucifera* L.) and subsequent regeneration of plantlets. Cryo Letters 13:117–126
33. Kim HH, Cha YS, Baek HJ, Cho EG, Chae YA, Engelmann F (2002) Cryopreservation of tea (*Camellia sinensis* L.) seeds and embryonic axes. Cryo Letters 23:209–216
34. Engelmann F (1992) Cryopreservation of embryos. In: Dattée Y, Dumas C, Gallais A (eds) Reproductive biology and plant breeding. Springer, Berlin, pp 281–290
35. Chandel KPS, Chaudhury R, Radhamani J, Malik SK (1995) Desiccation and freezing sensitivity in recalcitrant seeds of tea, cocoa and jackfruit. Ann Bot 76:443–450
36. Nadarajan J, Staines HJ, Benson EE, Marzalina M, Krishnapillay B, Harding K (2006) Optimization of cryopreservation protocol for *Sterculia cordata* zygotic embryos using Taguchi experiments. J Tropical Forest Sci 18:222–230
37. Normah MN, Marzalina M (1995) Achievements and prospects of in vitro conservation for tree germplasm. In: Normah MN, Narimah MK, Clyde MM (eds) In vitro conservation of plant genetic resources. Plant Biotechnology Laboratory, Universiti Kebangsaan, Kuala Lumpur, pp 253–261
38. Chai Y, Marzalina M, Krishnapillay B (1994) Effect of sucrose as a cryoprotective agent in the cryopreservation of some Dipterocarp species. In: Proceedings of VIth National Symposium, Kuala Lumpur
39. Hor YL, Stanwood PC, Chin HF (1990) Effects of dehydration on freezing characteristics and survival in liquid nitrogen of three recalcitrant seeds. Pertanika 13:309–314
40. Krishnapillay B (1989) Towards the development of a protocol for cryopreservation of embryos of a recalcitrant seed (*Artocarpus heterophyllus* Lam.). Ph.D. Thesis, Universiti Pertanian, Malaysia
41. Berjak P, Farrant JM, Mycock DJ, Pammenter NW (1989) Homoiohydrous (recalcitrant) seeds: the enigma of their desiccation sensitivity and the state of water in axes of *Landolphia kirkii* Dyer. Planta 186:249–261
42. Mroginski LA, Sansberro PA, Scocchi AM, Luna C, Rey HYA (2008) Cryopreservation protocol for immature zygotic embryos of species of *Ilex* (Aquifoliaceae). Biocell 32:33–39
43. Kim HM, Shin JH, Sohn JK (2004) Cryopreservation of zygotic embryos of herbaceous peony (*Paeonia lactiflora* Pall.) by encapsulation-dehydration. Korean J Crop Sci 49:354–357
44. Steinmacher DA, Saldanha CW, Clement CR, Guerra MP (2007) Cryopreservation of peach palm zygotic embryos. Cryo Letters 28:13–22
45. San-Jose MC, Jorquera L, Vidal N, Corredoira E, Sanchez C (2005) Cryopreservation of European chestnut germplasm. Acta Hortic 693:225–231
46. Sam YY, Hor YL (1999) Effects of vitrification solution on survival of zygotic embryos of rubber (Hevea brasiliensis) in liquid nitrogen. In: Marzalina M, Khoo KC, Jayanthi N, Tsan FY, Krishnapillay B (eds) IUFRO Seed Symposium 1998 "Recalcitrant seeds". Proceedings of the Conference, Kuala Lumpur, Malaysia, 12–15 October 1998, Forest Research Institute Malaysia, Kuala Lumpur, Malaysia, pp 146–152
47. Nadarajan J, Staines HJ, Benson EE, Marzalina M, Krishnapillay B, Harding K (2007) Optimization of cryopreservation for *Sterculia cordata* zygotic embryos using vitrification techniques. J Tropical Forest Sci 19:79–85
48. Cho EG, Hor YL, Kim HH, Rao VR, Engelmann F (2001) Cryopreservation of *Citrus madurensis* zygotic embryonic axes by vitrification: importance of pregrowth and preculture conditions. Cryo Letters 22:391–396
49. Yap LV, Hor YL, Normah MN (1999) Effects of sucrose preculture and subsequent desiccation on cryopreservation of alginate-encapsulated Hevea brasiliensis embryo. In: Marzalina M, Khoo KC, Jayanthi N, Tsan FY, Krishnapillay B (eds) IUFRO Seed Symposium 1998 "Recalcitrant seeds". Proceedings of the Conference, Kuala Lumpur, Malaysia, 12–15 October 1998, Forest Research Institute Malaysia, Kuala Lumpur, pp 140–145
50. Cho EG, Hor YL, Kim HH, Ramanatha Rao V, Engelmann F (2002) Cryopreservation of *Citrus madurensis* embryonic axes by encapsulation-dehydration. Cryo Letters 23:325–332
51. Wesley-Smith J, Vertucci CW, Berjak P, Pammenter NW, Crane J (1992) Cryopreservation of desiccation-sensitive axes of *Camellia sinensis* in relation to dehydration, freezing rate and the thermal properties of tissue water. J Plant Physiol 140:596–604

52. Chaudhury R, Radhamani J, Chandel KPS (1991) Preliminary observations on the cryopreservation of desiccated embryonic axes of tea (*Camellia sinensis* L., O. Kuntze) seeds for genetic conservation. Cryo Letters 12:31–36
53. Kaviani B (2009) Cryopreservation of embryonic axes of *Melia azedarach* L. and *Camellia sinensis* L. by encapsulation-dehydration. In: First International Symposium "Cryopreservation in Horticultural Species", Leuven, Belgium, 5–8 April 2009, p 95 (abstract)
54. Engelmann F, Dumet D, Chabrillange N, Abdelnour-Esquivel A, Assy-Bah B, Dereuddre J, Duval Y (1995) Factors affecting the cryopreservation of coffee, coconut and oil palm embryos. Plant Genet Res Newslett 103:27–31
55. Dussert S, Chabrillange N, Roquelin G, Engelmann F, Lopez M, Hamon S (2001) Tolerance of coffee (*Coffea* spp.) seeds to ultra-low temperature exposure in relation to calorimetric properties of tissue water, lipid composition and cooling procedure. Physiol Plant 112:495–505
56. Hor YC, Kim YJ, Ugap A, Chabrillange N, Sinniah UR, Engelmann F, Dussert S (2005) Optimal hydration status for cryopreservation of intermediate oily seeds: *Citrus* as a case study. Ann Bot 95:1153–1161
57. Pence VC (1995) Cryopreservation of recalcitrant seeds. In: Bajaj YPS (ed) Biotechnology in agriculture and forestry, vol 32, Cryopreservation of plant germplasm I. Springer, Berlin, pp 29–52
58. Mycock D (1999) Addition of calcium and magnesium to a glycerol and sucrose cryoprotectant solution improves the quality of plant embryo recovery from cryostorage. Cryo Letters 20:77–82
59. Uchendu EE, Muminova M, Traber MG, Reed BM (2009) Antioxidants improve regrowth of cryopreserved in vitro shoot tips. In: CRYO 2009, 46th Annual meeting of the Society for Cryobiology, 19–23 July 2009, Sapporo, Japan, p 53 (abstract)
60. Dussert S, Engelmann F (2006) New determinants of coffee (*Coffea arabica* L.) seed tolerance to liquid nitrogen exposure. Cryo Letters 27:169–178
61. Varghese D, Berjak P, Pammenter NW (2009) Cryopreservation of shoot tips of *Trichilia emetica,* a tropical recalcitrant-seeded species. Cryo Letters 30:280–290
62. Kioko JI, Berjak P, Pammenter NW, Watt P, Wesley-Smith J (1998) Desiccation and cryopreservation of embryonic axes of *Trichilia dregeana*. Cryo Letters 19:5–11
63. Florin B, Ducos JP, Firmin L, Meschine MC, Thierry C, Pétiard V, Deshayes A (1995) Preservation of coffee somatic embryos through desiccation and cryopreservation. In: Association Scinetique Inernationale du Café (ed) 16th Int. Sci. Colloquium Coffee, A.S.I.C., Paris, pp 542–547
64. Fang JY, Wetten A, Hadley P (2004) Cryopreservation of cocoa (*Theobroma cacao* L.) somatic embryos for long-term germplasm storage. Plant Sci 166:669–675
65. Kim HM, Shin JH, Sohn JK (2006) Cryopreservation of somatic embryos of the herbaceous peony (*Paeonia lactiflora* Pall.) by air drying. Cryobiology 53:69–74
66. Engelmann F, Dereuddre J (1988) Cryopreservation of oil palm somatic embryos: importance of the freezing process. Cryo Letters 9:220–235
67. Dereuddre J, Hassen N, Blandin S, Kaminski M (1991) Resistance of alginate-coated somatic embryos of carrot (*Daucus carota* L.) to desiccation and freezing in liquid nitrogen: 2 thermal analysis. Cryo Letters 12:135–148
68. de Boucaud MT, Brison M, Negrier P (1994) Cryopreservation of walnut somatic embryos. Cryo Letters 15:151–160
69. Tanino KK, Chen THH, Fughigami LH, Weiser CJ (1990) Metabolic alterations associated with abscisic acid-induced frost hardiness in bromegrass suspension culture cells. Plant Cell Physiol 31:505–511
70. Lee SP, Chen THH, Fuchigami LH (1991) Changes in the translatable RNA population during abscisis acid induced freezing tolerance in bromegrass suspension culture. Plant Cell Physiol 32:45–56
71. Tessereau H, Florin B, Meschine MC, Thierry C, Pétiard V (1994) Cryopreservation of somatic embryos: a tool for germplasm storage and commercial delivery of selected plants. Ann Bot 74:547–555
72. Bertrand-Desbrunais A, Fabre J, Engelmann F, Dereuddre J, Charrier A (1998) Adventitious embryogenesis recovery from coffee (*Coffea arabica* L.) somatic embryos after freezing in liquid nitrogen. C R Acad Sci Paris 307:795–801
73. Marin ML, Gogorcena Y, Ortiz J, Duran-Vila N (1993) Recovery of whole plants of sweet orange from somatic embryos subjected to freezing thawing treatments. Plant Cell Tiss Org Cult 34:27–33
74. Shimonishi K, Ishikawa M, Suzuki S, Oosawa K (2000) Cryopreservation of melon somatic embryos by desiccation method. In: Engelmann F, Takagi H (eds) Cryopreservation

of Tropical Plant Germplasm: Current Research Progress and Application. Proceedings of an International Workshop, Tsukuba, Japan, October 1998, International Plant Genetic Resources Institute (IPGRI), Rome, pp 167–171

75. Engelmann F, Duval Y, Dereuddre J (1985) Survie et prolifération d'embryons somatiques de palmier à huile (*Elaeis guineensis* Jacq.) après congélation dans l'azote liquide. C R Acad Sci Paris 301:111–116
76. Dumet D, Engelmann F, Chabrillange N, Duval Y (1993) Cryopreservation of oil palm (*Elaeis guineensis* Jacq.) somatic embryos involving a desiccation step. Plant Cell Rep 12:352–355
77. Gonzalez-Arnao MT, Juarez J, Ortega C, Navarro L, Duran-Villa N (2003) Cryopreservation of ovules and somatic embryos of citrus using the encapsulation-dehydration technique. Cryo Letters 24:85–94
78. Scocchi A, Vila S, Mroginski L, Engelmann F (2007) Cryopreservation of somatic embryos of paradise tree (*Melia azedarach* L.). Cryo Letters 28:281–290
79. Dereuddre J, Tannoury M, Hassen N, Kaminski M, Vintejoux C (1991) Application of the technique of encapsulation of somatic embryos of carrot (*Daucus carota* L.) when freezing in liquid nitrogen: cytological study. Bull Soc Bot Fr 139:15–33
80. Shibli RA, Al-Juboory KH (2000) Cryopreservation of 'Nabali' olive (*Olea europea* L.) somatic embryos by encapsulation-dehydration and encapsulation-vitrification. Cryo Letters 21:357–366
81. Fang JY, Wetten A, Adu-Gyamfi R, Wilkinson M, Rodriguez-Lopez C (2009) Use of secondary somatic embryos promotes genetic fidelity in cryopreservation of cocoa (*Theobroma cacao* L.). Agric Food Sci 18:152–159
82. Hatanaka T, Yasuda T, Yamaguchi T, Sakai A (1994) Direct regrowth of encapsulated somatic embryos of coffee (*Coffea canephora*) after cooling in liquid nitrogen. Cryo Letters 15:47–52
83. Turner SR, Tan B, Senaratna T, Bunn E, Dixon KW, Touchell D (2000) Cryopreservation of the Australian species *Macropidia fuliginosa* (Haemodoraceae) by vitrification. Cryo Letters 21:379–388
84. Corredoira E, San-José MC, Ballester A, Vieitez AM (2004) Cryopreservation of zygotic embryo axes and somatic embryos of European chestnut. Cryo Letters 25:33–42
85. Adu-Gyamfi R, Wetten A (2009) Cryopreservation of cocoa somatic embryos by vitrification. In: Application of Cryopreservation from Human Tissue to Plant Genebank Integration, Society for Low temperature Biology Annual meeting, University Hannover, Germany, 7–9 Sept 2009, p 34 (abstract)
86. Martinez-Montero ME, Martinez J, Engelmann F (2008) Cryopreservation of sugarcane somatic embryos. Cryo Letters 29:229–242
87. Bertrand-Desbrunais A (1991) La conservation des ressources génétiques des caféiers, Ph.D. Thesis, University Paris 6, France
88. Mycock DJ, Wesley-Smith J, Berjak P (1995) Cryopreservation of somatic embryos of four species with and without cryoprotectant pretreatment. Ann Bot 75:331–336
89. Florin B, Brulard B (2008) Examples of integration of cryopreservation in different plant biotechnology programmes. In: Laamanen J, Uosukainen M, Häggman H, Nukari A, Rantala S (eds) Agrifood Research Working papers 153. Proceedings of COST Action 871 meeting, Oulu, Finland, 20–23 Feb 2008
90. Dulloo ME, Ebert AW, Dussert S, Gotor E, Astorga C, Vasquez N, Rakotomalala JJ, Rabemiafara A, Eira M, Bellachew B, Omondi C, Engelmann F, Anthony F, Watts J, Qamar Z, Snook L (2009) Cost efficiency of cryopreservation as a long term conservation method for coffee genetic resources. Crop Sci 49:1–16
91. Pukacki PM, Jarzabech M, Pukacka M (2009) Characterization of cryoprotective activity of proteins in *Acer*, *Fagus* and *Quercus* embryonic axes. In: First International Symposium "Cryopreservation in Horticultural Species", Leuven, Belgium, 5–8 April 2009, p 117 (abstract)
92. Pence VC, Dresser BL (1988) Embryo cryostorage as a technique for germplasm preservation of several large-seeded tree species. In: Beltsville Symp in Agric Res XIII Biotic Diversity and Germplasm Preservation – Global Imperatives, 9–11 May 1988, p 24 (abstract)
93. Pence VC (1992) Desiccation and the survival of *Aesculus, Castanea*, and *Quercus* embryo axes through cryopreservation. Cryobiology 29:391–399
94. Sershen PNW, Berjak P, Wesley-Smith J (2007) Cryopreservation of selected Amaryllid species. Cryo Letters 28:387–399
95. Nery FC, Paiva R, Campos ACAL, Nogueira GF, Stein VC, Alvarenga AA (2009) Cryopreservation of *Anadenanthera colubrine*

(Vell.) Brenan embryonic axes. In: First International Symposium "Cryopreservation in Horticultural Species", Leuven, Belgium, 5–8 April 2009, p 67 (abstract)

96. Gagliardi RF, Pacheco GP, Valls JFM, Mansur E (2002) Cryopreservation of cultivated and wild *Arachis* species embryonic axes using desiccation and vitrification methods. Cryo Letters 23:61–68
97. Pritchard HW, Prendergast FG (1986) Effects of desiccation on the *in vitro* viability of embryos of the recalcitrant seed species *Araucaria hunstenii* K. Schum. J Plant Physiol 37:1388–1397
98. Thammasiri K (1999) Cryopreservation of embryonic axes of jackfruit. Cryo Letters 20:21–28
99. Berjak P, Dumet D (1996) Cryopreservation of seeds and isolated embryonic axes of neem (*Azadirachta indica*). Cryo Letters 17:99–104
100. Chandel KPS, Chaudhury R, Radhamani J (1996) Cryopreservation of embryos/ embryonic axes – a novel method for the long-term conservation of recalcitrant seed species. In: Normah MN, Narimah MK, Clyde MM (eds) In vitro conservation of plant genetic resources. Percetakan Watan Sdn Bhd, Kuala Lumpur, Malaysia, pp 53–71
101. Normah MN, Mainah G, Saraswathy R (2000) Cryopreservation of zygotic embryos of tropical fruit trees – a study on Lansium domesticum and baccaurea species. In: Engelmann F, Takagi H (eds) Cryopreservation of tropical plant germplasm: current research progress and application. Proceedings of an International Workshop, Tsukuba, Japan, October 1998, International Plant Genetic Resources Institute, Rome, pp 156–160
102. Ishikawa K, Harata K, Mii M, Sakai A, Yoshimatsu K, Shimomura K (1997) Cryopreservation of zygotic embryos of a Japanese terrestrial orchid (*Bletilla striata*) by vitrification. Plant Cell Rep 16:754–757
103. Withers LA (1982) The development of cryopreservation techniques for plant cell, tissue and organ culture. In: Fujiwara A (ed) Proceedings 5th International Congress Plant Cell Tissue Culture, Tokyo, Japan, pp 793–794
104. Nogueira GF, Paiva R, Carvalho MA de F, da Silva DPC, Silva LC, Nogueira RC (2009) Cryopreservation of *Byrsonima intermedia* A. Juss. embryos using different moisture contents. In: First International Symposium "Cryopreservation in Horticultural Species", Leuven, Belgium, 5–8 April 2009, p 76
105. Krishnapillay B, Marzalina M, Aziah MY (1992) Cryopreservation of excised embryos of rattan manau. In: Proceedings 3rd National Biotechnological Research Conference of National Universities of Malaysia, Kuala Lumpur, Malaysia, pp 325–330
106. Janeiro LV, Vieitez AM, Ballester A (1996) Cryopreservation of somatic embryos and embryonic axes of *Camellia japonica* L. Plant Cell Rep 15:699–703
107. Kim HH, Yoon JW, Park SU, Kim JH, Cho EG, Engelmann F (2005) Assessment of desiccation sensitivity of tea embryos for cryopreservation. Cryo Letters 26:269–276
108. Monnier M, Leddet C (1980) Action du saccharose sur la résistance au gel des embryons immatures de capselle. Bull Soc Bot Fr 127:71–77
109. Abou Taleb S, Yates I, Wood BW, Fouad MM (1992) Cryogenics and tissue culture for preserving pecan germplasm. HortScience 27:693
110. Cho EG, Normah MN, Kim HH, Ramanatha Rao V, Engelmann F (2002) Cryopreservation of *Citrus aurantifolia* seeds and embryonic axes using a desiccation protocol. Cryo Letters 23:309–316
111. Malik SK, Chaudhury R (2006) The cryopreservation of embryonic axes of two wild and endangered *Citrus* species. Plant Genet Res Charact Utiliz 4:204–209
112. Cho EG, Hor YL, Kim HH, Gwag JG, Ramanatha Rao V, Engelmann F (2003) Cryopreservation of *Citrus madurensis* zygotic embryos by encapsulation-dehydration. Korean J Breed 35:148–153
113. Santos IRI, Stushnoff C (2002) Cryopreservation of embryonic axes of *Citrus* species by encapsulation-dehydration. Plant Genet Res Newslett 131:36–41
114. Santos IRI, Stushnoff C (2003) Desiccation and freezing tolerance of embryonic axes from *Citrus sinensis* (L.) OSB. pretreated with sucrose. Cryo Letters 24:281–292
115. Makeen MA, Noor NM, Dussert S, Clyde MM (2006) Cryopreservation of whole seeds and excised embryonic axes of *Citrus suhuiensis* cv. limau langkat in accordance to their desiccation sensitivity. Cryo Letters 26:259–268
116. Chin HF, Krishnapillay B, Hor YL (1989) A note on the cryopreservation of embryos from young coconuts (*Cocos nucifera* var. Mawa). Pertanika 12:183–186

117. Assy-Bah B, Engelmann F (1992) Cryopreservation of immature embryos of coconut (*Cocos nucifera* L.). Cryo Letters 13:67–74
118. Sajini KK, Karun A, Kumaran PM (2006) Cryopreservation of coconut (*Cocos nucifera* L.) zygotic embryos after pre-growth desiccation. J Plantation Crops 34:576–581
119. Abdelnour A, Villalobos V, Engelmann F (1992) Cryopreservation of zygotic embryos of *Coffea* spp. Cryo Letters 13:297–302
120. Dussert S, Chabrillange N, Engelmann F, AnthonyF,HamonS(1997)Cryopreservation of coffee (*Coffea arabica* L.) seeds: importance of the precooling temperature. Cryo Letters 18:269–276
121. Dussert S, Chabrillange N, Engelmann F, Hamon S (1999) Quantitative estimation of seed desiccation sensitivity using a quantal response model: application to nine species of the genus *Coffea*. Seed Sci Res 9:135–144
122. Martinez M, Gonzalez-Arnao MT, Urra C, Rojas R, Cuba M, Garcia D (1996) Preliminary studies on the cryopreservation of zygotic embryos of *Coffea arabica* variety 972. Cultivos Tropicales 17:79–81
123. Normah MN, Vengadasalam M (1992) Effects of moisture content on cryopreservation of *Coffea* and *Vigna* seeds and embryos. Cryo Letters 13:199–208
124. Gonzales-Benito ME, Perez C (1994) Cryopreservation of embryonic axes of two cultivars of hazelnut (*Corylus avellana L.*). Cryo Letters 15:41–46
125. Reed BM, Normah MN, Yu XL (1994) Stratification is necessary for successful cryopreservation of axes from stored hazelnut seed. Cryo Letters 15:377–384
126. Peran R, Berjak P, Pammenter NW, Kioko JI (2006) Cryopreservation, encapsulation and promotion of shoot production of embryonic axes of a recalcitrant species *Ekebergia capensis* Sparrm. Cryo Letters 27:5–16
127. Grout BWW, Shelton K, Pritchard HW (1983) Orthodox behaviour of oil palm seed and cryopreservation of the excised embryo for genetic conservation. Ann Bot 52:381–384
128. Engelmann F, Chabrillange N, Dussert S, Duval Y (1995) Cryopreservation of zygotic embryos and kernels of oil palm (*Elaeis guineensis* Jacq.). Seed Sci Res 5:81–86
129. Villa AL, Jimenez PE, Valbuena RI, Bastidas S, Nunez VM (2007) Preliminary study of the establishment of cryoconservation protocol for oil palm (*Elaeis guineensis* Jacq.). Agronomia Colombiana 25:215–223
130. Zaimah NAN, Benson EE, Marzalina M (2007) Viability of *Elateriospermum tapos* (perah) embryo after storage in liquid nitrogen. J Tropical Forest Sci 19:1–5
131. Fu JR, Zhang BZ, Wang XP, Qiao YZ, Huang XL (1990) Physiological studies on desiccation, wet storage and cryopreservation of recalcitrant seeds of three fruit tree species and their excised embryonic axis. Seed Sci Technol 18:743–754
132. Brearley J, Henshaw GG, Davey C, Taylor NJ, Blakesley D (1997) Cryopreservation of *Fraxinus excelsior* L. zygotic embryos. Cryo Letters 16:215–218
133. Normah MN, Chin HF, Hor YL (1986) Desiccation and cryopreservation of embryonic axes of *Hevea brasiliensis*, Muell-Arg. Pertanika 9:99–303
134. Chin HF, Krishnapillay B, Alang ZC (1988) Cryopreservation of *Veitchia* and *Howea* palm embryo: non-development of the haustorium. Cryo Letters 9:372–379
135. Wen B, Song SQ (2007) Acquisition and loss of cryotolerance in *Livistonia chinensis* embryos during seed development. Cryo Letters 28:291–302
136. Marin ML, Mafla G, Roca WM, Withers LA (1990) Cryopreservation of cassava zygotic embryos and whole seeds in liquid nitrogen. Cryo Letters 11:257–264
137. Abdelnour-Esquivel A, Mora A, Villalobos V (1992) Cryopreservation of zygotic embryos of *Musa acuminata* (AA) and *M. balbisiana* (BB). Cryo Letters 13:159–164
138. Gonzalez-Rio F, Gurriaran MJ, Gonzalez-Benito E, Revilla MA (1994) Desiccation and cryopreservation of olive (*Olea europaea* L.) embryos. Cryo Letters 15:337–342
139. Zavala ME, Sussex IM (1986) Survival of developing wheat embryos and bean axes following cryoprotection and freezing in liquid nitrogen. J Plant Physiol 122:193–197
140. Hargreaves CL, Grace LJ, van der Maas SA, Menzies MI, Kumar S, Holden DG, Foggo MN, Low CB, Dibley MJ (2005) Comparative in vitro and early nursery performance of adventitious shoots from cryopreserved cotyledons and axillary shoots from epicotyls of the same zygotic embryo of control-pollinated *Pinus radiata*. Can J Forest Res 35:2629–2641
141. Wesley-Smith J, Walters C, Berjak P, Pammenter NW (2004) Non-equilibrium cooling of *Poncirus trifoliata* (L.) embryonic axes at various water contents. Cryo Letters 25:121–128

142. Wesley-Smith J, Walters C, Berjak P, Pammenter NW (2004) The influence of water content, cooling and warming rate upon survival of embryonic axes of *Poncirus trifoliata* (L.). Cryo Letters 25:129–138
143. Chaudhury R, Chandel KPS (1995) Cryopreservation of embryonic axes of almond (*Prunus amygdalus* Batsch.) seeds. Cryo Letters 16:51–56
144. de Boucaud MT, Helliot B, Brison M (1996) Desiccation and cryopreservation of embryonic axes of peach. Cryo Letters 17:379–390
145. Gonzales-Benito ME, Perez-Ruiz C (1992) Cryopreservation of *Quercus faginea* embryonic axes. Cryobiology 29:685–690
146. Gonzalez-Benito ME, Prieto RM, Herradon E, Martin C (2002) Cryopreservation of *Quercus suber* and *Quercus ilex* embryonic axes: in vitro culture, desiccation and cooling factor. Cryo Letters 23:283–290
147. Berjak P, Walker M, Mycock DJ, Wesley-Smith J, Watt P, Pammenter NW (2000) Cryopreservation of recalcitrant zygotic embryos. In: Engelmann F, Takagi H (eds) Cryopreservation of Tropical Plant Germplasm: Current Research Progress and Application. Proceedings of an International Workshop, Tsukuba, Japan, October 1998. International Plant Genetic Resources Institute, Rome, pp 140–155
148. Nogueira GF, Paiva R, Campos ACAL, Nery FC, Vargas DP, Paiva PD de O (2009) Cryopreservation of embryonic axes of castor bean cv Al-Guarani 2002. In: First International Symposium "Cryopreservation in Horticultural Species", Leuven, Belgium, 5–8 April 2009, p 108
149. Abdelnour-Esquivel A, Engelmann F (2002) Cryopreservation of chayote (*Sechium edule* Jacq. Sw.) zygotic embryos and shoot-tips from *in vitro* plantlets. Cryo Letters 23:299–308
150. Marzalina M (1995) Pemyimpanan biji benih mahogany (*Swietenia macrophylla* King.). Ph.D. Thesis, Universiti Kebangsaan Malaysia, Malaysia
151. Pence VC (1991) Cryopreservation of immature embryos of *Theobroma cacao*. Plant Cell Rep 10:144–147
152. Kendall EJ, Kartha KK, Qureshi JA, Chermak P (1993) Cryopreservation of immature spring wheat zygotic embryos using an abscisic acid pretreatment. Plant Cell Rep 12:89–94
153. Delvallée I, Guillaud J, Beckert M, Dumas C (1989) Cryopreservation of immature maize embryos after freeze-hardening in the ear and in vitro. Plant Sci 60:129–136
154. Wen B, Song SQ (2007) Acquisition of cryotolerance in maize embryos during seed development. Cryo Letters 28:109–118
155. Touchell D, Walters C (2000) Recovery of embryos of *Zizania palustris* following exposure to liquid nitrogen. Cryo Letters 21:261–270
156. Walters C, Touchell DH, Power P, Wesley-Smith J, Antolin MF (2002) A cryopreservation protocol for embryos of the endangered species *Zizania texana*. Cryo Letters 23:291–298
157. Misson JP, Druart P, Panis B, Watillon B (2006) Contribution to the study of the maintenance of somatic embryos of *Abies nordmanniana* LK: culture media and cryopreservation method. Propag Ornamental Plants 6:17–23
158. Jekkel Z, Gyulai G, Kiss J, Kiss E, Heszky LE (1998) Cryopreservation of horse-chestnut (*Aesculus hippocastanum* L.) somatic embryos using three different freezing methods. Plant Cell Tiss Org Cult 52:193–197
159. Uragami A, Sakai A, Nagai M, Takahashi T (1989) Survival of cultured cells and somatic embryos of *Asparagus officinalis* cryopreserved by vitrification. Plant Cell Rep 8:418–421
160. Uragami A, Sakai A, Nagai M, Takahashi T (1990) Cryopreservation by vitrification of cultured cells and somatic embryos from mesophyll tissue of asparagus. Acta Hortic 271:109–116
161. Li CR, Loh CS, Sun WQ (1999) An improved dehydration protocol for cryopreservation of *Brassica napus* somatic embryos. Cryo Letters 20:263–268
162. Deepu V, John MKM, Balasaravanan T, Pius PK, Kumar RR (2005) Influence of dehydration methods for cryopreservation of somatic embryos of tea. J Plantation Crops 33:156–159
163. Kumar S, Sharma S (2005) Somatic embryogenesis and cryopreservation of walnut (*Juglans regia* L.) and pecan (*Carya illinoensis* Koch). Acta Hortic 696:143–147
164. Oh SD (1997) The effect of prefreezing treatment and cryoprotectants on the survival of cryopreserved somatic embryos and plant regeneration in Korean native citrus species. Acta Hortic 447:499–505
165. Marin ML, Duran-Vila N (1988) Survival of somatic embryos and recovery of plants of sweet orange (*Citrus sinensis* (L.) Osb.) after immersion in liquid nitrogen. Plant Cell Tiss Org Cult 14:51–57
166. Nair DS, Reghunath BR (2007) Effective use of encapsulation-dehydration technique in cryopreserving somatic embryos of butterfly

pea (*Clitoria ternatea* L.). J Herbs Spices Med Plants 13:83–95
167. Cho HJ, Yang WY, Jeong MG, Kim HH, Ma KH (1998) Survival of somatic embryo of *Cnidium officinale* Makino cryopreserved by vitrification. RDA J Indust Crop Sci 40:41–46
168. Popova EV, Lu ZW, Hahn EJ, Paek KY (2009) Cryopreservation of *Coriandrum sativum* somatic embryos and plant regeneration. In: First International Symposium "Cryopreservation in Horticultural Species", Leuven, Belgium, 5–8 April 2009, p 116
169. Shimonishi K, Ishikawa M, Suzuki S, Oosawa K (1991) Cryopreservation of melon somatic embryos by desiccation method. Jap J Breed 41:347–351
170. Lecouteux C, Florin B, Tessereau H, Bollon H, Pétiard V (1991) Cryopreservation of carrot somatic embryos using a simplified freezing process. Cryo Letters 12:319–328
171. Thierry C, Florin B, Pétiard V (1999) Changes in protein metabolism during the acquisition of tolerance to cryopreservation of carrot somatic embryos. Plant Physiol Biochem 37:145–154
172. Dumet D, Engelmann F, Chabrillange N, Duval Y (1994) Effect of desiccation and storage temperature on the conservation of cultures of oil palm somatic embryos. Cryo Letters 15:85–90
173. Shibli RA (2000) Cryopreservation of black iris (*Iris nigricans*) somatic embryos by encapsulation-dehydration. Cryo Letters 21:39–46
174. Lee BC (1989) Survival and regeneration of somatic embryos of *Juglans regia* L. (English walnut) after being frozen in liquid nitrogen. Res Rep Inst For Genet 25:159–165
175. Danso KE, Ford-Lloyd BV (2002) Induction of high-frequency somatic embryos in cassava for cryopreservation. Plant Cell Rep 2:226–232
176. Deb CR (2002) Cryopreservation of somatic embryos and artificial seeds of *Melia azedarach* by vitrification. J Plant Biol 29:71–76
177. Bomal C, Tremblay FM (2000) Dried cryopreserved somatic embryos of two *Picea* species provide suitable material for direct plantlet regeneration and germplasm storage. Ann Bot 86:177–183
178. Percy REL, Livingston NJ, Moran JA, von Aderkas P (2001) Desiccation, cryopreservation and water relations parameters of white spruce (*Picea glauca*) and interior spruce (*Picea glauca x engelmannii* complex) somatic embryos. Tree Physiol 21:1303–1310
179. Ford CS, Jones NB, van Staden J (2000) Cryopreservation and plant regeneration from somatic embryos of *Pinus patula*. Plant Cell Rep 19:610–615
180. Gale S, John A, Harding K, Benson EE (2008) Developing cryopreservation for *Picea sitchensis* (sitka spruce) somatic embryos: a comparison of vitrification protocols. Cryo Letters 29:135–144
181. Fernandes P, Rodriguez E, Pinto G, Roldan-Ruiz I, Loose M, de Santos C (2008) Cryopreservation of *Quercus suber* somatic embryos by encapsulation-dehydration and evaluation of genetic stability. Tree Physiol 28:1841–1850
182. Fang JY, Sacandé M, Pritchard H, Wetten A (2009) Influence of freezable/non-freezable water and sucrose on the viability of *Theobroma cacao* somatic embryos following desiccation and freezing. Plant Cell Rep 28:883–889
183. Miaja ML, Gambino G, Vallania R, Gribaudo I (2004) Cryopreservation of *Vitis vinifera* L. somatic embryos by vitrification or encapsulation-dehydration. Acta Hort 663:599–603

Chapter 14

Cryogenic Technologies for the Long-Term Storage of *Citrus* Germplasm

Anna De Carlo, Maurizio Lambardi, and Elif Aylin Ozudogru

Abstract

With its beautiful trees, *Citrus* species have long been valued by humanity. The tasteful fruits, extensively used for nutrition, are also good for health due to the high content in vitamins, minerals, and dietary fibers. Like majority of the woody fruit plants, *Citrus* germplasm is conserved mainly as field collections in clonal orchards. However, such a traditional approach presents several difficulties, among which are the high cost, manual labor, and extensive land required to maintain the collections, as well as the necessity of a careful protection of plants from diseases and extreme environmental conditions. As many species in the genus have seeds recalcitrant to desiccation, conservation in seed banks is also inadequate. On the other hand, cryopreservation, i.e., the storage of specimens at ultra-low temperatures (usually in liquid nitrogen, at –196°C) where reactions within the cells are minimized, presents a unique alternative for the safe storage of such germplasm. The present contribution outlines the cryopreservation techniques applied to seeds, zygotic and somatic embryos, embryogenic callus cultures of *Citrus* spp. and provides sample protocols to be used for *Citrus* conservation.

Key words: *Citrus*, Cryopreservation, Desiccation, Embryogenic callus, Embryos, Encapsulation, Nucellar cells, Seeds, Slow-cooling

1. Introduction

Citrus, a member of the *Rutaceae* family, is one of the major fruit crops, produced in about 90 countries worldwide (1). True citrus fruits fall into three genera: *Citrus*, *Poncirus*, and *Fortunella*. Among these, the genus *Citrus* includes all the commercial citrus fruits, while *Poncirus* is a monotypic genus, represented only by the commercial rootstock *P. trifoliata*, and *Fortunella* includes some ornamental plants with relatively less commercial interest (2).

Like most vegetatively propagated species, citrus has traditionally been conserved as field collections. However, ancient – and

Trevor A. Thorpe and Edward C. Yeung (eds.), *Plant Embryo Culture: Methods and Protocols*, Methods in Molecular Biology, vol. 710, DOI 10.1007/978-1-61737-988-8_14,

often unique – citrus germplasm preserved this way is highly vulnerable to damage due to the action of pests and diseases (the control of which requires careful periodical monitoring of plants) and to the effects of late and hard frosts, particularly when collections are located outside the optimum climatic zone of the species (3, 4). Hence, much effort is being conducted today to develop alternative techniques to field collections for germplasm conservation, such as the conservation of in vitro cultures at above freezing temperatures (slow growth storage) and the cryopreservation, i.e., the storage of plant organs and tissues at the ultra-low temperature of liquid nitrogen (LN). The latter technique, in particular, is today attracting much attention for its potential in allowing a safe long-term storage of plant germplasm (5, 6).

Cryopreservation was initially developed in the 1960s for the storage of suspension and callus cultures; the experiment carried out by Quatrana (7) on flax cells being the first successful in vitro plant cryopreservation. Today, the technology is largely applied also to conserve reproductive organs, such as seeds, embryonic axes, and shoot tips (8, 9). The various procedures today available can be grouped within the ones requiring a controlled-rate freezing of specimens (slow cooling) and the ones allowing their direct immersion in LN (one-step freezing), such as "vitrification," "encapsulation-desiccation," "encapsulation-vitrification," and the "droplet method" (10).

Slow cooling is the traditional approach for the cryopreservation of embryogenic callus cultures (11). The principle of the method is the precooling (ranging from 0.1 to 5°C/min, more often 1°C/min) of embryogenic cultures to an intermediate temperature of –40°C (in the presence of suitable cryoprotectants) before being immersed in LN. Hence, the method is also known as two-step freezing. Determination of how fast the cooling rate should be is the key step for the prevention of cryoinjuries. Indeed, too slow cooling rates may lead to desiccation injury due to the loss of excess amount of water from the cells, while too fast cooling rates do not permit sufficiently the loss of freezable water and thus lead to the lethal intracellular ice formation (12). The method requires a programmable freezer for a gradual and precise drop rate of temperature; however, a cheap device ("Mr. Frosty®") is also available today (see 1.1.1).

The desiccation of seeds, excised embryos, or embryonic axes by air flow in a laminar flow cabinet, or by exposure to silica gel or saturated salt solutions, followed by the direct immersion of specimens in LN, is the simplest one-step freezing technique. The removal of the freezable water within the cells by desiccation prevents the formation of intracellular ice crystals during ultra-rapid freezing in LN (13).

As for the encapsulation methods, the inclusion of explants (generally, shoot tips) in alginate capsules allows them to tolerate the exposure to extreme treatments, such as preculturing with high sucrose concentrations and desiccation to relatively lower moisture contents (MC), which would be damaging or even lethal if applied to naked tissues (14). The gelation method developed by Redenbaugh et al. (15) is still today the most employed for the production of encapsulated explants (commonly called "synthetic seeds" or "synseeds"), firm enough to allow manipulation and handling of beads in cryopreservation procedures. The method involves the incubation of the explants in a sterile Na-alginate solution (ranging from 2 to 5%), and release of the solution in a drop-wise manner, each drop containing one explant, into a sterile solution of complexing agent (usually 100 mM $CaCl_2$). Capsule hardening occurs in 20–30 min at room temperature due to the ion-exchange reaction, where calcium replaces sodium (16).

Alternatively, cryoprotection of shoot tips, somatic embryos, and cultured cells can be induced by chemical desiccation through the use of highly concentrated vitrification solutions, PVS2 being largely the most used (17). Vitrification solutions, applied both to naked or encapsulated explants, provide cell protection during ultra-rapid freezing in LN by the physical process of transition of the aqueous solution of cell cytosol into an amorphous and glassy (i.e., noncrystalline) state, thus preventing the formation of intracellular ice crystals (18).

The present contribution focuses the attention on the application of cryopreservation techniques to seeds, embryos, and embryogenic/nucellar callus of *Citrus* spp., with the aim of the safe storage of valuable germplasm (see Table 1). Sample protocols, characterized by successful recovery of explants after the storage in LN, are also provided.

1.1. Practical Comments

1.1.1. Slow Cooling

An expensive apparatus, i.e., programmable freezer, was initially the only sound method for the slow cooling of the plant material to be cryopreserved. In time, use of Nalgene Freezing Containers (also known as "Mr. Frosty®", where a cooling rate of 1°C/min is achieved with the use of chilled isopropyl alcohol inside the freezer) or –70/–80°C freezers (usually for 1–24 h) became also widespread (11).

1.1.2. Encapsulation

Na-alginate is by far the most used gelling agent for capsule preparation. Excellent water solubility, moderate viscosity at room temperature, long-term storability, and the absence of any kind of toxicity are among the main advantages of the compound. Once mixed with the complexing agent (i.e., di- or tri-valent metal salt, such as calcium chloride or calcium nitrate), it easily turns into a hardened Ca-alginate gel by an ion-exchange reaction. Rigidity of the capsule is simply adjusted by optimizing the concentration of the calcium solution and/or gellation time (usually 20–30 min at room temperature) (16).

Table 1
Cryopreservation of *Citrus* spp. both by slow cooling and one-step freezing approaches

Species	Explant type	Cryopreservation procedure	Thawing (°C)	Recovery rate (%)	References
Slow cooling					
Citrus spp.	Embryogenic culture	0.5°C/min	37	100	(33)
Citrus spp.	Embryogenic callus	0.5°C/min	40	NR	(34)
C. deliciosa	Embryogenic culture	0.5°C/min	37	NR	(35)
C. grandis	Somatic embryos	−16 and −32°C respectively, in 24 h	37	80	(36)
C. hystrix	Embryonic axes	0.3°C/min	40	20	(37)
C. junos	Somatic embryos	−16 and −32°C respectively, in 24 h	37	82	(36)
C. nobilis x *C. deliciosa*	Somatic embryos	4°C, 30 min; 0°C, 24 h; slow cooling to −20°C	40	51	(38)
C. nobilis x *C. deliciosa*	Encapsulated somatic embryos	4°C, 30 min; 0°C, 24 h; slow cooling to −20°C	40	58	(38)
C. platymamma	Somatic embryos	−16°C and −32°C respectively, in 24 h	37	75	(36)
C. sinensis	Nucellar callus	0.5°C/min	40	70	(29)
C. sinensis	Somatic embryos	0.5°C/min	37	31	(39)
C. sinensis	Nucellar callus	0.5°C/min	37	NR	(40)
One-step freezing					
Citrus spp.	Embryogenic callus	Vitrification (PVS2, 3 min at 25°C)	40	100	(28)
Citrus spp.	Ovules and somatic embryos	Encapsulation-desiccation in LFC, 5 h (MC, 20–25%)	RT	16 and 100, respectively	(27)
Citrus spp.	Seeds	Desiccation in LFC, 10–36 h (MC, 15–25%)	37	27–100	(4, 9)

Citrus spp.	Seeds	Desiccation by equilibrium	37	NR	(19)
C. aurantifolia	Seeds	Desiccation in LFC, 4 h (MC, 13%)	40	50	(37)
C. aurantifolia	Embryonic axes	Desiccation in LFC, 3 h (MC, 9–11%)	40	100	(37)
C. aurantifolia	Seeds (without testa)	Desiccation on bench, 8 days (MC, 6.1%)	40	85	(25)
C. aurantifolia	Embryonic axes	Desiccation in LFC, 2 h (MC, 7.3%)	40	92.5	(41)
C. halimii	Seeds	Desiccation in LFC, 6 h (MC, 9.5%)	40	25	(37)
C. halimii	Embryonic axes	Desiccation in LFC, 1 h (MC, 16.6%)	40	100	(37)
C. hystrix	Embryonic axes	Desiccation in LFC, 2 h (MC, 11%)	40	60	(37)
C. madurensis	Embryonic axes	Encapsulation-desiccation on SG, 3 h (MC, 30%)	40	65	(41, 42)
C. madurensis	Embryonic axes	Vitrification (1/2 strength PVS2, 20 min; PVS2, 40 min, at 0°C)	40	85	(26, 43)
C. medica	Seeds	Desiccation on bench, 6 days (MC, 7.7%)	40	82	(44)
C. medica	Zygotic embryos	Desiccation in LFC, 2 h (MC, 6.7%)	40	90	(44)
C. sinensis	Nucellar tissue	Vitrification (PVS2, 3 min at 25°C)	RT	90	(30)
C. suhuiensis	Seeds	Desiccation by equilibrium	37	8.3	(45)
C. suhuiensis	Embryonic axes	Desiccation in LFC, 4 h	40	83.3	(45)
Poncirus trifoliata	Embryonic axes	Desiccation in LFC, 4 h (MC, 14%)	37	68	(46)

LFC laminar flow cabinet, *NR* not reported, *RT* room temperature, *SG* silica gel

1.1.3. Desiccation

Physical desiccation of the naked or encapsulated specimens is achieved by subjecting them either to the sterile air flow of a laminar flow cabinet or to activated silica gel. Desiccation time and the method to be used may depend on the type of the explant and its initial MC, and it ranges from minutes to hours. Alternatively, desiccation of the specimens can also be achieved by the use of saturated solutions (e.g., KOH, KNO_3, K_2CO_3, NH_4NO_3, $(NH_4)_2SO_4$, NH_4Cl, NaCl, KCl, $BaCl_2$). Here, specimens are considered desiccated when they reach equilibrium to the certain relative humidity of the salt solution at the given temperature (19).

1.1.4. Vitrification

Cryoprotectants, used both in slow cooling and one-step freezing approaches, belong to one of the two category; those that penetrate the plasma membrane (i.e., DMSO and glycerol), and those that do not penetrate (i.e., polyethylene glycol (PEG) and sucrose). Numerous authors use DMSO solution (at a concentration of 5–10%) as a single cryoprotectant, while many others prefer a combination of cryoprotectants at lower concentrations, considering the latter approach less toxic to the plant tissues (11). Indeed, there is a small amount of evidence indicating that DMSO may be involved in generating a variety of genetic/epigenetic alterations. For that reason, it is often applied for short intervals and at ice temperature, where both physical and chemical reactions are considerably slower (20). The most common cryoprotectant combination is called "Plant Vitrification Solution 2" (PVS2; (17)), consisting of 30% glycerol (w/v), 15% ethylene glycol (w/v), and 15% DMSO (w/v), prepared in liquid MS (21) medium containing 0.4 M sucrose. PVS2 does not permeate into the cytosol during the desiccation process. The solution easily supercools below −100°C and solidifies at −115°C (18). However, PVS2 may also exhibit some toxic effects to the tissues (due to the presence of DMSO), thus its incubation time and the temperature to be applied are fundamental parameters which must always be optimized (11). Toxic effect of the solution may be reduced or eliminated also by dehydrating the specimens in two steps. First step involves the incubation of the specimens in "loading solution" ((LS), 2 M glycerol and 0.4 M sucrose; (22)) at room temperature for 20 min, and the second step is the incubation in PVS2 at 0°C. During treatment with loading solution, the cells are considerably desiccated and plasmolyzed. However little or no permeation of glycerol into the cytosol was observed after a 20-min incubation. Thus, the protective effect of a brief incubation with the solution might be a result of the protective effect of plasmolysis (18).

2. Materials

1. All culture media are based on either the Murashige and Skoog (MS) (21) or Murashige and Tucker medium (MT) (23) with different concentrations of plant growth regulators, sucrose, and additives.
2. Plant growth regulators: N6-benzyladenine (BA), naphthalene acetic acid (NAA), gibberellic acid (GA_3), zeatin, and kinetin.
3. Media additives: yeast or malt extract, Difco bacto agar.
4. "Plant Vitrification Solution 2" (PVS2): 30% glycerol (w/v), 15% ethylene glycol (w/v), and 15% DMSO (w/v) liquid MS (21) medium containing 0.4 M sucrose.
5. Loading solution: 2 M glycerol and 0.4 M sucrose.
6. Other chemicals and solutions: glycerol, DMSO, sodium alginate, calcium chloride, silica gel, Tween 20, bleach, 2% (w/v) sodium hypochlorite solution, 70% ethanol, absolute ethanol.
7. Other materials: liquid nitrogen, water bath, cryovial, silica gel, Nalgene Freezing Containers.

3. Methods

3.1. Cryopreservation of Whole Seeds (4, 9)

3.1.1. Seed Decontamination

1. Harvest fresh ripe fruits and extract the seeds.
2. Surface sterilize the seeds by a 5-min immersion in 70% (v/v) ethanol, followed by a 15-min treatment in 2% (w/v) sodium hypochlorite solution.
3. Rinse the seeds three times (5 min each) in sterile distilled water and peel off the external integument in a laminar flow cabinet.
4. Store the seeds at 4°C in darkness until use in cryopreservation trials.

3.1.2. Cryopreservation by Desiccation

1. To determine the initial MC of the seeds (on a fresh weight basis), dry batch of seeds at 60°C for 48 h (see Note 1).
2. To desiccate the seeds, maintain them under the sterile air flow of a laminar flow cabinet. To bring the seeds to comparable levels of MC, the MC decrease in relation to the air-flow exposure time should be determined preliminarily (see Note 2).
3. Place the seeds in 2-mL cryovials (5 seeds/cryovial) and plunge directly into LN.

4. For recovering from LN, rewarm the seeds in a water bath at 37°C for 5 min.
5. For germination, place the seeds on semisolid MS medium, containing 146 mM sucrose and 500 mg/L malt extract (from (24)), at pH 5.7.
6. Maintain the seeds at 26°C in darkness for 1 week, then transfer them to plastic boxes (Magenta™ GA-7, Sigma Chem.), containing 50 mL of medium, under a 16-h photoperiod, provided by cool-white fluorescent tubes (60 µmol/m²/s).

3.2. Cryopreservation of Embryonic Axes (25, 26)

3.2.1. Excision of Embryonic Axes

1. Extract the seeds and surface sterilize them with absolute ethanol for 2 min. Then place them in 20% commercial bleach (0.54% active chlorine), with a few drops of Tween 20, for 20 min on a rotary shaker (see Note 3).
2. Rinse the seeds three to five times with sterile distilled water.
3. Remove the testa of the seeds and excise aseptically the 1–2 mm long embryonic axes from the cotyledons with a scalpel blade.

3.2.2. Cryopreservation by Desiccation

1. Preculture the embryonic axes on MS medium, containing 0.1 M sucrose, 7 g/L Difco agar, and 0.3 mg/L BA for 24 h.
2. Desiccate the embryonic axes in the laminar flow cabinet for 2 h.
3. Wrap them in aluminum foil envelops and freeze rapidly by direct immersion in LN.
4. After 24-h storage in LN, rewarm the embryonic axes in a water bath at 40°C for 5 min.
5. Place them on MS medium, containing 0.3 M sucrose for 1 day, then transfer them on MS medium, containing 0.1 M sucrose and 0.1 mg/L BA at 25°C under 16-h photoperiod and 25 µmol/m²/s of light intensity.

3.2.3. Cryopreservation by Encapsulation-Desiccation

1. Pregrow excised embryonic axes on semisolid MS medium, containing 0.1 mg/L BA, 0.1 mg/L NAA, and 0.1 mg/L GA_3 for 3 day.
2. For preculture, transfer the embryonic axes on semisolid MS medium, containing 0.1 mg/L BA, 0.1 mg/L NAA, and 0.1 mg/L GA_3 and 0.3 M sucrose for 1 day.
3. Encapsulate pretreated embryonic axes in 3% alginate beads with 100 mM $CaCl_2$ solution.
4. Dehydrate the beads in liquid MS medium, containing 0.6 M sucrose and 2 M glycerol for 60 min on a rotary shaker (200 rpm) at 25°C.
5. Dry them in a Petri dish sealed with silica gel (20 beads/40 g silica gel) for 3 h.

6. Place the dried beads in cryovials and plunge into LN.
7. After storage in LN, rewarm the beads in a water bath at 40°C for 5 min.
8. For embryonic axes development, place the beads on MS medium, containing 0.3 M sucrose for 1 day, then transfer them on MS medium, containing 0.1 mg/L BA, 0.1 mg/L NAA, and 0.1 mg/L GA_3 at 25°C under a 16-h photoperiod (25 μmol/m^2/s).

3.2.4. Cryopreservation by Vitrification

1. For preculture, place freshly excised embryonic axes on semi-solid MS medium, containing 0.1 mg/L BA, 0.1 mg/L NAA, 0.1 mg/L GA_3, and 0.1 M sucrose for 1 day, then transfer them on MS medium, containing 0.3 M sucrose and 0.5 M glycerol for an additional day.
2. Transfer the embryonic axes in 2-mL cryovials (20 axes/cryovial), containing 2 M glycerol and 04 M sucrose (loading solution) and incubate at 25°C for 20 min.
3. Replace the loading solution with half-strength PVS2 solution and incubate at 0°C for 20 min. Then, replace the half-strength solution with full-strength solution and treat the embryonic axes for additional 40 min.
4. At the end of vitrification treatment, renew the PVS2 solution and adjust the final volume inside the cryovial to 0.5 mL and rapidly immerse in LN.
5. After storage in LN, rewarm the beads in a water bath at 40°C for 5 min.
6. Remove the PVS2 solution and incubate the embryonic axes with 0.5 mL liquid medium, containing 1.2 M sucrose for 20 min.
7. Retrieve the embryonic axes, dry them on sterile filter paper, and place on MS medium, containing 0.1 mg/L BA, 0.1 mg/L NAA, and 0.1 mg/L GA_3 at 25°C under a 16-h photoperiod (25 μmol/m^2/s).

3.3. Cryopreservation of Somatic Embryos (22, 24, 27)

3.3.1. Cryopreservation by Slow Cooling

1. Harvest the young fruits 4–5 week after pollination and excise the immature ovules.
2. Dissect the ovules into halves and introduce in vitro the micropylar portion.
3. For direct embryogenesis, incubate the tissues on MT medium, containing 0.1 mg/L zeatin and 500 or 1,000 mg/L yeast or malt extract.
4. Add cryoprotectant solution, containing 10% DMSO (w/v) and 1.0 M sucrose, prepared in liquid MS medium, into 50-mL flask and immerse the somatic embryos into the solution at room temperature for 48 h.

5. Following pretreatment, transfer the somatic embryos to a 5-mL cryovial, loaded with 3 mL MS medium, containing one-third strength PVS2 solution, with sucrose concentration raised to 1.0 M.
6. After 90-min PVS2 treatment at 0°C, prefreeze the somatic embryos by placing the flask at −16 and −32°C, respectively, in a total of 24 h exposure, prior to immersion in LN.
7. Thaw the somatic embryos in water bath at 37°C for 5 min.
8. Remove the embryos from the vial and culture them on MT medium, containing 1.0 mg/L zeatin and 0.01 mg/L NAA at 25°C and 16 h photoperiod (160 μmol/m²/s).

3.3.2. Cryopreservation by Encapsulation-Slow Cooling

1. Culture the unfertilized ovules on MS medium, containing 9.29 μM (i.e., 2 mg/L) kinetin, 3% (w/v) sucrose and 0.8% (w/v) agar (pH 5.6).
2. Allow the somatic embryos to mature on the same medium and use embryos at heart stage to torpedo stage for encapsulation.
3. For encapsulation, mix the freshly developed embryos with 4% sodium alginate solution, prepared in liquid MS medium, containing 3% sucrose.
4. Drop the embryos mixed with sodium alginate solution into 75 mM $CaCl_2 \cdot 2H_2O$ using a wide glass dropper.
5. Keep the drops, each containing a single embryo, in $CaCl_2 \cdot 2H_2O$ solution for 45 min on a rotary shaker (70 rpm).
6. After 45 min, decant the solution to recover the beads, wash them with sterile distilled water and dry by spreading on sterile filter paper for 30 min in laminar flow cabinet.
7. For the storage in LN, transfer the encapsulated embryos into 10-mL cryovial (10 beads per cryovial), containing 10% (w/v) DMSO and 0.4 M sucrose in liquid MS medium.
8. Keep the cryovials at 4°C for 30 min and then at 0°C for 24 h.
9. Cool them slowly in an ultra-cooling bath to −20°C, and then transfer to LN.
10. For thawing, remove the cryovials from LN and rapidly warm in water bath at 40°C for 4 min.
11. Wash the beads with sterile distilled water and place them on MS medium, containing 9.29 μM kinetin.
12. Maintain the cultures at 26°C and 16-h photoperiod (40 μmol/m²/s).

3.3.3. Cryopreservation by Encapsulation-Desiccation

1. Collect the flowers before opening and disinfect them by soaking in a 2% (v/v) sodium hypochlorite solution, containing 0.1% (v/v) wetting agent (Tween 20), for 20 min.
2. Rinse the seeds three times in sterile distilled water.

3. Open the flowers under sterile conditions and dissect the whole pistils with a scalpel.
4. Cut the stigmas, styles, and ovaries perpendicularly to the longitudinal axis at a thickness of 0.4–0.5 mm.
5. For induction of somatic embryogenesis, place the thin cut layers, with the abaxial surface toward the medium, on MS medium, containing 7 g/L agar (pH 5.7).
6. Use the somatic embryos for cryopreservation when they become available at globular, torpedo, or heart-shaped stages.
7. Incubate the encapsulated somatic embryos in liquid medium, containing 0.75 M sucrose for 1 day.
8. After pregrowth in liquid medium, dehydrate the encapsulated somatic embryos in the air current of a laminar flow cabinet for 5 h, reaching the final MC of 20–25% (fresh weight basis).
9. Rapidly freeze the desiccated, encapsulated somatic embryos by direct immersion in LN.
10. After storage in LN, thaw the samples in the laminar air flow for 2–3 min.
11. Transfer them to standard culture medium for recovery (1 week in the dark, followed by standard light conditions).

3.4. Cryopreservation of Embryogenic Callus Cultures (28)

3.4.1. Maintenance of Embryogenic Callus Cultures

1. Maintain the callus cultures by subculturing in 2-week intervals on MT medium, containing 40 g/L sucrose and 7 g/L agar.
2. For suspension cultures to be used in cryopreservation treatments, transfer the calluses to liquid MT medium, containing 500 mg/L malt extract and 50 g/L sucrose, and maintain the culture vessels on a rotary shaker (120 rpm) at 27°C and on 14-h photoperiod (33 μmol/m^2/s).
3. Subculture the calluses every 2 weeks for at least two times before using it for cryopreservation.

3.4.2. Cryopreservation by Vitrification

1. After 8–10 days of culture, transfer 1.5 mL of cell suspension (containing about 0.1 mL packed cell volume) to a 2-mL cryovial and allow to settle.
2. Discard the supernatant and add 1.5 mL of PVS2 at room temperature.
3. Centrifuge the cell suspension at 100 × *g* for 20 s.
4. Discard the supernatant and add 1 mL of fresh PVS2.
5. Following 3-min PVS2 treatment at room temperature, plunge the cryovial into LN.
6. After storage, thaw rapidly the cell suspension in a 40°C water bath.

7. Replace the PVS2 solution by 1.5 mL of MT medium, containing 1.2 M sucrose, and incubate for 10 min.
8. For recovery, disperse the cells over double-layer sterile filter paper, placed on 20 mL MT medium, containing 50 g/L sucrose and 7 g/L agar in a 90-mm Petri dish.
9. After overnight equilibration, transfer the cells with the upper filter paper to another Petri dish containing the same medium.
10. Incubate the Petri dishes at 27°C and on 14-h photoperiod (33 μmol/m²/s).

3.5. Cryopreservation of Nucellar Cells and Callus (29, 30)

3.5.1. Induction and Maintenance of Nucellar Callus Cultures

1. For nucellar callus induction (see Note 4), excise the ovules (better if it is close to the flowering stage) and place on MT basal medium, containing 0.15 M sucrose, 10 mg/L BA, and 0.8% Difco Bacto agar (pH 5.7).
2. Incubate the cultures at 25°C under 16-h photoperiod, provided by cool-white fluorescent tubes (25 μmol/m²/s) and subculture every 2 weeks.
3. For suspension cultures to be used in cryopreservation treatments, inoculate about 1 g of nucellar callus into 50 mL of liquid MT medium, supplemented with 10 mg/L BA, and culture on a shaker (110 rpm) in the same environmental conditions mentioned above.

3.5.2. Cryopreservation by Slow Cooling

1. Transfer 10 mL of the cell suspension (containing 3 mL of packed cell volume) from 6-day-old cultures to 50 mL glass tubes.
2. Replace completely the medium by MT medium, containing 1.2 M sucrose.
3. Chill the glass tubes containing the cells in an ice bath.
4. Add gradually 2 mL of ice-cold MT medium, containing 1.2 M sucrose and 30% DMSO (w/v) in 1 h, reaching the final DMSO concentration of 5%.
5. Dispense 500 μL aliquots of cell suspensions (containing about 100 mg cells) into 2-mL cryovials.
6. Place the cryovials in the freezing chamber of a programmable freezer and cool at cooling rate of −0.5°C/min to −40°C prior to immersion in LN.
7. After storage, thaw rapidly the cryovials in a 40°C water bath.
8. Dispense the cell suspensions of 0.5 mL on double-layer sterile filter paper (Ø 50 mm), placed on 20 mL of MT medium, containing 5 mg/L BA and 0.8% agar in a Petri dish (Ø 90 mm).
9. After 4–5 h, transfer the filter paper with the cells to another Petri dish containing the same medium.

10. Incubate the Petri dish at 25°C and 16-h photoperiod (25 μmol/m²/s).

3.5.3. Cryopreservation by Vitrification

1. Transfer the cell suspensions (8/10-day-old cultured cells) into a 10-mL conical glass tube (110×15 mm) and allow to settle.
2. Discard the supernatant.
3. Add 4 mL of PVS2 at 25°C to 0.2 mL of packed cells.
4. Centrifuge the cells at 100×*g* for 20 s.
5. Discard the supernatant and add 2 mL of fresh PVS2.
6. After 3 min of PVS2 treatment at 25°C, load the cell suspensions into a 0.1-mL plastic straw and seal the top end of each straw by a heat sealer (see Note 5).
7. Plunge the straws into LN.
8. For thawing, place the straws in a water bath at 25°C and hold there for 10 min.
9. Expel the cell suspensions into 2 mL of MT medium, containing 1.2 M sucrose, and hold there for 10 min at 25°C.
10. Pour the cell suspensions on two pieces of sterile filter paper (Ø 50 mm), placed on 20 mL of MT medium, containing 5 mg/L BA and 0.8% agar in a Petri dish (Ø 90 mm).
11. After 4–5 h, transfer the filter paper with the cells to another Petri dish containing the same medium.
12. Incubate the Petri dishes at 25°C and 16-h photoperiod (25 μmol/m²/s).

4. Notes

1. Alternatively, seed MC can be determined by a moisture ana lyzer device (e.g., HG63 Halogen, Mettler Toledo).
2. In these studies, desiccation time ranged from 10 to 36 h, depending on the species and the initial MC. However, the time necessary to reduce seed MC can change consistently according to humidity conditions in the room and the flow rate of the cabinet.
3. Alternatively, seeds can be sterilized in 20% commercial Clorox for 20 min, followed by 70% ethanol for 5 min.
4. It is worth to mention that the nucellar embryos in polyembryonic *Citrus* spp. are ovular in origin, and thus somatic in nature (31). In accordance with that, cultured nucellar cells present high phenotypic stability (32).

5. The mean cooling rate of the straws is about 1,600°C/min between -30 and -150°C. When the experiment is repeated with 2-mL plastic cryovials, cooling rate is about 280°C/min.

Acknowledgment

The procedure under Subheading 3.1 was optimized in the frame of the Project CRIOGERM, financially supported by the Ente Cassa di Risparmio di Firenze, which is gratefully acknowledged.

References

1. Duran-Vila N (1995) Cryopreservation of germplasm of citrus. In: Bajaj YPS (ed) Biotechnology in agriculture and forestry, cryopreservation of plant germplasm I. Springer, Berlin, pp 70–86
2. Hodgson RW (1967) Horticultural varieties of citrus. In: Reuther W, Webber W, Batchelor LD (eds) The citrus industry. Division of Agricultural Science, University of California, Berkeley, pp 431–591
3. IBPGR (1982) Genetic resources of citrus. AGPG: IBPGR/82/7 Report, Rome, pp. 13
4. Lambardi M, Halmagyi A, Benelli C, De Carlo A, Vettori C (2007) Seed cryopreservation for conservation of ancient *Citrus* germplasm. Adv Hort Sci 21:198–202
5. Panis B, Lambardi M (2006) Status of cryopreservation technologies in plants (crops and forest trees). In: Ruane J, Sonnino A (eds) The role of biotechnology in exploring and protecting agricultural genetic resources. Food and Agriculture Organization of the United Nations (FAO), Rome, pp 61–78
6. Sakai A, Engelmann F (2007) Vitrification, encapsulation-vitrification and droplet-vitrification: a review. CryoLetters 28:151–172
7. Quatrana RS (1968) Freeze preservation of cultured flax cells utilizing dimethyl sulfoxide. Plant Physiol 43:2057–2061
8. Abdelnour-Esquivel A, Engelmann F (2002) Cryopreservation of chayote (*Sechium edule* JACQ. SW.) zygotic embryos and shoot tips from in vitro plantlets. CryoLetters 23:299–308
9. Lambardi M, De Carlo A, Biricolti S, Puglia AM, Lombardo G, Siragusa M, De Pasquale F (2004) Zygotic and nucellar embryo survival following desiccation/cryopreservation of *Citrus* intact seeds. CryoLetters 25:81–90
10. Reed BM (2008) Cryopreservation – Practical considerations. In: Reed BM (ed) Plant cryopreservation: a practical guide. Springer, Berlin, pp 3–13
11. Lambardi M, Ozudogru EA, Benelli C (2008) Cryopreservation of embryogenic cultures. In: Reed BM (ed) Plant cryopreservation: a practical guide. Springer, Berlin, pp 177–210
12. Meryman HT, Williams RJ (1985) Basic principles of freezing injury to plant cells: natural tolerance and approaches to cryopreservation. In: Kartha KK (ed) Cryopreservation of plant cells and organs. CRC Press, Boca Raton, pp 13–47
13. Normah MN, Makeen AM (2008) Cryopreservation of excised embryos and embryonic axes. In: Reed BM (ed) Plant cryopreservation: a practical guide. Springer, Berlin, pp 211–240
14. Engelmann F, Gonzalez-Arnao MT, Wu Y, Escobar R (2008) The development of encapsulation desiccation. In: Reed BM (ed) Plant cryopreservation: a practical guide. Springer, Berlin, pp 59–75
15. Redenbaugh K, Slade D, Viss P, Fujii JA (1987) Encapsulation of somatic embryos in synthetic seed coats. HortSci 22:803–809
16. Lambardi M, Benelli C, Ozudogru EA, Ozden-Tokatli Y (2006) Synthetic seed technology in ornamental plants. In: Teixeira da Silva JA (ed) Floriculture, ornamental and plant biotechnology: advances and topical issues. Global Science Books, UK, pp 347–354
17. Sakai A, Kobayashi S, Oiyama I (1990) Cryopreservation of nucellar cells of navel orange (*Citrus sinensis* Osb. var. *brasiliensis* Tanaka) by vitrification. Plant Cell Rep 9:30–33

18. Sakai A (2000) Development of cryopreservation techniques. In: Engelmann F, Takagi H (eds) Cryopreservation of tropical plant germplasm. current research progress and application. Japan International Research Center for Agricultural Sciences, International Plant Genetic Resources Institute, Tsukuba, Rome, pp 1–7
19. Hor YL, Kim YJ, Ugap A, Chabrillange N, Sinniah UR (2005) Optimal hydration status for cryopreservation of intermediate oily seeds: *Citrus* as a case study. Ann Bot 95:1153–1161
20. Finkle BJ, Zavala ME, Ulrich JM (1985) Cryoprotective compounds in the viable freezing of plant tissues. In: Kartha KK (ed) Cryopreservation of Plant Cells and Organs. CRC Press, Boca Raton, pp 75–113
21. Murashige T, Skoog F (1962) A revised medium for rapid growth and bioassays with tobacco tissue cultures. Physiol Plant 15:473–497
22. Matsumoto T, Sakai A, Yamada K (1994) Cryopreservation of *in vitro*-grown apical meristems of wasabi (*Wasabia japonica*) by vitrification and subsequent high plant regeneration. Plant Cell Rep 13:442–446
23. Murashige T, Tucker DPH (1969) Growth factor requirements of Citrus tissue cultures. In: Chapman HD (ed) Proc 1st Int Citrus Symp 3, University of California, Riverside, pp 1151–1161
24. Carimi F, Tortorici MC, De Pasquale F, Crescimanno FG (1998) Somatic embryogenesis and plant regeneration from undeveloped ovules and stigma/style explants of sweet orange navel group (*Citrus sinensis* (L.) Osb.). Plant Cell Tiss Org Cult 54:183–189
25. Cho EG, Noor NM, Kim HH, Rao VR, Engelmann F (2002) Cryopreservation of *Citrus aurantifolia* seeds and embryonic axes using a desiccation protocol. CryoLetters 23:309–316
26. Cho EG, Hor YL, Kim HH, Rao VR, Engelmann F (2001) Cryopreservation of *Citrus madurensis* zygotic embryonic axes by vitrification: importance of pregrowth and preculture conditions. CryoLetters 22:391–396
27. González-Arnao MT, Juárez J, Ortega C, Navarro L, Duran-Vila N (2003) Cryopreservation of ovules and somatic embryos of citrus using the encapsulation-desiccation technique. CryoLetters 24:85–94
28. Hao Y-J, You C-X, Deng X-X (2002) Effects of cryopreservation on developmental competency, cytological and molecular stability of citrus callus. CryoLetters 23:27–35
29. Kobayashi S, Sakai A, Oiyama I (1990) Cryopreservation in liquid nitrogen of cultured navel orange (*Citrus sinensis* Osb.) nucellar cells and subsequent plant regeneration. Plant Cell Tiss Org Cult 23:15–20
30. Sakai A, Kobayashi S, Oiyama I (1991) Survival by vitrification of nucellar cells of navel orange (*Citrus sinensis* var. *brasiliensis* Tanaka) cooled to −196°C. J Plant Physiol 137:465–470
31. Bajaj YPS (1985) Cryopreservation of embryos. In: Bajaj YPS (ed) Biotechnology in agriculture and forestry, cryopreservation of plant germplasm I. Springer, Berlin, pp 227–242
32. Kobayashi S (1987) Uniformity of plants regenerated from orange (*Citrus sinensis* Osb.) protoplasts. Theor Appl Genet 74:10–14
33. Pérez RM, Navarro L, Duran-Vila N (1997) Cryopreservation and storage of embryogenic callus cultures of several *Citrus* species and cultivars. Plant Cell Rep 17:44–49
34. Engelmann F, Dambier D, Ollitrault P (1994) Cryopreservation of cell suspensions and embryogenic calluses of *Citrus* using a simplified freezing process. CryoLetters 15: 53–58
35. Pérez RM, Mas O, Navarro L, Duran-Vila N (1999) Production and cryoconservation of embryogenic cultures of mandarin and mandarin hybrids. Plant Cell Tiss Org Cult 55:71–74
36. Oh S-D (1997) The effect of prefreezing treatment and cryoprotectants on the survival of cryopreserved somatic embryos and plant regeneration in Korean native citrus species. Acta Hortic 447:499–505
37. Normah MN, Serimala SD (1995) Cryopreservation of seeds and embryonic axes of several *Citrus* species. In: Ellis RH, Black M, Murdock AJ, Hong TD (eds) Basic and applied aspects of seed biology. Kluwer Academic, Dordrecht, pp 817–823
38. Singh B, Sharma S, Rani G, Virk GS, Zaidi AA, Nagpal A (2007) *In vitro* response of encapsulated and non-encapsulated somatic embryos of Kinnow mandarin (*Citrus nobilis* Lour x *C. deliciosa* Tenora). Plant Biotechnol Rep 1:101–107
39. Marin ML, Gogorcena Y, Ortiz J, Duran-Vila N (1993) Recovery of whole plants of sweet orange from somatic embryos subjected to freezing thawing treatments. Plant Cell Tiss Org Cult 34:27–33
40. Olivares-Fuster O, Asins MJ, Duran-Vila N, Navarro L (2000) Cryopreserved callus, a source of protoplasts for citrus improvement. J Hortic Sci Biotechnol 75:635–640

41. Cho EG, Hor YL, Kim HH, Rao VR, Engelmann F (2002) Cryopreservation of *Citrus madurensis* embryonic axes by encapsulation-desiccation. CryoLetters 23:325–332
42. Cho E-G, Hor Y-L, Kim H-H, Gwag J-G, Rao VR, Engelmann F (2003) Cryopreservation of *Citrus madurensis* zygotic embryos by modified encapsulation-desiccation. Korean J Breed 35:148–153
43. Cho EG, Hor YL, Kim HH, Rao VR, Engelmann F (2002) Cryopreservation of *Citrus madurensis* embryonic axes by vitrification: importance of loading and treatment with vitrification solution. CryoLetters 23:317–324
44. Cho E-G, Kim H-H, Baek H-J, Gwag J-G, Noor NM (2003) Cryopreservation of *Citrus medica* seeds. J Korean Soc Hortic Sci 44:565–568
45. Makeen MA, Noor NM, Dussert S, Clyde MM (2005) Cryopreservation of whole seeds and excised embryonic axes of *Citrus suhuiensis* cv. Limau Langkat in accordance to their desiccation sensitivity. CryoLetters 26:259–268
46. Radhamani J, Chandel KPS (1992) Cryopreservation of embryonic axes of trifoliate orange (*Poncirus trifoliata* (L.) RAF.). Plant Cell Rep 11:372–374

Chapter 15

Cryopreservation of Zygotic Embryonic Axes and Somatic Embryos of European Chestnut

Ana M. Vieitez, M. Carmen San-José, and Elena Corredoira

Abstract

For *Castanea sativa* (European chestnut), a species with recalcitrant seeds that is not easily propagated vegetatively, cryopreservation is one of the most promising techniques for maintaining genetic resource diversity and for conservation of selected germplasms. Long-term conservation of selected seeds and valuable embryogenic lines can be achieved through the cryopreservation of zygotic embryonic axes and somatic embryos, respectively. This chapter describes methods for the desiccation-based cryostorage of zygotic embryonic axes, and the vitrification-based cryopreservation of somatic embryos. For zygotic embryonic axes, the highest post-thaw survival and plantlet recovery rates are obtained by desiccation in a laminar flow hood to 20–25% moisture content, followed by direct immersion in liquid nitrogen. For somatic embryos, embryogenesis resumption rates of over 60% are achieved by preculture of embryo clumps for 3 days on solid medium containing 0.3 M sucrose, incubation in PVS2 vitrification solution for 60 min at 0°C, and direct immersion in liquid nitrogen. Plantlet recovery from cryostored embryogenic lines requires proliferation of the thawed embryos and subsequent maturation before germination and conversion into plantlets.

Key words: *Castanea sativa*, Chestnuts, Cryostorage, Cryopreservation, Embryo desiccation, European chestnut, Plant regeneration, Somatic embryogenesis, Vitrification, Zygotic embryos

1. Introduction

The genus *Castanea*, belonging to the family Fagaceae, is native to the temperate regions of the Northern hemisphere. European chestnut (*C. sativa* Mill.) is a hardwood species with a wide distribution and an important economic role in Europe. It probably originated in the Caucasus mountains, but is currently present in 25 European countries. European chestnut stands cover an area of over two million hectares (1), mainly in the Mediterranean region. Historically, chestnuts have been a major food source for many European mountain populations, while chestnut wood has

Trevor A. Thorpe and Edward C. Yeung (eds.), *Plant Embryo Culture: Methods and Protocols*,
Methods in Molecular Biology, vol. 710, DOI 10.1007/978-1-61737-988-8_15,

been used in furniture, for tannin production, and as a source of renewable energy (2). Although many chestnut stands have undergone degradation by disease and cultural abandonment since the end of the nineteenth century, the past 25 years have seen an upturn. In line with the sustainable agriculture policies followed in many countries, the chestnut has gained in value not only for timber and nut production, but also because of the contribution of European chestnut stands to wildlife conservation, protection from erosion, and recreational landscapes (3). Accordingly, there is increasing interest in safeguarding European chestnut from disease.

The two most devastating diseases of European chestnut – which also affect American chestnut (*C. dentata*) – are root rot or ink disease (caused by *Phytophthora* spp.) and chestnut blight (caused by *Cryphonectria parasitica*). Resistance to these diseases has been sought through conventional cross-breeding with resistant or tolerant Asian chestnut species (*C. crenata* and *C. mollisima*). Since such programmes require long periods of time, genetic transformation by biotechnological methods appears as an attractive and complementary alternative approach to the traditional breeding. Regardless of how resistant genotypes are produced, once obtained they must be propagated and conserved.

Chestnut trees have recalcitrant seeds, and because of strong maturation effects, they are also difficult to propagate vegetatively by conventional methods based on cuttings, grafting, or stooling. These difficulties, which complicate the conservation of high-value genotypes (4), could be overcome, at least partially, by micropropagation methods (5) including somatic embryogenesis (SE). Embryogenic cultures of European chestnut have been initiated from both immature zygotic embryos (6–9) and leaf sections obtained from shoots cultured in vitro (10). The proliferation of these somatic embryos by secondary embryogenesis has also been reported, as has the achievement of viable rates of germination and plant recovery (9, 11). These embryogenic systems have allowed the development of a relatively efficient genetic transformation protocol involving the culture of somatic embryos of *C. sativa* with different strains of *Agrobacterium tumefaciens* carrying marker genes (12–14).

Embryogenic lines obtained from selected elite individuals or genetically transformed embryos must be conserved. However, long-term maintenance by subculture involves a concomitant risk of contamination, somaclonal variation, and loss of embryogenic capability. Cryopreservation of these embryogenic cultures appears to offer the best prospects for long-term conservation and management of these valuable genotypes (15, 16). Moreover, the importance of cryopreservation as a link between breeding and subsequent mass propagation after clonal selection of field-tested seed progenies has also been emphasized for both conifers (17) and hardwood forest trees (18). Cryopreservation can in fact be applied to both somatic embryos and zygotic embryos or their embryonic axes (4), and in the case of *C. sativa*, both approaches have been investigated. Pence (19) applied a desiccation-based procedure to embryonic axes, but although surviving axes underwent root pole elongation, no plantlets developed. Later, however, a method likewise based on desiccation and rapid immersion in liquid nitrogen (LN) allowed Corredoira et al. (20) to achieve 90% survival and 63% plant recovery rates.

Both desiccation-based and vitrification-based procedures have been used to cryopreserve somatic embryos of European chestnut (20). The most efficient is the latter, which achieved an embryogenesis resumption rate of 68% as against 33% for the desiccation-based technique. The vitrification protocol also allows cryostorage of transgenic embryogenic lines, with post-thaw embryo recovery rates higher than 60%, and proven stability of the inserted foreign genes (13).

In this chapter, we provide details of viable methods for the cryopreservation of European chestnut based on (1) the desiccation of zygotic embryonic axes and (2) the vitrification of somatic embryos. In both cases, the protocol described covers the whole process, from the plant material of origin to the regenerated plantlet: (i) preparation of the original plant material by preculture, desiccation, or application of vitrification solutions; (ii) immersion in LN, thawing, and rewarming; and (iii) post-thaw culture and regeneration.

2. Materials

2.1. Cryopreservation of Zygotic Embryonic Axes

1. Plant material: mature chestnut seeds collected from open burs.
2. Material for nut surface sterilization: 70% ethanol, 5% chlorine solution (Millipore® chlorine tablets), Tween 80, and sterile distilled water.
3. Culture medium for recovery of embryonic axes after thawing: MS medium (21) with half-strength nitrates, 1 mL/L preservative for plant tissue culture media (PPM™, Plant Cell Technology, Inc., Washington, DC), 0.09 M sucrose, 6.0 g/L Vitro agar (Hispanlab, S.A., Spain), and the other organic additives listed in Table 1. The pH is adjusted to between 5.6 and 5.7 before autoclaving at 121°C for 20 min, after which the recovery medium is dispensed into 90-mm diameter Petri dishes (25 mL/dish) and 150×20 mm culture tubes (16 mL/tube).
4. Empty sterile Petri dishes and 2-mL cryovials.

2.2. Cryopreservation of Somatic Embryos

1. Plant material: stock chestnut embryogenic lines maintained by secondary embryogenesis.
2. Proliferation medium (for maintenance of embryogenic lines): MS mineral salts (with macronutrients at half strength) with 0.09 M sucrose, 7 g/L agar (Sigma), and the other organic additives listed in Table 1. The pH is adjusted to 5.6–5.7 before autoclaving at 121°C for 20 min, after which the

Table 1
Culture media used in the cryopreservation and recovery of zygotic embryonic axes and somatic embryos of European chestnut

	Embryonic axes	Somatic embryos		
Components	Recovery medium	Proliferation/ recovery medium	Preculture medium	Unloading medium
Macronutrients	MS (½ nitrates)	MS (half strength)	MS (half strength)	MS (half strength)
Micronutrients	MS	MS	MS	MS
Fe-EDTA	MS	MS	MS	MS
m-Inositol (mg/L)	100	100	100	100
Thiamine-HCl (mg/L)	1	1	1	1
Nicotinic acid (mg/L)	0.1	0.1	0.1	0.1
Pyridoxine-HCl (mg/L)	0.1	0.1	0.1	0.1
L-Glutamine (mg/L)	–	438	–	–
PPM (mL/L)	1	–	–	–
BA (μM)	0.88	0.44	–	–
NAA (μM)	–	0.54	–	–
Sucrose (M)	0.09	0.09	0.3	1.2
Agar (g/L) (Sigma A-1296)	–	7[a]	7	–
Vitro agar (g/L) (Hispanlab)	6[b]	–	–	–

BA 6-benzyladenine; *MS* medium of Murashige and Skoog (21); *NAA* naphthaleneacetic acid; *PPM* preservative for plant tissue culture media

[a]Reduced to 6 g/L in medium used for the first 24 h of recovery culture

[b]Reduced to 5 g/L in medium used for the first 24 h of recovery culture

medium is dispensed into 90-mm diameter Petri dishes (25 mL/dish).

3. Preculture medium: proliferation medium devoid of glutamine and plant growth regulators, and with sucrose concentration increased to 0.3 M (Table 1); dispensed in Petri dishes (25 mL/dish).
4. Plant Vitrification Solution 2 (PVS2): 30% w/v glycerol, 15% w/v dimethylsulfoxide, and 15% w/v ethylene glycol in liquid MS medium containing 0.4 M sucrose (22).
5. Unloading medium: liquid preculture medium devoid of agar, with sucrose concentration increased to 1.2 M (see Table 1); for unloading embryo samples after thawing.
6. Sterile filter paper discs (Whatman no. 1) and cryovials.
7. Sterile pipettes for replacing solutions from cryovials.
8. Recovery medium: same as the proliferation medium, except that agar concentration is reduced to 6 g/L in the medium used for the first 24 h of recovery culture (Table 1); dispensed in Petri dishes.

3. Methods

3.1. Cryopreservation of Zygotic Embryonic Axes

A reliable method for successful cryopreservation of zygotic embryonic axes is described that is based on partial desiccation. Desiccation-based methods, consisting in dehydrating explants following induction of dehydration tolerance, and rapid cooling by direct immersion in LN, are simpler and more practical than classical slow-cooling protocols involving freeze-induced dehydration (20, 23). In the protocol described below, plantlets are regenerated directly in the recovery medium.

3.1.1. Desiccation-Based Cryopreservation

1. Collect chestnuts of selected trees from mature open burs, generally in October, and store them in paper bags at 4°C until use, which should occur within 4–6 weeks of nut collection (see Note 1).
2. Remove the external seed coat, leaving the inner coat intact, and surface sterilize by successive immersion in (1) 70% (v/v) ethanol for 2 min and (2) a 5% solution of free chlorine with three or four drops of Tween 80/500 mL for 30 min (stir gently) (see Note 2).
3. Drain off the chlorine solution and rinse the decoated seeds three times in sterile distilled water; the first water bath should last just a few seconds and the other two 10 min each. The seeds are then transferred to a fourth water bath pending axis excision.

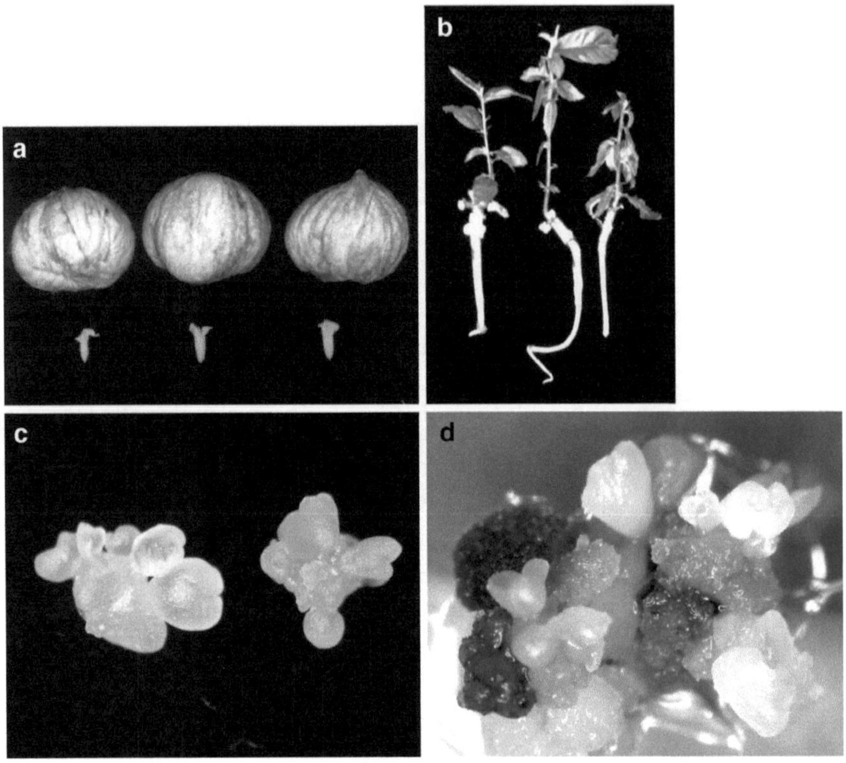

Fig. 1. Plant material of *Castanea sativa* before and after application of cryopreservation methods. (**a**) Freshly isolated zygotic embryonic axes prior to being subjected to the desiccation-based procedure. (**b**) Plantlet development from chestnut embryonic axes after 8 weeks of culture on recovery medium following a 4-h desiccation period (24% moisture content) and rapid cooling in liquid nitrogen. (**c**) Somatic embryo clumps prior to being subjected to the vitrification-based procedure. (**d**) Somatic embryo formation from a cryopreserved somatic embryo clump after 6 weeks of culture on recovery medium following preculture in 0.3 M sucrose medium and exposure to PVS2 solution for 60 min and rapid cooling in liquid nitrogen.

4. Dissect the embryonic axes aseptically from the surrounding cotyledons and transfer them to empty Petri dishes (20–25 axes/dish) (Fig. 1a).
5. Desiccate the embryonic axes in the open Petri dishes in a laminar flow hood until their moisture content has been reduced to 20–24% of fresh weight (see Note 3).
6. Transfer the desiccated axes to 2-mL cryovials (five axes per vial), place the vials in a cryostorage cane or cryobox, and plunge them rapidly into LN.
7. For thawing, immerse the vials for 2 min in a water bath at 40°C.

3.1.2. Plantlet Recovery

1. To rehydrate embryonic axes, transfer them to Petri dishes containing recovery medium (Table 1) solidified with 5 g/L agar (10 axes to a dish) and culture in the dark for 24 h at 24°C.

2. Transfer the axes to culture tubes containing fresh recovery medium solidified with 6 g/L agar (place each axis upright in its own individual tube) and keep in the dark at 24°C for 2 weeks (see Notes 4 and 5).
3. Subject the cultures for 5–6 weeks to a 16-h photoperiod with a photon flux density of 50–60 μmol/m²/s provided by white fluorescent lights, with light and dark temperatures of 25 and 20°C, respectively.
4. Seven to eight weeks after thawing, plantlet recovery may be evaluated as the percentage of whole plantlets (plantlets with both root and shoot growth, Fig. 1b) that has developed directly from embryonic axes (see Note 6).
5. The plantlets recovered from cryostorage may be used either directly for plant regeneration (by transfer to substrate and acclimatization in a greenhouse for further growth under nonsterile conditions), or as a source of clonal micropropagation cultures.

3.2. Cryopreservation of Somatic Embryos

The protocol for reliable cryopreservation of chestnut somatic embryos that is described below is based on vitrification, i.e., the transition of water directly from the liquid phase into an amorphous phase or glass, while avoiding ice crystallization (24). Vitrification involves treatment of samples with cryoprotective substances, dehydration in a highly concentrated vitrification solution, and rapid cooling. In the present case, somatic embryos recovered from cryopreservation must be proliferated and matured before plantlets can be obtained by germination.

3.2.1. Vitrification-Based Cryopreservation

1. As the source of somatic embryos, use stock embryogenic lines maintained by secondary embryogenesis with sequential subculture at 5–6-week intervals onto solidified embryo proliferation medium (Table 1) and incubation under a 16-h photoperiod of 50–60 μmol/m²/s photon flux density, with 25°C light/20°C dark temperatures (standard conditions) (see Note 7).
2. Isolate 6–8 mg of clumps of globular and/or heart-shaped embryos (Fig. 1c) from stock embryogenic cultures 3–4 weeks after the last subculture (see Note 8).
3. Place the embryo clumps in Petri dishes containing solid preculture medium (Table 1; use 1 dish per 10 clumps) and incubate for 3 days under the standard conditions defined in step 1.
4. After preculture, place the embryo clumps in 2-mL cryovials (10 clumps to a vial), and to each vial, add 1.8 mL of ice-cold PVS2 vitrification solution and leave for 60 min at 0°C (see Notes 9 and 10).

5. Resuspend the embryo clumps in 0.6 mL of fresh PVS2, place the cryovials in a cryostorage cane or cryobox, and immerse rapidly in LN.
6. To thaw, proceed as in step 7 under Subheading 3.1.1.

3.2.2. Embryo Recovery

1. Drain off the PVS2 solution, replace it with unloading medium (Table 1), leave for 10 min, replace the medium with fresh unloading medium, and leave for a further 10 min.
2. Transfer the embryo clumps in each cryovial to a filter paper disc placed on recovery medium in a Petri dish (use medium with 0.6 g/L agar; see Table 1). Culture for 24 h under the standard light and temperature conditions described in step 1 under Subheading 3.2.1.
3. Transfer the embryo clumps to fresh recovery medium gelled with 0.7 g/L agar in Petri dishes without any filter paper disc. Culture for 6 weeks under standard conditions (see Note 11).
4. Six weeks after thawing, the embryo recovery frequency may be evaluated as the proportion of clumps showing resumption of embryogenesis (Fig. 1d). The number of new embryos per embryogenic clump should also be considered in evaluating the efficiency of the cryopreservation process (see Notes 12 and 13).

3.2.3. Maintenance and Multiplication of Recovered Cryostored Embryos

1. Isolate somatic embryos or embryo clumps from embryogenic cultures retrieved from LN and grown on recovery medium, and transfer to Petri dishes containing 25 mL of embryo proliferation medium (10 explants to a dish). Culture under standard growth conditions, as in step 1 under Subheading 3.2.1, to multiply embryos by secondary embryogenesis.
2. Subculture onto fresh multiplication medium at 5–6-week intervals to maintain or increase the number of embryogenic cultures (see Note 14).

3.2.4. Embryo Maturation

1. Isolate opaque white cotyledonary somatic embryos 4–8-mm long from proliferating embryogenic cultures and transfer to Petri dishes (10–12 embryos to a dish) containing 25 mL of maturation medium, defined as basal medium (proliferation medium devoid of phytohormones) with 0.08 M maltose instead of 0.09 M sucrose. Culture for 4 weeks under standard growth conditions.
2. Transfer the embryos from maturation medium to dishes containing basal medium and keep in the dark at 4°C for 2 months (see Note 15).
3. Transfer the cold-treated matured embryos to sterile open Petri dishes, 12 embryos to a dish, and dehydrate in a laminar flow

cabinet at room temperature to a moisture content of 54–58% (approximately 2 h under our conditions) (see Note 16).

3.2.5. Embryo Germination and Conversion to Plantlets

1. Place the partially desiccated embryos horizontally in Petri dishes containing germination medium consisting of basal medium supplemented with 0.44 μM 6-benzylaminopurine (BA) and 0.49 μM 3-indolebutyric acid (IBA) and with glutamine concentration reduced to 200 mg/L (for the definition of basal medium, see step 1 under Subheading 3.2.4).
2. Incubate for 6–8 weeks under standard light and temperature conditions as in step 1 under Subheading 3.2.1.
3. Germination response may be evaluated as the proportion of embryos that have developed into plantlets (conversion into somatic seedlings). For evaluation of potential plant recovery, the proportion of embryos developing only shoots ("shoot germination") should also be taken into account (see Notes 17 and 18).
4. Isolate shoots from embryos that have grown shoots but not roots, and use micropropagation procedures developed for chestnut (5).
5. Place plantlets obtained in steps 3 and 4 in pots containing a 3:1 mixture of commercial substrate (Pinot®) and perlite. Keep under a 16-h photoperiod (95 μmol/m^2/s from cool-white fluorescent lamps) in a growth chamber at 25°C and 85–90% relative humidity until resumption of growth within 6–12 weeks of transplantation. Move the plantlets to greenhouse conditions for further growth before planting out.

4. Notes

1. Chestnut seeds require cold storage to germinate. Storage of chestnuts in paper bags for several weeks is preferred to the alternative of cold stratification in sand, which has proved to reduce the efficiency of sterilization. Contamination rates never exceed 5–10% among seeds stored in paper bags (25).
2. To avoid contamination problems during culture, use only healthy chestnuts without any signs of deterioration of seed covers or cotyledon tissues (signs include any kind of stain or blot, which will probably have been due to attack by fungi or microorganisms). It is important to perform surface sterilization of the seeds before the excision of embryo axes, rather than surface sterilization of isolated axes, which may negatively act along with freezing stress.

3. The success of embryonic axis cryopreservation depends on procuring moisture levels that minimize both desiccation damage and freezing damage. During desiccation, samples of axes should be used for periodic monitoring of water content, at least during method development. By way of orientation, freshly excised zygotic axes (Fig. 1a) have a water content of approximately 66%, and the optimal moisture content defined in this protocol (20–24%) is achieved by 4–5 h of desiccation in a laminar flow cabinet at room temperature (20).
4. In method development, in order to be able to optimize cryopreservation procedures, it is first necessary to identify conditions that are appropriate for the culture of embryonic axes in recovery medium (25). In optimizing the BA content of the recovery medium used in this protocol, it was found that varying the concentration between 0.22 and 2.22 μM had little effect on the germination and development of non-cryostored axes, but omission of BA or its inclusion at a concentration of 4.44 μM or higher resulted in reduced plant recovery and/or was detrimental for plantlet quality (20).
5. When this protocol is followed, 70–100% of non-cryostored embryo axes develop as whole plantlets within 8 weeks of culture, regardless of whether or not they had previously been desiccated to a moisture content of 20% (20).
6. Although the efficiency of the protocol must evidently be measured on a whole plantlet basis, it may also be of interest to evaluate survival, i.e. the percentage of cryopreserved axes exhibiting any kind of growth (including not only normal development, but also development of the root pole only and callus formation). Under the conditions of this protocol, survival increased with decreasing moisture content from 0% for non-desiccated axes to 100% for axes in which water content had been reduced to approximately 20% by 5-h desiccation; with this treatment, 63% of surviving axes achieved plantlet recovery and 37% only root growth.
7. Embryogenic lines may be initiated from immature zygotic embryos or from leaf explants excised from axillary shoot multiplication cultures (9). Although somatic embryogenesis induction rates from original explants are relatively low, mass production of embryos may be achieved by secondary embryogenesis.
8. The developmental stage of somatic embryos used for cryopreservation did influence post-cryostorage recovery rates in chestnut and other related species such as oaks (26). Clumps of 2–3 globular or heart-shaped embryos withstand storage in LN better than more differentiated, cotyledonary stage embryos. In the globular and early torpedo stages, somatic embryos have a greater number of active embryogenic cells

than at the cotyledonary stage, when cells exhibit higher levels of vacuolization and differentiation (27).

9. An adequate response to dehydration in vitrification solution is essential for successful vitrification. PVS2 solution is well tolerated by chestnut embryogenic cultures, the recovery rate of PVS2-treated and non-cryopreserved embryos generally exceeding 85% even after exposure times of up to 120 min.
10. Using PVS2 solution at room temperature (24°C) instead of 0°C significantly reduces the recovery of cryostored embryo clumps treated for 60 min.
11. Following rewarming, cryopreserved embryo clumps turn brown-black, but 2–3 weeks later, their surviving cells begin to produce cream-coloured globular-stage embryos. In our experience, whole original embryos do not survive cryopreservation.
12. As in the case of cryopreserved zygotic embryonic axes (see Note 6), survival (evidenced by any sign of growth, including callus formation) must be distinguished from embryo recovery (evidenced by the production of torpedo or cotyledonary stage embryos). The protocol described here (20) has achieved post-cryostorage embryogenesis resumption rates of 60–70% when applied to a variety of chestnut lines, including genetically transformed embryogenic lines (13), with numbers of new embryos per clump ranging from 1.4 to 3.4.
13. The post-thaw survival rate of European chestnut achieved by the present vitrification-based protocol is similar to that of American chestnut embryogenic cultures cryopreserved by a cryoprotectant/slow-freezing method (28), and the present protocol is both less expensive and simpler than the slow-freezing procedure.
14. Repetitive embryogenesis is mainly sustained through the development of secondary embryos on the root-hypocotyl zone of torpedo or early cotyledonary stage primary embryos. In addition, nodular proembryogenic masses (PEMs) may also be developed from cotyledons of primary embryos (10).
15. A 2–3-month cold treatment before germination has been found to be necessary to obtain plantlets of both European chestnut and hybrids (9), and also for the germination of American chestnut somatic embryos (29).
16. During desiccation, samples of embryos should be used for periodic monitoring of water content as a percentage of fresh weight, at least during method development. Under our conditions, water content was on average 85% before desiccation treatment, and fell to 54–58% after 2-h desiccation by laminar flow (11). Desiccation treatment is not essential for conversion to plantlets, but it does enhance both potential plant

recovery (by increasing the number of germinating embryos exhibiting only shoot development; see Note 17) and the quality of regenerated plantlets.

17. Although rates of direct conversion into plantlets are relatively low (18–22%), total plant recovery can be increased to 40–50% by making use of shoots produced by embryos that develop shoots but not roots (10, 11). These shoots can be multiplied by axillary shoot proliferation, rooted, and acclimatized (5).
18. The genotype is an important factor influencing not only the embryo proliferation ability but also the germination and plantlet recovery of chestnut somatic embryos.

References

1. Conedera M, Manetti MC, Giudici F, Amorini E (2004) Distribution and economic potential of the sweet chestnut (*Castanea sativa* Mill.) in Europe. Ecol Medit 30:47–61
2. Bellini E (2005) The chestnut and its resources: images and considerations. Acta Hort 693:85–96
3. Bounous G (2005) The chestnut: a multipurpose resource for the new millennium. Acta Hort 693:33–40
4. Pence VC (1995) Cryopreservation of recalcitrant seeds. In: Bajaj YPS (ed) Biotechnology in agriculture and forestry, vol 32, Cryopreservation of plant germplasm I. Springer, Berlin, pp 29–50
5. Vieitez AM, Sánchez MC, García-Nimo ML, Ballester A (2007) Protocol for micropropagation of *Castanea sativa*. In: Jain SM, Häggman H (eds) Protocols for micropropagation of woody trees and fruits. Springer, Dordrecht, pp 299–312
6. Vieitez FJ, San-José MC, Ballester A, Vieitez AM (1990) Somatic embryogenesis in cultured immature zygotic embryos in chestnut. J Plant Physiol 136:253–256
7. Vieitez FJ (1995) Somatic embryogenesis in chestnut. In: Jain MS, Gupta PK, Newton RJ (eds) Somatic embryogenesis in woody plants, vol 2. Kluwer Academic, Dordrecht, pp 375–407
8. Sauer U, Wilhelm E (2005) Somatic embryogenesis from ovaries, developing ovules and immature zygotic embryos, and improved embryo development of *Castanea sativa*. Biol Plant 49:1–6
9. Corredoira E, Ballester A, Vieitez AM (2006) Somatic embryogenesis in chestnut. In: Mujib S, Samaj J (eds) Somatic embryogenesis, plant cell monograph 2. Springer, Heidelberg, pp 177–199
10. Corredoira E, Ballester A, Vieitez AM (2003) Proliferation, maturation and germination of *Castanea sativa* Mill. somatic embryos originated from leaf explants. Ann Bot 92:129–136
11. Corredoira E, Valladares S, Vieitez AM, Ballester A (2008) Improved germination of somatic embryos and plant recovery of European chestnut. In Vitro Cell Dev Biol Plant 44:307–315
12. Corredoira E, Montenegro D, San-José MC, Vieitez AM, Ballester A (2004) *Agrobacterium*-mediated transformation of European chestnut embryogenic cultures. Plant Cell Rep 23:311–318
13. Corredoira E, San-José MC, Vieitez AM, Ballester A (2007) Improving genetic transformation of European chestnut and cryopreservation of transgenic lines. Plant Cell Tissue Organ Cult 91:281–288
14. Maynard CA, Powell WA, Polin-McGuigan LD, Vieitez AM, Ballester A, Corredoira E, Merkle SA, Andrade GM (2008) Chestnut. In: Kole C, Hall TC (eds) Compendium of transgenic crop plants: transgenic forest tree species. Blackwell, Oxford, pp 169–192
15. Engelmann F (2000) Importance of cryopreservation for the conservation of plant genetic resources. In: Engelmann F, Takagi H (eds) Cryopreservation of tropical plant germplasm – current research progress and application. JIRCAS, Tsukuba/IPGRI, Rome, pp 8–20
16. Lambardi M, De Carlo A (2003) Application of tissue culture to the germplasm conservation of temperate broad-leaf trees. In: Jain SM, Ishii K (eds) Micropropagation of woody

trees and fruits. Kluwer Academic, Dordrecht, pp 815–840
17. Klimaszewska K, Cyr DR (2002) Conifer somatic embryogenesis: I. Development. Dendrobiology 48:31–39
18. Vendrame WA, Holliday CP, Montello PM, Smith DR, Merkle SA (2001) Cryopreservation of yellow-poplar and sweetgum embryogenic cultures. New Forests 21:283–292
19. Pence VC (1992) Desiccation and the survival of *Aesculus, Castanea* and *Quercus* embryo axes through cryopreservation. Cryobiology 29:391–399
20. Corredoira E, San-José MC, Ballester A, Vieitez AM (2004) Cryopreservation of zygotic embryo axes and somatic embryos of European chestnut. CryoLetters 25:33–42
21. Murashige T, Skoog F (1962) A revised medium for rapid growth and bioassay with tobacco tissue culture. Physiol Plant 15:473–497
22. Sakai A, Kobayashi S, Oiyama I (1990) Cryopreservation of nucellar cells of navel orange (*Citrus sinensis* Osb. Var. *brasiliensis* Tanaka) by vitrification. Plant Cell Rep 9:30–33
23. Poulsen KM (1992) Sensitivity to desiccation and low temperatures (-196°C) of embryo axes from acorns of the pedunculate oak (*Quercus robur* L.). CryoLetters 13:75–82
24. Sakai A (1995) Cryopreservation of germplasm of woody plants. In: Bajaj YPS (ed) Biotechnology in agriculture and forestry, vol 32, Cryopreservation of plant germplasm I. Springer, Berlin, pp 53–69
25. San-José MC, Ballester A, Vieitez AM (2001) Effect of thidiazuron on multiple shoot induction and plant regeneration from cotyledonary nodes of chestnut. J Hortic Sci Biotech 76:588–595
26. Valladares S, Toribio M, Celestino C, Vieitez AM (2004) Cryopreservation of embryogenic cultures from mature *Quercus suber* trees using vitrification. CryoLetters 25:177–186
27. Yeung EC (1995) Structural and developmental patterns in somatic embryogenesis. In: Thorpe TA (ed) In vitro embryogenesis in plants. Kluwer Academic, Dordrecht, pp 205–247
28. Holliday C, Merkle S (2000) Preservation of American chestnut germplasm by cryostorage of embryogenic cultures. J Am Chestnut Found 14:46–52
29. Andrade GM, Merkle SA (2005) Enhancement of American chestnut somatic seedling production. Plant Cell Rep 24:326–334

Chapter 16

Cryopreservation of *Ilex* Immature Zygotic Embryos

Luis Mroginski, Natalia Dolce, Pedro Sansberro, Claudia Luna, Ana Gonzalez, and Hebe Rey

Abstract

Tropical *Ilex* species have recalcitrant seeds. This chapter describes protocols for long-term conservation of *Ilex brasiliensis*, *I. brevicuspis*, *I. dumosa*, *I. microdonta*, *I. integerrima*, *I. paraguariensis*, *I. pseudoboxus*, *I. taubertiana*, and *I. theezans* through cryopreservation of zygotic rudimentary embryos at the heart developmental stage. The embryos are aseptically removed from the seeds and precultured (7 days) in the dark at 27 ± 2°C on solidified quarter-strength Murashige and Skoog medium with 3% sucrose and 0.1 mg/L zeatin. The embryos are then encapsulated in 3% calcium alginate beads and pretreated at 24-h intervals in liquid medium supplemented with progressively increasing sucrose concentrations (0.5, 0.75, and 1 M). The beads are dehydrated for 5 h with silica gel to 25% water content (fresh weight basis) and then placed in sterile 5-mL cryovials. Then the beads are either plunged rapidly in liquid nitrogen where they are kept for 1 h (rapid cooling), or cooled at 1°C/min to −30°C and then immersed in liquid nitrogen for 1 h (slow cooling). After cryopreservation, the beads are rewarmed by immersion of the cryovials for 1 min in a water bath at 30°C. Finally, the beads are transferred onto culture medium (1/4MS, 3% sucrose, and 0.1 mg/L zeatin, solidified with 0.8% agar) and incubated in a growth room at 27 ± 2°C under a 14-h light (116 μmol/m²/s) and 10-h dark photoperiod. Maximum recovery percentages between 15 and 83% (depending on the species and the treatment) were obtained with the cryopreserved embryos.

Key words: Cryopreservation, Embryo culture, Encapsulation dehydration, Fruit cryopreservation, Germplasm preservation, *Ilex spp*, Liquid nitrogen, Plant, Regeneration

1. Introduction

The genus *Ilex* is the largest of the family Aquifoliaceae. The systematics of the genus presents some difficulties and the total number of species is still uncertain, probably due to the lack of appropriate collections considering that *Ilex* is usually comprised of woody dioecious plants (1). Giberti (2) has mentioned 400 species widely distributed in the world. However, more recent taxonomic studies recognized at least 600 species (3, 4). Most species of this genus are deciduous or evergreen shrubs or small

Trevor A. Thorpe and Edward C. Yeung (eds.), *Plant Embryo Culture: Methods and Protocols*, Methods in Molecular Biology, vol. 710, DOI 10.1007/978-1-61737-988-8_16,

trees, but in the tropics the genus also includes some very large trees and, in some cases, climbers. Despite that the distribution is predominantly in subtropical–tropical regions of both hemispheres, few species grow in temperate areas. *Ilex* species extend north to 64° and south to 33° (3).

The *Ilex* genus is comprised of several species of economic importance. Some of them, commonly named "hollies," such as "English holly" (*Ilex aquifolium* L.), "Chinese holly" (*I. cornuta* Lindl. Hitchcock), "American holly" (*I. opaca* Ait.Farage), and "Japanese holly" (*Ilex crenata* Thunb.), have long been symbolic of Christmas and also have been cultivated by nurserymen in the United States for landscaping, and various institutions and commercial breeders are developing hybrids with improved tolerance to winter and with more foliage (5). In South America, *I. paraguariensis* St. Hil. (named "maté tree," "yerba mate," or "ka'a y") is a species that grows typically in acidic soil with its natural distribution restricted to 3% of the South American territory (6). This species, a tree or sometimes a shrub as much as 16-m tall, is a perennial crop which is an important source of income in some regions of northeastern Argentina, Paraguay, northern Uruguay, and southern Brazil (2, 7). Of these regions, Argentina (Provinces of Misiones and Corrientes) is the largest producer with around 152,000 hectares. The overall value of "mate tree" production around the world was estimated in 2004 at one billion USD (8). Its dried leaves and twigs are used for making a stimulating drink named "mate," "chimarrao," "mate tea," and "tereré," which has been consumed for centuries and are very appreciated by people of this region. Actually, it is well known that its leaves contain caffeine, which can explain this physiological effect (9–11). This plant has many other uses in canned drinks, soluble teas, cosmetics, colourings, and medicines (11). Mate tea has been mentioned as a central nervous system stimulant, diuretic (12), antioxidant (13), and cardioprotective agent (14) and as having other health benefits (8, 15). Other species of *Ilex*, such as *I. brevicuspis* (16) and *I. dumosa* (17), have been recently mentioned for their health benefits. The leaves of *I. dumosa* have similar components to *I. paraguariensis* while they have less caffeine and saponins content. Other members of the genus *Ilex*, such as "Yaupon Holly" (*I. vomitoria*), "guayusa" (*I. guayusa*), and *I. tarapotina*, are also used in infusions (18).

Regarding the possibilities for germ plasm conservation, most of the *Ilex* species present two major constraints: (A) They usually have seeds with rudimentary embryos that remain in the immature heart-shaped stage for a long time after the fruits reach maturity (5, 19, 20). When fruits of *I. paraguariensis* are ripe, only about 1% of the seeds (pyrenes) contain mature embryos and 99% of the seeds have embryos either in the heart stage (70%) or in the torpedo stage (29%) (21). In the same sense, it has been reported that when fruits of *I. dumosa* and *I. brevicuspis* are mature, only 7.22 and 1.90% of the embryos are at the cotyledonary stage, respectively (22). Similar results were reported in 14 species of *Ilex* which occur in Hong Kong (3). As a result, seed germination is delayed, and in *I. paraguariensis,* a minimum of 5–9 months under appropriate environmental conditions is required for embryo maturation (23). In the case of *I. opaca*, germination in nature requires 1–3 years, as the percentage of germination is about one in ten million (19). The technique of embryo rescue has been employed successfully in many crops to solve this problem (24–26). This technique permits the acceleration of the maturation of rudimentary embryos, resulting in the highest germination rates. It was used in numerous species of *Ilex* (Table 1).

Table 1
Plant regeneration by in vitro culture of rudimentary embryos of 21 species of *Ilex*

Species	Maximum percentage of embryos converted to seedlings	References
I. aquifolium	86.13	(5, 20, 47)
I. argentina	97	(43, 48)
I. brasiliensis	56–82	(36, 38)
I. brevicuspis	8–94	(36, 38, 48)
I. cassine	1.37–62.5	(5, 20)
I. cornuta	97.51	(5, 20, 47)
I. crenata	96.40	(5, 20, 47)
I. dumosa	40–74	(36, 38, 48, 49)
I. integerrima	27–61	(36, 38, 48)
I. glabra	84.89	(5, 20, 47)
I. longipes	93.10	(20, 47)
I. microdonta	43–89	(36, 48)
I. opaca	28.96–75.0	(5, 20, 43, 44, 47)
I. pedunculosa	87.77	(5, 20)
I. pernyi	15.73–91.7	(5, 20, 47)
I. paraguariensis	40–60	(36–38, 48, 50, 51)
I. pseudoboxus	23–93	(36, 38, 48)
I. serrata	60.81	(20, 47)
I. taubertiana	10	(38)
I. theezans	33–67	(36, 38, 48)
I. verticillata	89.89	(5, 20, 47)

(B) They have seeds (especially the subtropical and tropical species) which are highly sensitive to desiccation and which cannot be stored at low temperatures. In other words, according to Roberts (27), they are recalcitrant seeds and, therefore, are not suitable for long-term preservation using conventional seed storage methods. Thus, the germ plasm of *Ilex* spp. is maintained in the field as ex situ gene banks (28). Under this storage method, the genetic resources are exposed to diseases, pests, and natural hazards. In addition, labor costs and trained personnel requirements are very important (29–31).

Currently cryopreservation appears, at least in some species, to be an ideal procedure for long-term conservation of plant germ plasm. It consists of bringing the plant material to a metabolically inactive state through its immersion in liquid nitrogen (−196°C). Cryopreservation protocols have been developed for seeds and several explants of various plant species (32–34).

Two strategies for in vitro cryopreservation of germ plasm of *Ilex* species using zygotic embryos were tested: (a) embryos cryopreserved in fruits and (b) cryopreservation of isolated embryos (Fig. 1) (28). The protocols for both procedures require several steps which depend on the procedure chosen.

2. Materials

1. Nine species of *Ilex* (see Table 2).
2. Surface sterilization of fruits and embryos: 70% ethanol and 1.8% sodium hypochlorite, with two drops of Triton X-100® (Merck, Darmstadt, Germany).
3. Pretreatment of fruits: plastic bags and refrigerator (4°C).
4. Solution for cryopreservation: 2.5 mL of cryoprotectant solution (50% sucrose and 50% glycerol) in 5-cm^3 polypropylene cryotubes.
5. Controlled Rate Freezing System (Gordinier Electronics, Inc., USA, Model 9000).
6. Liquid nitrogen.
7. Water bath (30°C) for thawing.
8. For isolation of embryos: laminar flow hood and sterile tools (scalpel, sterile glass Petri plate, needle, razor, blade, and forceps) (see Note 1).
9. For embryo culture: Murashige and Skoog (MS) medium (35) with agar (Sigma Chemical Co. A-1296) and zeatin (6-[4-hydroxy-3-methylbut-2-enylamino] purine) (Sigma Chemical Co.) in 11-cm^3 glass tubes.

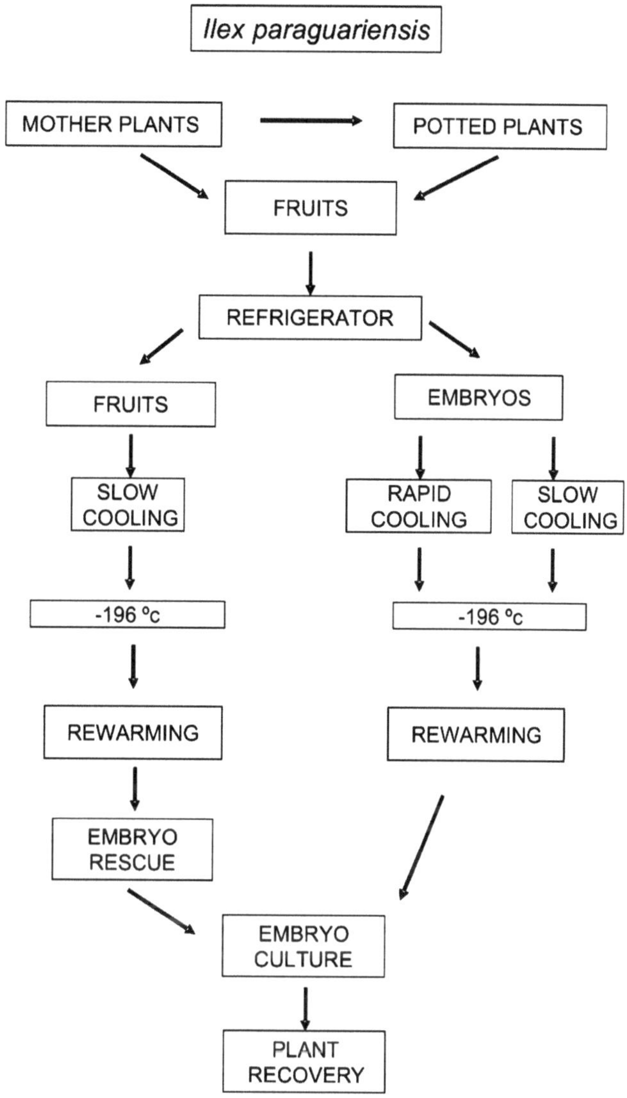

Fig. 1. Strategies for cryopreservation of germ plasm of *Ilex* spp.

10. Aluminum foil and Resinite AF® (Casco SAIC Company) for cover glass tubes.
11. Autoclave.
12. Silica gel (Riedel-de Haën, AG-D-30926, Seelze, Germany).
13. Growth room at 27 ± 2°C with a 14-h photoperiod (116 μmol/m²/s).
14. For transplanting seedlings: pots containing fine vermiculite.

Table 2
In vitro germination (%) of cryopreserved zygotic embryos of *Ilex* species

Maximum percentage of embryos

Species converted to seedlings by using	Embryos cryopreserved in fruits (36)[a]	Isolated cryopreserved embryos (38)[a]
I. brasiliensis	3	67
I. brevicuspis	0	3
I. dumosa	13	30
I. integerrima	0	40
I. microdonta	3	–
I. paraguariensis	10	67
I. pseudoboxus	23	10
I. taubertiana	–	10
I. theezans	0	40

[a]Ref. (36, 38) – not tested

3. Methods

3.1. Cryopreservation of Embryos in Fruits

This protocol (36) using vitrification method consists of the following:

1. Preparation of plant materials. Fruits will be chosen either from plants growing in the field or from potted plants growing in greenhouse (Fig. 1). This condition is generally preferable, since the fruits are more free of superficial contamination with bacteria or fungi. It is highly desirable that in both cases the mother plants have good sanitary conditions.

   Immature light green fruits (drupes) of *Ilex* spp. (Fig. 2a) should be harvested during summer (2–3 months after hand pollination) and superficially sterilized in 70% ethanol for 5 min, followed by immersion for 30 min in 1.8% sodium hypochlorite, with two drops of Triton X-100®. Subsequently, the fruits should be rinsed three times with autoclaved distilled water. They can be stored in a refrigerator (4°C) in plastic bags until use (see Notes 2–4).
2. Pretreatment. The superficially sterilized fruits should be cold-pretreated (for 1 month at 4°C) and hermetically sealed in 5-cm^3 polypropylene cryotubes (ten fruits per tube)

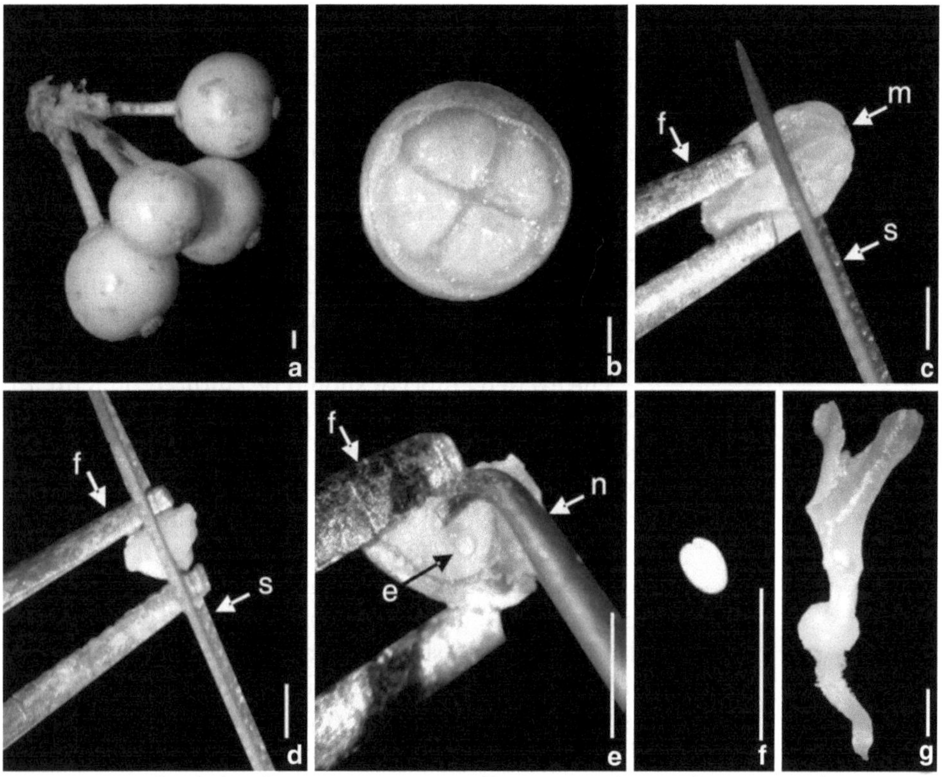

Fig. 2. Embryo cryopreservation of *Ilex* spp. (*Vertical bars* represent 1 mm) (**a**) Fruits. (**b**) Seeds. (**c–e**) Embryo excision procedure. (**f**) Embryo at the heart development stage. (**g**) Seedling obtained by embryo culture.

containing 2.5 cm^3 of cryoprotectant solution composed of 50% sucrose (v/v) and 50% glycerol (v/v).

3. Freezing and storage. The cryotubes containing fruits immersed in the cryoprotectant solution should be frozen slowly, by cooling from 25°C (room temperature) to −40°C at 1°C/min before storage in liquid nitrogen (−196°C) using a Controlled Rate Freezing System (Gordinier Electronics, Inc., USA, Model 9000) (see Note 5).
4. Thawing. For thawing, the cryotubes containing the fruits should be placed in a 30°C water bath for 1 min.
5. Viability assessment. After freezing, the rudimentary embryos at the heart-shaped stage (Fig. 2f) should be excised and cultured in vitro for survival assessment. For isolation of embryos, firstly, the fruits should be surface sterilized by soaking in 70% ethanol (5 min) followed by immersion in 1.8% hypochlorite and two drops of Triton X-100® (30 min), rinsing three times with sterile distilled water, and maintaining in the final rinse until embryo excision. After that the embryos should be

separated from pulp under aseptic condition, working with the aid of a stereomicroscope in a laminar flow hood, using a sterile glass Petri plate. The dissecting tools should be disinfected frequently by dipping into 70% ethanol and drying them on a sterile glass Petri plate. Since the embryos are minute (0.16–0.35 mm in length) and easily damaged during the manipulation of the isolation, it is necessary to know exactly the place of the seed in which they are located. Normally, the embryos appear at the end of the seeds, close to the micropylar plug, and it possible to separate of the rest of the tissues of the seeds with a scalpel (Fig. 2b–e) (see Notes 6–8).

The excised embryos should be cultured on 3 mL of culture medium in 11-mL glass tubes (one embryo per tube). The tubes are sealed with Resinite AF® (Casco SAIC Company) and incubated in darkness at a constant temperature of 27 ± 2°C. The culture medium was that reported by Sansberro et al. (37) for embryo culture of *I. paraguariensis* and consists of quarter-strength MS medium (35) with 3% sucrose, 0.65% agar, and 0.1 mg/L zeatin (see Note 9). The cultured embryos show the same developmental sequences as they pass in situ, and the first seedlings (Fig. 2g) can be observed after 14–35 days (depending upon the species) of culture. These seedlings can be successfully transplanted to pots containing vermiculite in a growth room at 27 ± 2°C with a 14-h photoperiod (116 μmol/m^2/s). Relative humidity should be maintained at 95–100% during the first week and then decreased gradually during the establishment in a greenhouse.

3.2. Cryopreservation of Isolated Embryos

This protocol (38), using encapsulation–dehydration method, was tested for eight species of *Ilex*, Table 2.

1. Preparation of plant material. This step is the same as the one described in Subheading 3.1 step 1, and for the isolation of rudimentary embryos, the procedure is described in Subheading 3.1 step 5.
2. Pretreatment. Excised embryos (Fig. 2a–f) should be precultured for 7 days in the dark at 27 ± 2°C on solidified (0.8% agar) quarter-strength MS medium (35) with 3% sucrose and 0.1 mg/L zeatin. After preculture, the embryos should be encapsulated in 3% calcium alginate (see Note 10) and then (beads of approximately 4–5 mm in diameter) transferred at 27 ± 2°C with 24-h intervals in liquid medium supplemented with progressively increasing sucrose concentration (0.5, 0.75, and 1 M). The beads should then be dehydrated for 5 h with silica gel to 25% water content (fresh weight basis). Finally, the dried beads should be placed in sterile 5-mL cryotubes (ten beads/cryotube) (see Notes 11–13).

3. Storage. Cryotubes containing dried beads can be plunged rapidly in liquid nitrogen (rapid cooling) or cooled at 1°C/min to −30°C (using a Controller Rate Freezing System, Gordiner Electronics, Inc., USA) and then immersed in liquid nitrogen (slow cooling).
4. Thawing and viability assessment. The beads should be rewarmed by immersing cryotubes in a 30°C water bath for 1 min. Finally, the beads should be transferred to the same culture medium described under Subheading 3.1 step 5 and incubated under a 14-h light (116 μmol/m^2/s)/10-h dark photoperiod. Survival of the embryos should be evaluated after freezing by counting the number of embryos that develop plantlets.

4. Notes

1. For tissue culture laboratory facilities, media preparation, equipment, and handling, see refs. (39–42).
2. For seed preparation, embryo excision, culture, and incubation, see ref. (5).
3. In addition to immature fruits, mature red fruits can be a source of embryos.
4. Embryos cryopreserved in fruits using rapid cooling do not germinate.
5. In the case of slow freezing, a programmable freezing apparatus will be necessary in order to obtain precise and reproducible freezing conditions.
6. Seeds that float should be discarded.
7. Take care that the pressure does not injure the fragile embryonic tissue.
8. A drop of sterile water can be added to the seed during dissection in order to avoid dehydration injury of immature embryos.
9. Excised embryos of most of *Ilex* spp. are sensitive to light during the initial period of incubation. Thus, dark incubation is recommended for at least the first week (5, 43, 44).
10. The encapsulation of embryos should be done individually by using the classical procedure described for somatic embryos (45), where the explants are suspended in liquid culture medium and 3% sodium alginate. This mixture, with the explants, is dispensed with a 2-mL sterile Pasteur pipette by dropping in a 0.1-M calcium chloride solution at room temperature. After 30 min, the resulting beads (about 4–5 mm in

diameter) containing one embryo are removed from the liquid medium and dried on filter paper (46).

11. As an alternative to the drying treatment with silica gel, embryos can be dried in a laminar flow hood.
12. Dehydration was carried out by setting the beads on an aluminum net at 15 mm from 30 g silica gel in a hermetically closed sterile plastic container (50 mL capacity)(46).
13. The period of dehydration was obtained from dehydration curves by determining gravimetrically the water content (fresh weight basis) at 1-h intervals between 0 and 6 h (46).

References

1. Del Pero Martínez MA, Pelotto JP, Basualdo N (1997) Distribution of flavonoid aglycones in *Ilex* species (Aquifoliaceae). Biochem Sys Ecol 25:619–622
2. Giberti GC (1994) Aquifoliaceae. In: Spichiger R, Ramella L (eds) Flora de Paraguay, Editions des Conservatoire et Jardin botaniques de la Ville de Genève and Missori Botanical Garden. p 32
3. Tsang ACW, Corlett RT (2005) Reproductive biology of the *Ilex* species (Aquifoliaceae) in Hong Kong, China. Can J Bot 83:1545–1554
4. Loizeau PA, Spichiger R (2004) Aquifoliaceae. In: Smith N, Mori SA, Henderson A, Stevenson DW, Heald SV (eds) Flowering plants of the neotropics. Princeton University Press, Princeton, pp 26–28
5. Hu CY (1989) Holly (*Ilex spp*). In: Bajaj YPS (ed) Biotechnology in agriculture and forestry, Trees II, vol 5. Springer, Berlin, pp 412–427
6. Reissmann CB, Radomski MI, Bianchini de Quadros RM (1999) Chemical composition of *Ilex paraguariensis* St.Hil. under different management conditions is seven localities of Paraná state. Braz Arch Biol Technol 42:187–194
7. Grondona EM (1954) Historia de la yerba mate. II. Sinonimia, cariología y distribución geográfica. Rev Arg Agron 21:9–24
8. Heck CI, De Mejía EG (2007) Yerba mate tea (*Ilex paraguariensis*): a comprehensive review on chemistry, health implications, and technological considerations. J Food Sci 72:138–151
9. Alikaridis F (1987) Natural constituents of *Ilex* species. J Ethnopharmacol 20:121–144
10. Mazzafera P (1997) Maté drinking: caffeine and phenolic acid intact. Food Chem 1:67–71
11. Gauer L, Cavalli-Molina S (1999) Genetic variation in natural populations of maté (*Ilex paraguariensis* A. St.-Hil., Aquifoliaceae). Heredity 84:647–656
12. Gonzalez A, Ferreira F, Moyna P, Paz EA (1993) Biological screening of Uruguayan medicinal-plants. J Ethnopharmacol 39: 217–220
13. Filip R, Lotito SB, Ferraro G, Fraga GC (2000) Antioxidant activity of *Ilex paraguariensis* and related species. Nutr Res 20:1437–1446
14. Schinella G, Fantinelli JC, Mosca SM (2005) Cardioprotective effects of *Ilex paraguariensis* extract: evidence for a nitric oxide-dependent mechanism. Clin Nutr 24:360–366
15. Carducci CN, Dabas C, Muse JO (2000) Determination of inorganic cations by capillary ion electrophoresis of *Ilex paraguariensis* (St.H.), a plant used to prepare tea in South America. J AOAC Int 83:1167–1173
16. Filip R, Ferraro GE (2003) Researching on new species of mate: *Ilex brevicuspis*: phytochemical and pharmacology study. Eur J Nutr 42:50–54
17. Filip R, Lopez PG, Ferraro GE (1999) Phytochemical study of *Ilex dumosa*. Acta Hortic 501:333–336
18. Loizeau PA (1994) Las Aquifoliaceas péruviennes. Boissiera 48:1–306
19. Ives SA (1923) Maturation and germination of seeds of *Ilex opaca*. Bot Gaz 76:60–77
20. Hu CY (1975) In vitro culture of rudimentary embryos of eleven *Ilex* species. J Am Soc Hortic Sci 100:221–225
21. Niklas CO (1987) Estudios embriológicos y citológicos en la yerba mate *Ilex paraguariensis* (Aquifoliaceae). Bonplandia 6:45–56
22. Dolce NR, Gonzalo AG, Rey HY (2007) Embriología de *Ilex dumosa* e *I.brevicuspis* (Aquifoliaceae). Bol Soc Arg Bot 42(suppl): 40–41
23. Fontana HP, Prat Krikum SD, Belinghieri LD (1990) Estudios sobre la germinación y conservación de semillas de yerba mate (*Ilex paraguariensis* St.Hil.). *Informe Técnico*,

EEA INTA Cerro Azul (Misiones, Argentina) 52:1–12

24. Hu CY, Wang PJ (1986) Embryo culture: technique and application. In: Evans DA, Sharp WR, Ammirato PB (eds) Handbook of plant cell culture, vol 4. Macmillan, New York, pp 43–96
25. Sharma DR, Kaur R, Kumar K (1996) Embryo rescue in plants: a review. Euphytica 89: 325–337
26. Raghavan V (2003) One hundred years of zygotic embryo culture investigations. In Vitro Cell Dev Biol Plant 39:437–442
27. Roberts EH (1973) Predicting the storage life to seeds. Seeds Sci Technol 1:499–514
28. Mroginski L, Rey H (2007) Cryopreservation of plant germplasm in Argentina. Adv Hortic Sci 21:270–273
29. Engelmann F (1991) In vitro conservation of tropical germplasm- a review. Euphytica 57: 227–243
30. Withers LA, Engelmann F (1998) In vitro conservation of plant genetic resources. In: Altman A (ed) Agricultural biotechnology. Marcel Dekker, New York, pp 57–88
31. Sarasan VC, Ramsay MM, Atherton C, McMichen M, Prendergast G, Rowntree JK (2006) Conservation *in vitro* of threatened plants – progress in the past decade. In Vitro Cell Dev Biol-Plant 42:206–214
32. Panis B, Lambardi M (2006) Status of cryopreservation technologies in plants (crops and forest trees). In: Ruoane J, Sonnino A (eds) The role of biotechnology in exploring and protecting agricultural genetic resources. FAO, Rome, pp 61–78
33. Kameswara Rao N (2004) Plant genetic resources: advancing conservation and use through biotechnology. Afr J Biotechnol 3:136–145
34. Engelmann F (2000) Importance of cryopreservation for the conservation of plant genetic resources. In: Engelmann F, Takagi H (eds) Cryopreservation of tropical plant germplasm. IPGRI, Rome, pp 8–20
35. Murashige T, Skoog F (1962) A revised medium for rapid growth and bioassays with tobacco tissue cultures. Physiol Plant 15: 473–497
36. Mroginski LA, Sansberro PA, Scocchi AM, Luna CV, Rey HY (2006) Effect of fruit cryopreservation on in vitro germination of zygotic embryos of several species of *Ilex*. Acta Hortic 725:417–419
37. Sansberro PA, Rey HY, Mroginski LA, Collavino MM (1998) In vitro culture of rudimentary embryos of *Ilex paraguariensis*: Responses to exogenous cytokinins. J Plant Growth Regul 17:101–105
38. Mroginski LA, Sansberro PA, Scocchi AM, Luna CV, Rey HY (2008) A cryopreservation protocol for immature zygotic embryos of species of *Ilex* (Aquifoliaceae). Biocell 32:33–39
39. Gamborg OL, Phillips GC (1995) Laboratory facilities, operation, and management. In: Gamborg OL, Phillips GC (eds) Plant cell, tissue and organ culture. Springer, Berlin, pp 3–20
40. Gamborg OL, Phillips GC (1995) Media preparation and handling. In: Gamborg OL, Phillips GC (eds) Plant cell, tissue and organ culture. Springer, Berlin, pp 21–34 and appendix pp 301–307
41. Brown DCW, Thorpe TA (1984) Organization of a plant tissue culture laboratory. In: Vasil IK (ed) Plant cell culture and somatic cell genetics of plants. Laboratory procedures and their applications, vol 1. Academic, New York, pp 1–12
42. Harry IS, Thorpe TA (1994) In vitro methods for forest trees. In: Vasil IK, Thorpe TA (eds) Plant cell and tissue culture. Kluwer Academic Publishers, Dordrecht, pp 539–560
43. Ferreira AG, Hu CY (1989) Light-mediated inhibition of in vitro late embryogeny of *Ilex*. J Am Soc Hortic Sci 114:819–823
44. Hu CY (1976) Light-mediated inhibition of in vitro development of rudimentary embryos of *Ilex opaca*. Am J Bot 63:651–656
45. Redenbaugh KA, Paasch BD, Nichol JW, Kessler ME, Viss PE, Walker KA (1986) Somatic seeds: encapsulation of asexual plant embryos. Biotechnology 4:797–801
46. Rey HY, Faloci M, Medina R, Dolce N, Mroginski L, Engelmann F (2009) Cryopreservation of in vitro-grown shoot tips and apical meristems of the forage legume *Arachis pintoi*. Cryoletters 30:347–358
47. Hu CY, Rogalski F, Ward C (1979) Factors maintaining *Ilex* rudimentary embryos in the quiescent state and the ultrastructural changes during in vitro activation. Bot Gaz 140:272 279
48. Sansberro PA, Rey HY, Luna CV, Mroginski L (2001) Influence of gelling agents on *Ilex paraguariensis* tissue culture. Acta Hortic 560:453–456
49. Dolce NR, Rey HY, Mroginski LA (2009) Cryoconservation of *Ilex dumosa* (Aquifoliaceae) germplasm. Acta Hortic (submitted)
50. Ferreira AG, Cunha GG, da Silveira TS, Hu CY (1991) In vitro germination of immature embryos of *Ilex paraguariensis* St. Hil. Phyton 52:27–32
51. Sansberro PA, Rey HY, Mroginski L (2000) Efectos de bajas temperatures en la germinación *in vitro* de embriones inmaduros de yerba mate, in *Proc Congreso sul Americano da Erva Mate*. Río Grande do Sul, Brazil, pp 443–446

Part IV

The Use of Zygotic Embryos as Explants for Somatic Embryogenesis and/or Organogenesis

Chapter 17

The Use of Zygotic Embryos as Explants for In Vitro Propagation: An Overview

Mohamed Elhiti and Claudio Stasolla

Abstract

Plant propagation in vitro via somatic embryogenesis or organogenesis is a complicated process requiring the proper execution of several steps, which are affected by culture conditions and environment. A key element for a successful outcome is the choice of the explants. Several studies have shown that factors such as age, ontogenic and physiological conditions, and degree of differentiation affect the response of the explants to culture conditions. As a general rule, younger tissues, such as zygotic embryos, are the preferred choice for tissue culturists as they have better potential and competence to produce embryos and organs compared to more differentiated and mature tissues. This chapter focuses on how competence and commitment to regenerate embryos and organs in cultures are acquired by somatic cells and why zygotic embryos are so often utilized for propagation practices.

Key words: Embryo, Organogenesis, Plant growth regulators, Propagation, Somatic embryogenesis

1. Introduction

The utilization of in vitro techniques to regenerate plants in vitro has been largely used as a "propagation" tool as well as a model system to understand basic aspects of plant development. Plant regeneration in culture can be achieved either through embryogenesis, that is the formation of bipolar structures, i.e., embryos, or through organogenesis, the generation of a specific organs, i.e., shoots or roots. Regeneration through either process can occur directly or indirectly, depending on whether an intervening proliferation of undifferentiated tissue (callus) is required.

1.1. Embryogenesis

Embryogenesis represents an important event in the plant life cycle which is initiated with double fertilization, followed by the formation of embryos which are characterized by lateral, radial,

Trevor A. Thorpe and Edward C. Yeung (eds.), *Plant Embryo Culture: Methods and Protocols*,
Methods in Molecular Biology, vol. 710, DOI 10.1007/978-1-61737-988-8_17,

and longitudinal growth. Embryonic growth, which has been well documented in both angiosperms and gymnosperms, results in precise morphological changes which in angiosperms define the globular, heart, and torpedo stages of development. The final phases of embryogenesis are marked by the accumulation of storage products and, at least for the majority of species, acquisition of dormancy. Proper execution of these events is ensured by an intrigued genetic network contributing to the expression and interaction of factors which regulate histodifferentiation and tissue patterning. Perturbations of this signaling often results in aberrant growth and ultimately embryo abortion. In higher plants, embryogenesis can also occur through apomixis, which involves the asexual formation of embryos from the maternal tissues. This process, executed without intervening meiotic and fertilization phases, has been described in a variety (more than 400) of species included in 40 different families (1). As suggested by Feher (2) the apomictic process suggests that fertilization is not a determinant factor for embryogenesis and that cells other than gametes have the potential to regain embryogenic potential and form embryos. Both notions have been demonstrated in vivo by the early realizations that embryos can often arise from the leaf margins of several species, including *Bryophyllum* or *Malaxis* (3, 4).

Applications of these concepts have been extrapolated in vitro where somatic embryogenesis, the process whereby somatic cells, i.e., cells other than gametes, change their developmental fate and embark on an embryogenic pathway culminating in the formation of somatic embryos, has been implemented successfully in many species. The history of somatic embryogenesis starts with the pioneering work of Levine (5) who reported the regeneration of carrot embryos from tissue culture in the presence of low levels of auxins. This work was later followed by Steward et al. (6) who documented embryo-like structures originating from cultured carrot cells. Since then these studies have been replicated in other systems resulting in numerous species able to form somatic embryos in culture. For a historical perspective of in vitro embryogenesis the readers are encouraged to consult a recent review (7). The process of somatic embryogenesis is successfully carried out by a proper selection of the explant, medium, growth substances, and the control of the physical culture environment. Thorpe and Stasolla (8) proposed that the limited production of somatic embryos for several species is ascribed to the failure to achieve the proper balance and order of the above factors.

Embryo development in vitro is generally divided into an induction phase, in which somatic cells must be reprogrammed to embark into a new fate, followed by a developmental phase characterized by the growth of the embryo. A lot of information concerning the latter phase is available in literature as many structural, physiological, and molecular studies have been conducted during the different phases of embryo development (8). However, the inductive phase has not been investigated in details given the difficulties in identifying cells within the explant undergoing somatic-embryogenic transition.

1.2. Organogenesis

Besides embryogenesis, plantlet formation from cultured cells can occur through the formation of primordia, which subsequently undergo organogenesis. In many instances, shoot primordia are formed first followed by leafy vegetative shoots, which are then rooted via root organogenesis. The organogenic process was first documented by White (9) who obtained shoots from tobacco hybrids and Nobecourt (10) who observed root formation from carrot callus. During the following years several other plant species were shown to form de novo shoots and roots from callus, thanks to the finding of Skoog and Miller (11) who identified the auxin/cytokinin balance as the main regulatory mechanism controlling organogenesis. In addition to phytohormones, other metabolites have shown to stimulate organogenesis in different species. These metabolites include adenine, amino acids, uracil, uridine, nicotine, and phenolic acids (12). However the interactions among all these compounds are in agreement with the notion postulated by Skoog and Miller (11); a concept which still leads the majority of research dealing with organogenesis. An updated review on the physiological and molecular events occurring during the organogenic process is provided in (13).

This chapter provides an overview on the use of zygotic embryos to initiate somatic embryogenesis and organogenesis in vitro. Factors regulating embryogenic/organogenic potential and competence are first discussed in order to appreciate why immature and mature zygotic embryos are the preferred choices of explants for many species.

2. The Embryogenic/Organogenic Pathway from Somatic Cells

Induction of embryos and/or organs from somatic cells within an explant is a complex process, which is exemplified as consisting of three conditions (2, 14). First, the explant must have the "potential" to produce embryos or organs. Second, some cells within the explant must be "competent" to respond to endogenous or exogenous signals. Third, these competent cells must be "induced" by specific signals and become "committed" to initiate the embryogenic/organogenic pathway.

2.1. Acquisition of Embryogenic/Organogenic Potential

Acquisition of potential is determined at several levels with the genotype being the most important factor. This concept is easily demonstrated by the variation in embryogenic response, which is often observed among different genotypes within the same species.

Several studies in alfalfa have also reinforced this notion (15). In an elegant experiment (15), it was possible to express embryogenic potential in a nonregenerative alfalfa clone through the introgression of "favorable" alleles isolated from an embryogenic clone. Although the genetic makeup of the explant plays a determinant role on embryogenic potential, it is also true that the inability to initiate the embryogenic process might be due to suboptimal culture conditions. Therefore, a recalcitrant clone might indeed have the genetic potential to regenerate embryos in culture but would fail to do so because of unsuitable in vitro conditions. Feher (2) further elaborated on the importance of the genotype suggesting that "genetic determinants" specify the temporal and spatial ability of the explant to express competence to regenerate embryos, which is, however, affected by both developmental and environmental cues. This statement implies that not all the tissues of a genotype with embryogenic potential are able to respond in culture and thus the choice of the explant has to be carefully considered.

In the case of embryogenesis, various plant tissues have been employed in culture to regenerate embryos. For some species, including alfalfa, somatic embryo formation can be initiated from all organs within the seedling, including hypocotyls and cotyledons (16). Excised cotyledons of soybean cultured with their abaxial epidermis in contact with the medium produce a large number of embryos (17). For other species, however, the explant is limited to a specific ontogenic stage. In grasses, for example, only zygotic embryos, inflorescences, and leaves can be used to generate somatic embryos, and all these explants contain meristematic cells that can be maintained and propagated in culture by applications of exogenous auxin (18). In pea, only the embryonic axis, and not the cotyledons from seed embryo explants, is embryogenic (19). Ewans et al. (20) reported that in about 40% of crop species undergoing somatic embryogenesis, the cultures were derived from either hypocotyls or zygotic embryos. For many species the explant of choice is the zygotic embryo which, according to Neumann (21) is at the top of the hierarchical gradient of tissues responding to embryogenesis (embryo > hypocotyl > petiole > leaf lamina > root). It is not surprising that the embryogenic potential is highest in zygotic embryos since they consist of cells already possessing embryogenic fate and therefore would respond better in culture. Other more differentiated plant tissues can, however, be induced to acquire embryogenic or organogenic potential through a reprogramming into the embryonic state.

An important factor which would advance the field of plant propagation would be understanding differences existing between embryogenic and nonembryogenic phenotypes. Physiological studies revealed that alfalfa embryogenic genotypes are more sensitive to auxin compared to nonembryogenic genotypes (22).

In the former, the expression of auxin-responsive genes was affected by low concentrations of auxin which did not have any affect on the nonembryogenic genotypes. In the same study, it was shown that auxin concentrations which arrested callus cell proliferation and induced embryo development in embryogenic genotypes promoted callus growth in the nonembryogenic genotype. These studies clearly indicate that possession/acquisition of embryogenic or organogenic competence make cells and tissues more receptive to environmental stimuli required to initiate the embryogenic process.

2.2. Acquisition of Competence and Induction

Expression of the embryogenic/organogenic potential only occurs if cells within the explant are "competent" or responsive to specific culture cues which allow them to differentiate into embryos or organs. In general, zygotic embryos can be considered competent explants, whereas others require an inductive signal to acquire a competent status. This concept is further complicated by the notion that it is often difficult to discriminate between cues promoting morphogenesis in competent cells and those inducing competence. Over the years several systems have been developed to study the nature of competent cells. Using carrot cultures, Nomura and Komamine (23) isolated small embryogenic cells able to differentiate into embryos in response to changes in auxin levels. Tracking experiments revealed that these cells have unique structural features which make them easily distinguishable from other cells. They tend to be small, highly cytoplasmic, generally contain a few small vacuoles and undergo asymmetric cell divisions (24). Cytological studies in maize further documented that competent cells have characteristic patterns of microtubule arrangements (25). These structural features are accompanied by a unique physiology. Independent studies (22, 26) indicated that competent leaf-protoplast-derived cells have a faster rate of DNA biosynthesis resulting in faster cell division. These events were accelerated by an experimental acidification of the medium which also promoted the formation of embryogenic cells in the presence of low 2,4-D levels, which under normal conditions prevent the process. In the same line, embryogenic cell formation was abrogated if the pH of the cultured was maintained high (2, 26). For more detailed information about comparative physiological studies between embryogenic and nonembryogenic cells the reader can consult Feher (2). A key element for the initiation of the embryogenic or organogenic process is represented by the inductive signals which trigger competent cells to form embryos or organs. These signals include stress and hormones.

2.2.1. Stress-Induced Competence

Cell fate acquisition and maintenance is determined by positional information cues existing within the organism's body. Excision of cells, tissues, or organs alters these cues and induces inevitable stress conditions. These events are observed in culture where the excised and wounded tissue is exposed to media providing suboptimal levels of nutrients and plant growth regulators. As suggested by Feher in (2), the response to stress conditions is dependent upon two key factors: the level of stress, which does not have to exceed the tolerance of the explant or cells will die, and the "physiological state" of the explant which mediates cellular responses. The imposed stress is important for promoting dedifferentiation of cells within the explant (2). The first notion is exemplified by the formation of callus observed when tissue is removed from its original in vivo environment and placed in artificial growth conditions. Grosset et al. (27) showed that the majority of genes expressed by tobacco mesophyll protoplasts are indeed induced by wounding. Similarly, in leaf-protoplast-derived cells the imposition of oxidative stress initiates the differentiation process, as estimated by the acidification of the culture medium (26). Besides dedifferentiation, wounding is also required for the activation of the embryogenic program. Extreme examples include the improved somatic embryogenesis observed in *Quercus suber* cells cultured by alternating between proliferation and starvation-stress media (28), and suspension cultures produced from mature explants of *Phoenix dactylifera* grown in media without sucrose (29). Inclusions of heavy metal stress were also used to induce somatic embryogenesis (without the applications of growth regulators normally needed to promote the process) from apical shoots of carrot seedlings (30). Although these experiments show the effects of extreme stress conditions, it must be mentioned that the simple dissecting of tissues/organs, including zygotic embryos, induces tremendous stresses which per se might be enough to elicit a response.

The mechanism(s) whereby stress affects embryogenesis and organogenesis has not been fully elucidated, although there is evidence that multiple cellular responses are triggered. Krishnaraj and Thorpe (31) showed that salt stress might be required for altering cellular energy status and reducing power which increase the rate of the pentose phosphate pathway. Activation of this pathway has been associated with in vitro morphogenesis. Stress has also been implicated in the production of arabinogalactan proteins, which have been found to promote somatic embryogenesis. In a study (32), it was shown how nonembryogenic cultures can acquire embryogenic competence if exposed to arabinogalactan proteins produced from embryogenic cells. A similar promotive effect was also observed by chitinases, which also tend to accumulate in response to stress conditions (33).

Oxidative stress plays a key role in the acquisition of embryogenic competence. Feher et al. (34) showed that compounds promoting oxidative stress are able to produce embryogenic cells from alfalfa leaf protoplasts. Although the inducting mechanisms remain elusive, it has been suggested that oxidative stress might be connected to mitogen-activated protein kinase (MPAK) phosphorylation cascade which plays a central role in cell division processes (2).

2.2.2. Hormone-Induced Competence

Competent cells can produce embryos or organs in response to internal and external stimuli often represented by changes in plant growth regulators. For embryogenesis the most effective inducer is generally auxin, the level of which increases during the early stages of carrot fertilization in vivo (35). As indicated in (2, 34), it is difficult to discern whether this growth regulator is only needed for the acquisition of cell competence or also for the initiation of the embryogenic process. Experiments conducted by Kitamiya et al. (36) on *Medicago* cells suggest that auxin is important in promoting the embryogenic program, which however can proceed in the absence of growth regulators. 2,4-D is the most commonly used auxin and its exact mode of action still remains elusive. It has been suggested that the direct effect of 2,4-D is less significant compared to the effect that this synthetic growth regulator has on the endogenous IAA content (2).

Inclusions of 2,4-D in the culture medium stimulate native IAA production, which has been associated to increased embryogenic response (37). A peak of endogenous IAA levels has been observed in immature zygotic embryos during the induction of somatic embryogenesis (38), as well as in alfalfa leaf protoplasts able to generate embryos in culture (26). The role of endogenous IAA as an inductive signal for the initiation of the embryogenic process was further confirmed by localization studies conducted during the induction of somatic embryos from cultured sunflower embryos (39).

Besides IAA, applications of 2,4-D increase the expression levels of ethylene and ABA biosynthetic genes: 1-aminocyclopropane-1-carboxylic acid synthase and 9-cis-epoxycarotenoid dehydrogenase, respectively (40). The fact that both ethylene and ABA are considered "stress" hormones, reinforces the importance of stress for the initiation of the embryogenic process and also suggests a link between auxin and stress.

The auxin-increase of ABA levels might per se be responsible for promoting embryogenesis. It is well established that ABA induces somatic embryogenesis in many systems. Charriere et al. (38) showed that applications of this hormone to sunflower zygotic embryos stimulate somatic embryo initiation under levels of sucrose not suitable for embryogenesis. A developmental arrest of somatic embryos as a result of ABA depletion was observed in *Nicotiana plumbaginifolia*; this arrest was overcome if ABA was reapplied in the culture medium (41). It is worth mentioning that besides auxin, stress can increase the endogenous ABA content.

Therefore, the level of this plant regulator also is altered as explants are dissected and placed on culture media.

Exposure to auxin increases DNA methylation level, which is involved in gene silencing (42). As reviewed by Feher (2) alterations in methylation level occur during the initiation of somatic embryogenesis. The author suggested that it may be possible to start the embryogenic process of recalcitrant species by experimental modifications of the methylation pattern. In the carrot system, the removal of 2,4-D results in a rapid drop in methylation and this pattern is reversed as the embryos develop (43). The regulation of methylation level by auxin appears to be mediated by S-adenosylmethionine (SAM) and S-adenosylcysteine (SAH). As reviewed by von Aderkas and Bonga (44), a reduction of auxin results in a reduction in ethylene production and this increases SAM levels, thereby favoring a high SAM/SAH ratio which in turn promotes methylation. Therefore a depletion of 2,4-D from the culture medium, which in many systems such as carrot promotes embryogenesis, increases the methylation level. These conditions are required for triggering embryogenesis. The role of methylation during embryogenesis was also observed during tobacco androgenesis. The methylation level in the degenerative cells decreased to values lower than those observed in the vegetative cells which are the only ones able to generate haploid embryos (45). Besides promoting the acquisition of the embryogenic potential SAM might also be required for the progression of embryo development (46). Since preferential synthesis of auxin in plants occurs in young and developing tissues and organs, it is assumed that zygotic embryos have higher levels of this growth regulator compared to more mature explants. This physiological condition would therefore render zygotic embryos more responsive to the culture environment and enable them to initiate somatic embryos at higher frequency.

3. Morphology and Physiology of Embryogenic/Organogenic Cells

As indicated in the previous section, studies by Toonen et al. (24) on carrot cultures revealed that several cell types can generate embryos in culture, although maximal embryogenic frequency was observed for small, highly cytoplasmic cells. Structure of embryogenic carrot cells was further investigated by Nomura and Komamine (23) who using fractionation studies identified single cells (state 0) which were able to form small cellular aggregates and developed into embryos upon removal of auxin. State 0 cells were also small and highly cytoplasmic, thereby confirming previous observations. Accurate identification of competent cells was rendered possible by the development of cellular markers, including the JIM8 cell wall epitope and somatic embryogenesis receptor kinase (SERK) (47, 48). The JIM 8 epitope was located on the wall of single carrot cells and specific to embryogenic cultures (47). The authors suggested that JIM 8-targeted cells represent a transitional state. These cells can undergo two distinct developmental pathways: either elongate and eventually die, or divide and form the initial cell of the somatic embryos. Cells expressing the JIM-8 epitope were also small and cytoplasmic rich. A different morphology of competent cells was reported by Schmidt et al. (48) who employed "SERK" as a marker to identify embryogenic cells (discussed below). Using hypocotyl explants of carrots as a model system, the authors demonstrated that cells competent for embryogenesis were large and elongated. A similar conclusion was also reached by Somleva et al. (49) who using a similar approach in *Dactylis glomerata* showed that the SERK marker was also expressed by a group of large, vacuolated cells; although only small, cytoplasmic cells were able to produce embryos.

Despite these isolated studies, it is accepted that embryogenic cells are generally round, small and highly cytoplasmic, whereas their nonembryogenic counterparts are elongated and contain large vacuoles. This notion is supported by several independent studies using angiosperm and gymnosperm systems (50). Another feature of embryogenic cells is the accumulation of starch, the presence of small vacuoles characterized by low transparency, as well as the high levels of proteins, possibly storage, as estimated by the strong staining with toluidine blue O (51). The small size and the highly cytoplasmic nature observed in the embryogenic cells are also featured by cells committed to undergo shoot organogenesis (12). These cells undergo high mitotic activity forming meristemoid centers which will subsequently form unipolar primordia.

Differences in pH and calcium level were also observed between embryogenic and nonembryogenic cells. Compared to their nonembryogenic counterparts, the symplast of embryogenic cells is characterized by an higher pH (52). This notion was experimentally proven by culturing carrot zygotic embryos in the presence of NH_4Cl, which increased the cellular pH. These conditions guaranteed a continuous production of somatic embryos (53). It was postulated (2) that the difference in pH observed between embryogenic and nonembryogenic cells is related to vacuolar function, with a storage role in the former and a lytic function in the latter cells. Alkalinization of competent cells can also be linked to the initiation of cell divisions, leading to somatic embryo formation. Despite the paucity of information regarding the functional role of pH changes in plant cells, Pichon and Desbiez (54) showed that cellular alkalinization induced cell divisions in meristematic and hypocotyl *Arabidopsis* cells, whereas cellular acidification had opposite effects. This notion is not new since the role of high pHs in promoting cell cycle progression in yeast cells has been well documented (55).

As reviewed in (2), competent cells respond to embryogenic inductive conditions by increasing the endogenous Ca^{++} level. The role of calcium as a key regulator for a variety of physiological events in plant cells is well recognized. Calcium level generally increased during the initial phases of fertilization in both animal and plant zygotes (56). In carrot cultures an experimental increase in Ca^{++} level in an auxin-free environment increased somatic embryo yield, whereas a decrease in Ca^{++} effected by ionophores and channel blockers had an opposite effect (57). A similar dependence of embryo initiation and calcium level was documented during sandalwood somatic embryogenesis. As reviewed in (2) calcium signal might be mediated by calmodulin and other Ca^{++}-dependent protein kinases and results in the establishment of cell polarity which is one of the first events in embryo initiation.

4. Molecular Events Related to Embryogenic/ Organogenic Competence

Extensive reprogramming of gene expression accompanies the transition from somatic cells into embryogenic competent cells in response to inductive signals. Extensive effort has been focused on the identification of "master" genes required for this transition although it is now apparent that the induction of the embryogenic pathway is not governed by a single gene, but it is under the control of an intrigued genetic network. It was documented (58) that ectopic expression of SERK resulted in a fourfold increase in embryogenic production from Arabidopsis seedlings. The expression of this gene, which is generally higher in cell cultures with enhanced embryogenic capabilities (59), is unique to cells showing a rapid response to hormonal signals and competent to produce somatic embryos (58). Two other genes involved in the somatic-embryogenic transition encode the transcription factors Leafy Cotyledons 1 (LEC1) and Baby Boom (BBM) (60, 61). Overexpression of both genes is sufficient to induce embryo development from *Arabidopsis* vegetative tissue.

As suggested by Feher (2), embryogenic competence might not be necessarily due to an induction of genetic events, but rather to release from a suppression state. This notion is supported by studies conducted on *pickle* mutants in which embryos form from root meristems. This gene encodes for a chromatin-remodeling ATPase which is required to suppress the expression of several embryogenesis-related genes, including LEC1, in somatic cells (62). Therefore, embryogenic competence might be acquired from a release of specific factors from a silencing condition mediated by the organization of chromatin (2). This notion is not new as in eukaryotes the overall gene expression pattern has been shown to be controlled by the presence of compact or loose regions within the chromatin (63).

Information on the genetics related to the acquisition of competence for shoot organogenesis is reviewed in (13). A candidate gene during the early dedifferentiation phases is CYCD3, whereas competence acquisition and shoot formation are related to changes in expression of CRE1 and CKI1 (involved in cytokinin reception and perception) as well as shoot meristem genes, such as *SHOOTMERISTEMLESS* and *WUSCHEL*.

5. Zygotic Embryos as the Preferred Explant for In Vitro Propagation

Based on the above studies it emerges that the majority of structural and physiological features needed for inducing somatic embryogenesis and organogenesis in culture are present in zygotic embryos. Zygotic embryo cells already express the "embryogenic potential" with many of the genes required for the induction process already expressed. Therefore, their fate is already committed and does not need to be redirected toward a new developmental path. This is why in many species embryogenic tissue can be readily obtained using immature or mature zygotic embryos. Of interest, a degree of response in culture is also related to the developmental stage of the zygotic embryos. He et al. (64) divided wheat embryos in several developmental stages and showed that the higher yield of embryogenic tissue was obtained using stage II and III embryos. A similar specificity was also observed in conifers where immature zygotic embryos are more responsive then their fully mature counterparts (7). Over the last few years the number of species regenerated in culture using somatic embryogenesis or organogenesis from zygotic embryos has increased and includes both conifers and angiosperms (Tables 1 and 2).

5.1. The Use of Zygotic Embryos for the Initiation of Somatic Embryogenesis

Several reports describe the use of zygotic embryos as initial explants for inducing somatic embryogenesis in both conifers and angiosperms. In the majority of the species, the generation of somatic embryos comprises five steps: induction, maintenance, development, maturation, and conversion. During the induction phase, embryogenic tissue is generated from zygotic embryos (immature or mature), and this step usually requires high levels of auxins and cytokinins, as well as high osmoticum. In white spruce, for example, BA and 2,4-D are used at a concentration of 5 and 10 μM, respectively (65). These requirements are also needed during the induction process of other species (8), although auxin alone is often sufficient (66, 67). It is not clear which regions of

Table 1
Samples of species exhibiting somatic embryogenesis from zygotic embryos over the past decade

Family	Species	Source	IM	MM	M/GM	References
Apocynaceae	*Catharanthus roseus*	Immature zygotic embryos	MS + 4.52 μM 2,4-D + 3% sucrose + 100 mg/L myo-inositol + 0.4 mg/L thiamine-HCl	MS + 4.52 μM 2,4-D + 3% sucrose	MS basal salt	(73)
Anacardiaceae	*Buchanania lanzan Spreng*	Immature zygotic embryos	MS + 4.53 m*M* 2,4-D + 5.32 m*M* NAA + 4.48 mM BA + 100 mg/L + 100 mg/L casein hydro-lysate + 100 mg/L adenine sulfate + 100 mg/L L-glutamine	–	MS + 2.24 μ*M* BA + 15 μ*M* ABA	(74)
	Mangijera indica	Cotyledons of immature zygotic embryos	MS basal + 5 mg/L 2,4-D + 5 mg/L GA3 + 500 mg/L glutamine + 300 mg/L casein hydrolyzate + 20% coconut water + 1 g/L AC + 30 g/L sucrose	–	B5 macroelements + MS microelements + 100 mg/L myo-inositol + 2 mg/L glycine + 0.4 mg/L thiamine-HCl + 0.5 mg/L nicotinic acid + 0.5 mg/L pyridoxine-HCl + 5 mg/L kinetin + 500 mg/L glutamine + 100 mg/L casein hydrolyzate + 20% coconut water + 1 g/L AC + 40 g/L sucrose	(75)

Arecaceae	*Bactris gasipaes*	Mature zygotic embryos	MS + 10 µM Picloram + 1 µM $AgNO_3$ + 3% sucrose	MS + 40 µM 2,4-D + 10 µM 2iP + 1.5 g/L AC + 1 g/L glutamine + 0.5 g/L hydrolyzed casein + 3% sucrose	MS + 24.6 µM 2-iP + 0.44 µM NAA	(76)
Araliaceae	*Kalopanax pictus*	Radical tips of immature zygotic embryos	MS + 4.4 µM 2,4-D + 0.5 µM TDZ + 3% sucrose	–	MS free hormones → ½ MS + 14.4 µM GA3 + 2% sucrose	(77)
Asteraceae (Compositae)	*Helianthus annuus*	Immature zygotic embryos	MS-B5 + 12% sucrose + 1 g/L casein hydro-lysate + 100 mg/L myo-inositol + 500 mg/L MES + 6.6 µM BAP	–	–	(78)
Bixaceae	*Bixa orellana*	Immature zygotic embryos	MS basal + B5 vitamins + 2.26 µM 2,4-D + 4.52 µM Kinetin + 100 mg/L myo-inositol + 87.6 m*M* sucrose + 1 g/L AC	MS basal + B5 vitamins + 2.26 µM 2,4-D + 4.52 µM Kinetin + 100 mg/L myo-inositol + 87.6 mM sucrose + 1 g/L AC	–	(79)
Cupressaceae	*Cryptomeria japonica*	Immature zygotic embryos	SMSG + 0.01% (w/v) myo-inositol + 0.15% glutamine + 3.2 m*M* 2,4-D + 1.8 m*M* BA + 3% sucrose	SMSG + 0.01% (w/v) myo-inositol + 0.15% glutamine + 3.2 m*M* 2,4-D + 1.8 m*M* BA + 3% sucrose	LEMM + EMM vitamins + EMM amino acids + 5% polyethylene glycol 4,000 + 3% maltose + 100 m*M* ABA. → 3/5-strength EMM vitamins + 400 mg/L glutamine + 260 mg/L arginine + 20 mg/L praline + 0.2% activated charcoal + 10 m*M* GA3	(80)

(continued)

Table 1 (continued)

Family	Species	Source	IM	MM	M/GM	References
Euophorbiaceae	*Manihot esculenta*	Mature and immature zygotic embryos	1/3 MS + 0.01 mg/L NAA + 0.01 mg/L GA3 + 1.0 mg/L thiamine-HCl + 100 mg/L inositol + 2% sucrose + 25 mg/L commercial fertilizer containing: N 10, P 52. K 10	–	–	(81)
Fagaceae	*Quercus robur*	Mature and immature zygotic embryos	WPM or MS + 200 mg/L glutamine + 500 mg/L casein hydrolysate + 2% sucrose + 2 mg/L IBA + 1 mg/L BA	WPM or MS + 0.1 mg/L BA + 0.1 mg/L IBA	WPM or ½ MS + 8% sucrose + 3 mg/L ABA → WPM or ½ MS + 0.1 mg/L BA + 2% sucrose	(82)
Icacinaceae	*Nothapodytesfo etida*	Torpedo stage of zygotic embryos	MS + 9.05 μM 2,4 D + 4.44 μM BA + 2.32 μM Kinetin + 20 g/L sucrose	MS + 9.05 μM 2,4-D + 4.44 μM BA + 2.32 μM Kinetin + 20 g/L sucrose	MS free hormones	(83)
Malvaceae	*Tilia cordata*	Cotyledons of immature embryos	MS + 0.56 m*M* myo-inositol + WPM vitamins + 87.6 m*M* sucrose + 4.5 μM 2,4D	MS + 1 μM IBA → MS + 131.5 m*M* sucrose + 5 μM ABA	–	(84)
Myrtaceae	*Eucalyptus globulus*	Mature zygotic embryos	MS + 30 g/L sucrose + 3 mg/L NAA + 100 mg/L ascorbic acid	–	MS + 30 g/L sucrose	(85)
Meliaceae	*Melia azedarach*	Immature zygotic embryos	MS + 13.62 μM TDZ + 3% sucrose + 2 mg/L AC	–	–	(86)

Poaceae (Gramineae)	*Triticuma estivum*	Scutella of immature zygotic embryos	MS + 9.96 µM 2,4-D + 10 mg/L casamino acids + 3% sucrose	–	–	(87)
	Paspalum scrobiculatum	Mature zygotic embryos	N6 + 100 µM 2,4-D + 2% sucrose	MS free hormones		(88)
Proteaceae	*Protea cynaroides*	Mature zygotic embryos	MS + 3% sucrose	–	½ MS + 3% sucrose + 0.3 µM GA3	(89)
Pinaceae	*Abies concolor*	Immature zygotic embryos	SH + 1 mg/L ABP	SH + 1 mg/L BAP + 1,000 mg/L casein hydrolysate + 500 mg/L l-glutamine	SH+10 mg/L ABA+30 g/L maltose+75 g/L PEG+500 mg/L casein hydrolysate+ 500 mg/L l-glutamine → ½ MS macro+MS micro+ 2.5 mg/L thiamine-HCl+ 20 g/L sucrose	(90)
	Larix sibirica	Mature and immature zygotic embryos	MSG + 1.45 g/L l-glutamine + 0.1 g/L mesoinositol + 2 mg/L 2,4-D + 1 mg/L BAP + 3% sucrose	MSG + 1.45 g/L l-glutamine + 0.1 g/L mesoinositol + 2 mg/L 2,4-D + 0.5 mg/L BAP + 1.5% sucrose	–	(91)
	Pinus pinea	Zygotic embryos	MLV + 9 µM 2,4-D + 4.5 µM BAP + 0.5 g/L l-glutamine + 1 g/L casein hydrolysate + 10 g/L sucrose	MLV + 9 µM 2,4-D + 4.5 µM BAP + 0.5 g/L l-glutamine + 1 g/L casein hydrolysate + 10 g/L sucrose	½ MLV + 60 g/L sucrose + 10 g/L AC + 80 µM → AFC + 30 g/L sucrose	(92)

(continued)

Table 1 (continued)

Family	Species	Source	IM	MM	M/GM	References
Rosaceae	*Prunus avium*	Immature zygotic embryos	MS basal + 100 mg/L myo-inositol + full – strength morel vitamins + 250 mg/L glutamine + 2 mg/L glycine + 500 mg/L casein hydrolysate + 88 m*M* sucrose + 0.54 μ*M* NAA + 0.46 μ*M* kinetin + 0.44 μ*M* BA	The same induction	MS basal + 100 mg/L myo-inositol + full – strength morel vitamins + 250 mg/L glutamine + 2 mg/L glycine + 500 mg/L casein hydrolysate + 88 m*M* maltose + 10 μ*M* ABA → WPM + 2 mg/L glycine + 2 mg/L glutamine + 44 m*M* sucrose	(93)
Schisandraceae	*Schisandra chinensis*	zygotic embryos	WV5 medium + 50 μM 2,4-D + 2% sucrose + 2 m*M* glutamine	WV5 medium + 4 μM BA + 10 μM 2,4-D + 2% sucrose + 2 m*M* glutamine	WV5 + 3% of polyethylene + glycol 4,000 (PEG) + 3% sucrose + 30 μM (±)-ABA → WV5 + 0.1% activated charcoal (AC) → WV5 + 1.5% sucrose + 0.1% AC + 0.05 μM IBA → ½ MS + 1.5% sucrose + 0.1% AC	(94)
Solanaceae	*Capsicum annuum*	Immature zygotic embryos	MS + 2 mg/L 2,4-D + 10% CW + 10% sucrose		MS + 2% sucrose	(95)
Vitaceae	*Vitis vinifera*	Immature zygotic embryos	NN + 1.0 mg/L 2,4-D	NN + 1.0 mg/L NAA + 0.5 mg/L BA	NN + 0.03 mg/L NAA + 0.5 mg/L BA	(96)

For previous years see Thorpe (97)
IM induction medium, *MM* maturation medium, *M/GM* maturation/germination medium

Table 2
Samples of species exhibiting shoot organogenesis from zygotic embryos over the past decade

Family	Species	Source	CIM	SIM	RIM	References
Asteraceae (Compositae)	*Helianthus annuus*	Cotyledons of zygotic embryos	–	MS + 100 mg/L myo-inositol + 2 mg/L glycine + 1.5 mg/L nicotinic acid + 0.5 mg/L pyridoxine HCI + 0.5 mg/L thia-mine HCl + 0.5 mg/L folic acid + 0.25 mg/L biotin	–	(98)
		Immature zygotic embryos	–	MS-B5 + 3% sucrose + 1 g/L casein hydro-lysate + 100 mg/L myo-inosi-tol + 500 mg/L MES + 6.6 μM BAP	–	(78)
Alliaceae	*Allium cepa*	The embryonic shoot apex and the lower part of the cotyledon	MS + 30 g/L sucrose + 1 mg/L 2,4-D	MS + 30 g/L sucrose	–	(99)
	Allium ampeloprasum	Mature zygotic embryos	MS + 30 g/L sucrose + 1 mg/L 2,4-D	MS + 1 mg/L kine-tin + 30 g/L sucrose	–	(100)

(continued)

Table 2 (continued)

Family	Species	Source	CIM	SIM	RIM	References
Cycadaceae	*Cycas revoluta*	Zygotic embryos	–	SH + 3% sucrose + 9 μM BA + 0.04 μM 2,4-D	½ SH + 5.4 μM NAA	(101)
Cupressaceae	*Thuja occidentalis*	Mature embryos	½ QP + 1 μM BA + 3% sucrose + 100 mg/L asparagin + 100 mg/L myo-inositol + 5 mg/L nicotinic acid + 5 mg/L pyrodoxine-HCl + 5 mg/L thiamine-HCl	½ QP + 10 μM zeatin	–	(102)
Fabaceae (Leguminosae)	*Arachis hypogaea*	Mature zygotic embryos	MS + 30 g/L sucrose + 4 mg/L NAA + 1 mg/L BAP	MS + 30 g/L sucrose + 0.5 mg/L BAP + 0.5 mg/L kinetin	–	(103)
	Acacia mangium	Embryo axes and cotyledons of mature zygotic embryos	MS + 9.05 μM 2,4-D + 13.95 μM KT + 100 mg/L casein enzymatic hydrolysate + 100 mg/L ascorbic acid + 150 mg/L glutamine + 150 mg/L asparagine + 150 mg/L proline + 30 g/L sucrose	MS + 4.55 μM TDZ + 1.43 μM IAA + 100 mg/L casein enzymatic hydrolysate + 100 mg/L ascorbic acid + 150 mg/L glutamine + 150 mg/L asparagine + 150 mg/L proline + 30 g/L sucrose	MS + 10.75 μM NAA + 2.33 μM KT + 100 mg/L casein enzymatic hydrolysate + 100 mg/L ascorbic acid + 150 mg/L glutamine + 150 mg/L asparagine + 150 mg/L proline + 30 g/L sucrose	(104)
	Cajanus cajan	Mature zygotic embryos	MS + 10 μM TDZ + 30 g/L sucrose	MS + 0.05 μM TDZ + 30 g/L sucrose	–	(105)

Lamiaceae (Labiatae)	*Salvia sclarea*	Cotyledons of immature embryos	MS + 9.05 μM 2,4-D + 30 g/L sucrose	MS + 7.22 μM GA3 + 4.44 μM BA + 2.69 μM NAA + 30 g/L sucrose	MS + 0.57 μM IAA + 0.54 μM NAA + 20 g/L sucrose	(106)
Oleaceae	*Fraxinus angustifolia*	Mature embryos	½ MS microelements + MS microelements and organic + 20 g/L sucrose + 4.4 μM BA + 0.44 μM 2,4-D	DKW + 20 g/L sucrose + 4.4 μM BA	DKW + 20 g/L sucrose + 4.4 μM BA + 0.4 μM 2,4-D	(107)
Pinaceae	*Larixo ccidentali*	Mature zygotic embryos	½ QP + organic SH (LVSH) + 2% sucrose + 10 μM BA	–	Aqueous solution of 0.1% IAA and IBA	(108)
	Pinus massoniana	Mature zygotic embryos	–	DCR + 0.5 mg/L BA + 0.05 mg/L IBA + 3% sucrose	1/2 GD + 2% sucrose + 2 mg/L IBA + 0.05 mg/L BA	(109)
	Pinus strobus	Embryonic cotyledons and hypocotyls	TE + 5 μM IAA + 3 μM IBA + 3 μM → TE + 3 μM IAA + 6 μM BA + 6 μM TDZ + 0.4% phytagel	PS + 8 μM TDZ + 0.01 μM IAA + 500 mg/L CH + 600 mg/L glutamine	PS + 0.5 μM BA + 0.01 μM IAA	(110)
	Pinus elliottii	Mature zygotic embryos	TE + 12 μM NAA + 15 μM 2,4D + 6 μM 2iP + 30 g/L sucrose + 500 mg/L Glutamine + 500 mg/L myo-Inositol	TE + 2 μM IBA + 3 μM BA + 9 μM TDZ + 30g/L sucrose + 500 mg/L Glutamine + 500 mg/L myo-Inositol	TE + 0.01 μM IAA + 0.01 μM IBA + 400 mg/L Glutamine + 250 mg/L myo-Inositol	(111)
Taxaceae	*Taxus wallichiana*	Zygotic embryos	½ WPMSH + 0.5 mg/L BA + 1 mg/L 2,4-D + 30 g/L sucrose	½ WPMSH + 2.5 mg/L BA + 30 g/L sucrose	1/5 MS	(112)

(continued)

Table 2 (continued)

Family	Species	Source	CIM	SIM	RIM	References
Umbeliferae	*Cuminum cyminum*	Mature embryos	B5 + 30 g/L sucrose + 0.2 μM BAP + 0.2 μM NAA	B5 + 30 g/L sucrose + 1 μM BAP + 0.2 μM NAA + 0.4 μM IAA	–	(113)
Zamiaceae	*Ceratozamia mexicana*	Immature embryo	–	B5 macroelements + MS microele-ments + 400 mg glutamine + 100 mg ascorbic acid + 100 mg casein hydrolys-tate + 100 mg argin-ine + 100 mg asparagine + 60 g sucrose + 9 μM 2,4-D	B5 macroelements + MS microele-ments + 400 mg glutamine + 100 mg ascorbic acid + 100 mg casein hydrolystate + 100 mg arginine + 100 mg asparagine + 60 g sucrose + 9 μM 2,4-D + 13.9 μM Kinetin	(114)
Alstroemeriaceae	*Alstroemeria (A. pelegrina* x, *A. psittacina)*	Mature zygotic embryos	MS-B5 + 3% sucrose + 40 μM NAA + 20 μM kinetin	MS-B5 + 3% sucrose + 20 μM NAA + 20 μM kinetin	–	(115)
Poaceae (Gramineae)	*Pennisetum glaucum*	Immature zygotic embryos	L3 + 2.5 mg/L 2,4-D + 30 g/L maltose	L3 + 2.5 mg/L 2,4-D + 0.1 mg/L BAP + 30 g/L maltose	–	(116)

For previous years see Thorpe (97)
CIM callus induction medium, *SIM* shoot induction medium, *RIM* root induction medium

the zygotic embryos are responsive to the induction process, although several studies indicate that embryogenic tissue formation is initiated from the upper hypocotyl (8). As indicated previously, the level of maturity of the zygotic embryos is a very important factor since differences in induction frequencies are often observed. As a general rule, immature (early cotyledonary) embryos are more responsive than their fully mature counterparts. Embryogenic tissue is generally easily recognizable from nonembryogenic tissue. In conifers, for example, embryogenic tissue is translucent and composed of immature, filamentous-shaped embryos characterized by a well-developed suspensor region and an embryo proper, which is formed by small highly cytoplasmic cells. This is in contrast to the nonembryogenic tissue which has a green/brown coloration, and it is mainly composed of parenchyma cells without any recognizable structure (65). Embryogenic tissue produced from zygotic embryos can be maintained on solid or liquid medium for a long period of time. Often the maintenance step also requires some auxins or cytokinins, although at lower concentrations to those used for the induction phase. In some instances, however, the hormonal requirements for induction and maintenance are identical (68). The time of subculturing during maintenance is strictly genotype dependent with some cell lines requiring a higher subculture frequency. Growth of the somatic embryos is encouraged on development medium, the composition of which is species dependent. In some species, including maize, embryo development can be achieved in the absence of plant growth regulators (68), whereas for others auxins or abscisic acid are required (65, 66). Physiological maturation is not necessarily a prerequisite found in fully developed embryos. In some instances, a desiccation period is required to terminate the developmental program and initiate the germination process. This maturation step is very common in conifers where embryos must experience a water stress prior to postembryonic growth. In spruce this can be achieved using a partial treatment, in which embryos are slowly dried in a high relative humidity environment. Inclusions of osmoticum agents in the development medium are also used to increase the tolerance to desiccation (65). It must be mentioned, however, that for the majority of species this desiccation step is not required and embryos can be transferred directly from the developmental medium onto the germination medium (68). The final step of the somatic embryogenic process is germination, in which embryos start their postembryonic growth and regenerate viable plants. This step is usually achieved in a medium devoid of plant growth regulators (65, 68).

5.2. The Use of Zygotic Embryos for Inducing Organogenesis

A general shoot organogenic process comprises three distinct steps: callus induction, shoot induction, and root induction. The last two steps are collectively called regeneration.

Induction of callus can be initiated using both mature and immature embryos (69, 70) and is usually achieved on media containing either auxins and cytokinins (69) or auxin alone (69, 70). The incubation time is species dependent and the callus responsive to organogenesis can be often recognized by its non-embryogenic counterpart. In rice, for example, responsive callus is creamy and become visible after only 2 weeks of incubation in the dark. Generation of shoots from the callus is achieved in the presence of cytokinins which stimulate the formation of meristemoids, i.e., organized cellular aggregates which will further develop into shoots. Compared to somatic embryogenesis, studies on the initial phases of the organogenic process are scarce in literature. Once developed, shoots are transferred onto rooting medium, which is generally devoid of plant growth regulators (71, 72), or it may contains low levels of auxins (70).

6. Conclusions

Plant regeneration in vitro via embryogenesis or organogenesis is a multistep process, the success of which is dependent upon the proper execution of each step. A key element for both propagation techniques is the choice of the explant. Overall young tissues, especially immature and mature embryos, represent the preferred choice since they are composed of cells early in their developmental path and able to dedifferentiate and embark on a new fate.

References

1. Bicknell RA, Koltunow AM (2004) Understanding apomixis: recent advances and remaining conundrums. Plant Cell 1:228–245
2. Feher A (2005) Why somatic plant cells start to form embryos? Plant Cell Monogr 2:85–101
3. Yarbrough JA (1932) Anatomical and developmental studies of the foliar embryos of *Bryophyllum calcynum*. Am J Bot 42:215–223
4. Taylor RL (1967) The foliar embryos of *Malaxis paludosa*. Can J Bot 45:1553–1556
5. Levine M (1947) Differentiation of carrot root tissue grown in culture. Bull Torrey Bot Club 74:321–323
6. Steward F, Mapes M, Mears K (1958) Growth and organized development of cultured cells. II Organization in cultures grown from freely suspended cells. Am J Bot 45:705–708
7. Stasolla C, Thorpe TA (2010) Tissue culture: historical perspective and applications. In: Kumar A, Sopory SK (eds) Applications of plant biotechnology: in vitro propagation, plant transformations and secondary metabolite production. I K International Publishers, New Delhi, pp 1–39
8. Thorpe TA, Stasolla C (2001) Somatic embryogenesis. In: Bhojwani SS, Soh WY (eds) Current trends in the embryology of angiosperms. Kluwer Academic Publishers, Dordrecht, pp 279–336
9. White PR (1939) Controlled differentiation in a plant tissue culture. Bull Torrey Bot Club 66:507–513
10. Nobecourt P (1939) Sur les radicelles naissant des cultures de tissues vegetaux. Compte Rendus Des Seances de la Societe de biologie et de ses Filiales 130:1271–1272
11. Skoog F, Miller CO (1957) Chemical regulation of growth and organ formation in plant

tissue cultures in vitro. Sym Soc Exp Biol 11:118–131
12. Thorpe TA (1980) Organogenesis in vitro: structural, physiological and molecular aspects. Int Rev Cytol Suppl 11A:71–111
13. Sugiyama M (1999) Organogenesis in vitro. Curr Opin Plant Biol 2:61–64
14. Feher A, Taras P, Pasternak P, Dudis D (2003) Transition of somatic plant cells to an embryogenic state. Plant Cell Tissue Organ Cult 74:201–228
15. Moltrasio R, Robredo CG, Gomez MC, Diaz Paleo AD, Diaz DG, Rios RD, Franzone PM (2004) Alfalfa (*Medicago sativa*) somatic embryogenesis: genetic control and introduction of favorable alleles into elite Argentinian germplasm. Plant Cell Tissue Organ Cult 77:119–124
16. Brown DCW, Atanassov A (1985) Role of genetic background in somatic embryogenesis in *Medicago*. Plant Cell Tissue Organ Cult 4:111–114
17. Hartweck LM, Lazzeri PA, Cui D, Collins GB, Williams EG (1988) Auxin orientation effects on somatic embryogenesis from immature soybean cotyledons. In Vitro Cell Dev Biol Plant 24:821–828
18. Vasil IK (1988) Progress in the regeneration and genetic manipulation of cereal crops. Biotechnology 6:397–402
19. Kysely W, Jacobsen H-J (1990) Somatic embryogenesis from pea embryos and shoot apices. Plant Cell Tissue Organ Cult 20:7–10
20. Ewans DE, Sharp WR, Flick CE (1981) Growth and behavior of cell cultures: embryogenesis and organogenesis. In: Thorpe TA (ed) Plant tissue culture. Academic, New York, pp 45–113
21. Neumann K H (2000) Some studies on somatic embryogenesis: a tool in plant biotechnology. http://geb.uni-giessen.de/geb/volltexte/2000/321/
22. Bogre L, Stefanov I, Abraham M, Somogyi I, Dudis D (1990) Differences in response to 2, 4-dichlorophenoxyacetic acid treatment between embryogenic and non embryogenic lines of alfalfa. In: Nijkamp HJJ, van der Plaas LHW, Van Aartrijk J (eds) Progress in plant cellular and molecular biology. Kluwer Academic Publishers, Dordrecht, pp 427–436
23. Nomura K, Komamine A (1985) Identification and isolation of single cells that produce somatic embryos at high frequency in a carrot cell suspension culture. Plant Physiol 79:988–991
24. Toonen MAJ, Hendriks T, Schmidt EDL, Verhoeven HA, van Kammen A, de Vries SC (1994) Description of somatic-embryo forming single cells in carrot suspension cultures employing video cell tracking. Planta 194:565–572
25. Samaj J, Baluska F, Pretova A, Volkmann D (2003) Auxin deprivation induces a developmental switch in maize somatic embryogenesis involving redistribution of microtubules and actin filaments from endoplasmic to cortical cytoskeletal arrays. Plant Cell Rep 21:940–945
26. Pasternak TP, Prinsen E, Ayaydin F, Miskolczi P, Potters G, Asard H, Van Onckelen HA, Dudis D, Feher A (2002) The role of auxin, pH, and stress in the activation of embryogenic cell division in leaf protoplast-derived cells of alfalfa. Plant Physiol 129:1807–1819
27. Grosset J, Marty I, Chartier Y, Meyer Y (1990) mRNAs newly synthesized by tobacco mesophyll protoplasts are wound-inducible. Plant Mol Biol 15:485–496
28. Fernandez-Guijarro B, Celestino C, Toribio M (1995) Influence of external factors on secondary embryogenesis and germination in somatic embryos from leaves of *Quercus suber*. Plant Cell Tissue Organ Cult 41:99–106
29. Veramendi J, Navarro L (1996) Influence of physical conditions of nutrient medium and sucrose on somatic embryogenesis of date palm. Plant Cell Tissue Organ Cult 45:159–164
30. Kiyosue T, Takano T, Kamada H, Harada H (1990) Induction of somatic embryogenesis in carrot by heavy metal ions. Can J Bot 68:2301–2303
31. Krishnaraj S, Thorpe TA (1996) Salinity stress effects on [^{14}C-1] and [^{14}C-6]-glucose metabolism of a salt tolerant and salt susceptible variety of wheat. Int J Plant Sci 157:110–117
32. McCabe PF, Valentine TA, Forsberg LS, Pennell RI (1997) Soluble signals from cells identified at the cell wall establish a developmental pathway in carrot. Plant Cell 9:2225–2241
33. De Jong AJ, Cordewener F, Lo Schiavo M, Terzi M, Vandekerckhove A, Van Kammeren A, de Vries SC (1992) A carrot somatic embryo mutant is rescued by chitinase. Plant Cell 4:425–433
34. Feher A, Pasternak T, Otvos K, Miskolczi P, Dudis D (2002) Induction of embryogenic competence in somatic plant cells: a review. Biologia 57:5–12
35. Ribnicky DM, Cohen JD, Hu WS, Cooke TJ (2001) An auxin surge following fertilization in carrots: a mechanism for regulating plant totipotency. Planta 45:345–467

36. Kitamiya E, Suzuki S, Sano T, Nagata T (2000) Isolation of two genes that were induced upon the initiation of somatic embryogenesis on carrot hypocotyls by high concentrations of 2, 4-D. Plant Cell Rep 19:551–557
37. Rajasekaran K, Hein MB, Davis GC, Carnes MG, Vasil IK (1987) Exogenous growth regulators in leaves and tissue cultures of *Pennisetum purpureum*. J Plant Physiol 130:13–25
38. Charriere F, Sotta B, Miginiac E, Hahne G (1999) induction of adventitious and somatic embryos on in vitro cultured zygotic embryos of *Helianthus annuus*: variation of endogenous hormone levels. Plant Physiol Biochem 37:751–757
39. Thomas C, Bronner R, Molinier J, Prinsen E, Van Onckelen H, Hahne G (2002) Immunocytochemical localization of indole-3-acetic acid during induction of somatic embryogenesis in cultured sunflower embryos. Planta 215:577–583
40. Hansen H, Grossmann K (2000) Auxin-induced ethylene triggers abscisic acid biosynthesis and growth inhibition. Plant Physiol 124:1437–1448
41. Senger S, Mock HP, Conrad U, Manteuffel R (2001) Immunomodulation of ABA function affects early events in somatic embryo development. Plant Cell Rep 20:112–120
42. Holliday R (1984) The biological significance of meiosis, in *Controlling events in meiosis*, (Evans, C. W. and Dickinson, H. G., eds.). Symp Soc Exp Biol 38:381–395
43. Lo Schiavo F, Pitto L, Giuliano G, Torti G, Nuti-Ronchi V, Marazziti D, Vergara S, Orselli S, Terzi M (1989) DNA methylation of embryogenic carrot cell cultures and its variations as caused by mutation, differentiation, hormones, and hypermethylating drugs. Theor Appl Genet 77:325–331
44. von Aderkas P, Bonga JM (2000) Influencing micropropagation and somatic emrbyogenesis in mature trees by manipulation of phase change, stress, and culture environment. Tree Physiol 20:921–928
45. Oakeley EJ, Podesta A, Jost JP (1997) Developmental changes in DNA methylation of the two tobacco pollen nuclei during maturation. Proc Nat Acad Sci (USA) 94:11721–11725
46. Munksgaard D, Mattsson O, Okkels FT (1995) Somatic embryo development in carrot is associated with an increase I levels of S-adenosylmethionine, S-adenosylhomocysteine and DNA methylation. Physiol Plant 93:5–10
47. Pennell RJ, Janniche L, Scofield GN, Booij H, de Vries SC, Roberts K (1992) Identification of a transitional state in the developmental pathway to carrot somatic embryogenesis. J Cell Biol 119:1371–1380
48. Schmidt ED, Guzzo F, Toonen MA, de Vries SC (1997) A leucine-rich repeat containing receptor-like kinase marks somatic plant cells competent to form embryos. Development 124:2049–2062
49. Somleva MN, Schmidt ED, de Vries SC (2000) Embryogenic cells in *Dactylis glomerata* L. explants identified by cell tracking and by SERK expression. Plant Cell Rep 19:718–726
50. Egertsdotter U, von Arnold S (1998) Development of somatic embryos in Norway spruce. J Exp Bot 49:155–162
51. Stasolla C (2006) Somatic embryogenesis in *Picea* suspension cultures. In: Loyola-Vargas VM, Vazquez-Flota F (eds) Plant cell culture protocols- methods in molecular biology. Humana Press, Totowa, pp 87–99
52. Schaefer J (1985) Regeneration in alfalfa tissue culture. Plant Physiol 79:584–588
53. Smith DL, Krikorian AD (1990) pH control of carrot somatic embryogenesis. In: Nijkamp HJJ, Van der Plas LHW, Van Aartrijk J (eds) Progress in plant cellular and molecular biology. Kluwer Academic Publishers, Dordrecht, pp 449–453
54. Pichon O, Desbiez MO (1994) Is cytoplasmic pH involved in the regulation of cell cycle in plants? Physiol Plant 92:261–265
55. Frelin C, Vigne P, Ladoux A, Lazdunski M (1988) The regulation of intercellular pH in cells from vertebrates. Eur J Biochem 174:3–14
56. Stricker SA (1999) Comparative biology of calcium signaling during fertilization and egg activation in animals. Dev Biol 211:157–176
57. Jansen MAK, Booij H, Schel JHN, de Vries SC (1990) Calcium increases the yield of somatic embryos in carrot embryogenic suspension cultures. Plant Cell Rep 9:221–223
58. Hecht V, Vielle-Calzada JP, Hartog MV, Schmidt ED, Boutilier K, Grossniklaus U, de Vries SC (2001) The *Arabidopsis* somatic embryogenesis receptor kinase 1 gene is expressed in developing ovules and embryos and enhances embryogenic competence in culture. Plant Physiol 127:803–816
59. Mordhorst P, Voerman KJ, Hhartog MV, Meijer EA, van Went J, Koornneef M, de Vries S (1998) Somatic embryogenesis in *Arabidopsis thaliana* is facilitated by mutations in genes repressing meristematic cell division. Genetics 149:549–563
60. Lotan T, Ohto M, Yee KM, West MAL, Lo R, Kwong RW, Yamagishi K, Fischer RL,

Goldberg RB, Harada JJ (1998) *Arabidopsis* LEAFY COTYLEDON 1 is sufficient to induce embryo development in vegetative cells. Cell 93:1195–1205
61. Boutilier K, Offringa R, Fukuoda H, Sharma V, Kieft H, van Lammeren AAM, Ouellett T, van Lookeren C (2000) Ectopic expression of the *Brassica napus* baby boom gene triggers a homeotic conversion of vegetative tissue into embryos and cotyledons. Plant Mol Biol Rep 18:s11–s14
62. Ogas J, Kaufmann S, Henderson J, Somerville C (1999) PICKLE is a CHD3 chromatin-remodeling factor that regulates the transition from embryonic to vegetative development in *Arabidopsis*. Proc Natl Acad Sci (USA) 96:13839–13844
63. Li G, Hall TC, Holmes-Davis R (2002) Plant chromatin: development and gene control. Bioessays 24:234–243
64. He GG, Tanner G, Scott KJ (1986) Somatic embryogenesis and morphogenesis in callus derived from the epiblast of immature embryos of wheat. Plant Sci 45:119–124
65. Stasolla C, Kong L, Yeung EC, Thorpe TA (2002) Maturation of somatic embryos in conifers: morphogenesis, physiology, biochemistry, and molecular biology. In Vitro Cell Dev Biol Plant 38:93–105
66. Lazzeri PA, Hildebrand DF, Collins GB (1985) A procedure for plant regeneration from immature cotyledon tissue of soybean. Plant Mol Biol Rep 3:160–167
67. Gonzalez ML, Vitietez AM, Vitietez E (1985) Somatic embryogenesis from chestnut cotyledon tissue cultured in vitro. Sci Hortic 27:97–103
68. Register JC, Peterson DJ, Bell PJ, Bullock WP, Evans IJ, Frame B, Greenland AJ, Higgs NS, Jepson I, Jiao S, Lewnau CJ, Sillick JM, Wilson HM (1994) Structure and function of selectable and non-selectable transgenes in maize after introduction by particle bombardment. Plant Mol Biol 25:951–961
69. Lee K, Jeon H, Kim M (2002) Optimization of a mature embryo-based in vitro culture system for high-frequency somatic embryogenic callus induction and plant regeneration from *japonica* rice cultivars. Plant Cell Tissue Organ Cult 71:237–244
70. Agrawal V, Sardar PR (2006) In vitro propagation of *Cassia angustifolia* through leaflet and cotyledon derived calli. Biol Plant 50:118–122
71. Elshanshoury A, Elshafy E, Hamoud M, Elhiti M (2005) Molecular studies on regenerated salt tolerant cell lines of rice. Egypt J Exp Bot 1:11–20
72. Datta MM, Jha AMS (2006) Organogenesis and plant regeneration in *Taxus wallichiana* (Zucc.). Plant Cell Rep 25:11–18
73. Kim SW, In DS, Choi PS, Liu JR (2004) Plant regeneration from immature zygotic embryo-derived embryogenic calluses and cell suspension cultures of *Catharanthus roseus*. Plant Cell Tissue Organ Cult 76:131–135
74. Sharma P, Koche V, Quraishi A, Mishra SK (2005) Somatic embryogenesis in *Buchanania lanzan* spreng. In Vitro Cell Dev Biol Plant 41:645–647
75. Xiao JN, Huang XL, Wu YJ, Li XJ, Zhou MD, Engelmann F (2004) Direct somatic embryogenesis induced from cotyledons of mango immature zygotic embryos. In Vitro Cell Dev Biol Plant 40:196–199
76. Steinmacher DA, Cangahuala-Inocente GC (2007) Somatic embryogenesis from peach palm zygotic embryos. In Vitro Cell Dev Biol Plant 43:124–132
77. Moon HK, Kim YW, Lee JS, Choi YE (2005) Micropropagation of *Kalopanax pictus* tree via somatic embryogenesis. In Vitro Cell Dev Biol Plant 41:303–306
78. Charrière F, Hahne G (1998) Induction of embryogenesis versus caulogenesis on in vitro cultured sunflower (*Helianthus annuus* L.) immature zygotic embryos: role of plant growth regulators. Plant Sci 137:63–71
79. Neto VBP, Botelho MN, Aguíar R, Silva EAM, Otoni WC (2003) Somatic embryogenesis from immature zygotic embryos of annatto (*Bixa orellana* L.). In Vitro Cell Dev Biol Plant 39:629–634
80. Igasaki T, Sato T, Akashi N, Mohri T, Maruyama E, Kinoshita I, Walter C, Shinohara K (2003) Somatic embryogenesis and plant regeneration from immature zygotic embryos of *Cryptomeria japonica* D. Don. Plant Cell Rep 22:239–243
81. Fregene M, Ospina JA, Roca W (1999) Recovery of cassava (*Manihot esculenta* Crantz) plants from culture of immature zygotic embryos. Plant Cell Tissue Organ Cult 55:39–43
82. Valladares S, Sánchez C, Martínez MT, Ballester SA, Vieitez AM (2006) Plant regeneration through somatic embryogenesis from tissues of mature oak trees: true-to-type conformity of plantlets by RAPD analysis. Plant Cell Rep 9:879–886
83. Fulzele DP, Satdive RK (2003) Somatic embryogenesis, plant regeneration, and the evaluation of camptothecin content in *Nothapodytes foetida*. In Vitro Cell Dev Biol Plant 45:212–216
84. Kärkönen A (2000) Anatomical study of zygotic and somatic embryos of *Tilia cordata*. Plant Cell Tissue Organ Cult 61:205–214
85. Pinto G, Silva S, Park YS (2008) Factors influencing somatic embryogenesis induction in *Eucalyptus globulus* Labill: basal

medium and anti-browning agents. Plant Cell Tissue Organ Cult 95:79–88
86. Vila S, Gonzalez ANA, Rey H, Mroginski L (2003) Somatic embryogenesis and plant regeneration from immature zygotic embryos of *Melia azedarach* (Meliaceae). In Vitro Cell Dev Biol Plant 39:283–287
87. Li B, Caswell K, Leung N, Chibbar RN (2003) Wheat (*Triticum aestivum* L.) somatic embryogenesis from isolated scutellum: days post anthesis, days of spike storage, and sucrose concentration affect efficiency. In Vitro Cell Dev Biol Plant 39:20–23
88. Vikrant A, Rashid A (2003) Somatic embryogenesis or shoot formation following high 2, 4 D pulse-treatment of Paspalum scrobiculatum. Biol Plant 46:297–300
89. Wu HC, du Toit ES, Reinhardt CF (2007) A protocol for direct somatic embryogenesis of *Protea cynaroides* L. using zygotic embryos and cotyledon tissues. Plant Cell Tissue Organ Cult 89:217–224
90. Kormuťák A, Salaj T, Vooková B (2006) Storage protein dynamics in zygotic and somatic embryos of white fir. Biol Bratislava 61:479–485
91. Belorussova AS, Tret'yakova IN (2008) Patterns of somatic embryo formation in Siberian larch: embryological aspects. Russ J Dev Biol 39:83–91
92. Carneros E, Celestino C, Klimaszewska K, Park YS, Toribio M, Bonga JM (2009) Plant regeneration in Stone pine (*Pinus pinea* L.) by somatic embryogenesis. Plant Cell Tissue Organ Cult 98:165–178
93. Talleux LR, Diemer F, Sourdioux M, Chapelain K, De March GG (1999) Improvement of somatic embryogenesis in wild cherry (*Prunus avium*). Effect of maltose and ABA supplements. Plant Cell Tissue Organ Cult 55:199–209
94. Smiskova A, Lasinova H, Havel L (2005) Somatic embryogenesis from zygotic embryos of *Schisandra chinensis*. Biol Plant 49:451–454
95. Kaparakis G, Alderson PG (2008) Role for cytokinins in somatic embryogenesis of pepper (*Capsicum annuum* L.). J Plant Growth Regul 27:110–114
96. Yang XM, Cao ZY, An LZ, Wang YM, Fang XW (2006) In vitro tetraploid induction via colchicine treatment from diploid somatic embryos in grapevine (*Vitis vinifera* L.). Euphytica 152:217–224
97. Thorpe TA (ed) (1995) In Vitro Embryogenesis. Kluwer Academic Publishers, Dordrecht
98. Power CJ (1987) Organogenesis from *Helianthus annuus* inbreds and hybrids from the cotyledons of zygotic embryos. Am J Bot 74:497–503
99. Zheng S, Henken B, Sofiari E, Jacobsen E, Krens FA, Kik C (1998) Factors influencing induction, propagation and regeneration of mature zygotic embryo-derived callus from *Allium cepa*. Plant Cell Tissue Organ Cult 53:99–105
100. Rinaldi LMR (1999) Factors affecting shoot regeneration from zygotic embryo and seedling explants of *Cycas revoluta* Thunb. In Vitro Cell Dev Biol Plant 35:25–28
101. Nour KA, Yeung EC, Thorpe TA (1993) Shoot bud histogenesis from mature embryos and shoots of eastern white cedar (*Thuja occidentalis* L.) cultured in vitro. Int J Plant Sci 154:378–385
102. Chengalrayan K, Hazra S, Meagher MG (2001) Histological analysis of somatic embryogenesis and organogenesis induced from mature zygotic embryo-derived leaflets of peanut (*Arachis hypogaea* L.). Plant Sci 161:415–421
103. Xie D, Hong Y (2001) In vitro regeneration of Acacia mangium via organogenesis. Plant Cell Tissue Organ Cult 66:167–173
104. Liu W, Chilcott CE, Reich RC, Hellmann GM (2000) Regeneration of *Salvia sclarea* via organogenesis. In Vitro Cell Dev Biol Plant 36:201–206
105. Harry IS, Thompson MR, Thorpe TA (1991) Regeneration of plantlets from mature embryos of western larch. In Vitro Cell Dev Biol Plant 27P:89–98
106. Zhang Y, Wei ZM, Xi MI, Shi JS (2006) Direct organogenesis and plantlet regeneration from mature zygotic embryos of masson pine (*Pinus massoniana* L.). Plant Cell Tissue Organ Cult 84:119–123
107. Ebrahimie E, Habashi AA, Ghareyazie B, Ghannadha M, Mohammadie M (2003) A rapid and efficient method for regeneration of plantlets from embryo explants of cumin (*Cuminum cyminum*). Plant Cell Tissue Organ Cult 75:19–25
108. Chavez VM, Litze E, Norstog K (1992) In vitro morphogenesis of *Ceratozamia hildae* and *C. mexicana* from megagametophytes and zygotic embryos. Plant Cell Tissue Organ Cult 30:93–98
109. Tang W, Newton RJ (2005) Peroxidase and catalase activities are involved in direct adventitious shoot formation induced by thidiazuron in eastern white pine (*Pinus strobus* L.) zygotic embryos. Plant Physiol Biochem 43:760–769
110. Singh ND, Sahoo L, Sarin NB, Jaiwal PK (2003) The effect of TDZ on organogenesis and somatic embryogenesis in pigeonpea (*Cajanus cajan* L. Millsp). Plant Sci 164:341–347

111. Mihaljević S, Vršek I (2009) In vitro shoot regeneration from immature seeds of *Epimedium alpinum* induced by thidiazuron and CPPU. Sci Hortic 120:406–410
112. Tang W, Newton RJ, Charles TM (2005) Plant regeneration through multiple adventitious shoot differentiation from callus cultures of slash pine (*Pinus elliottii*). J Plant Physiol 163:98–101
113. Datta MM, Majumder A, Jha S (2006) Organogenesis and plant regeneration in *Taxus wallichiana* (Zucc.). Plant Cell Rep 25:11–18
114. Buiteveld J, Valk PVD, Jansen J, Creemers-Molenaar JC, Hooymans CMC (1993) Callus induction and plant regeneration from explants of commercial cultivars of leek (*Allium ampeloprasum* var. porrum L.). Plant Cell Rep 12:431–434
115. Hutchinson MJ, Tsujita JM, Saxena PK (1994) Callus induction and plant regeneration from mature zygotic embryos of a tetraploid *Alstroemeria* (*A. pelegrina* x *A. psittacina*). Plant Cell Rep 14:184–187
116. Oldach KH, Morgenstern A, Rother S, Girgi M, O'Kennedy M, Lörz H (2001) Efficient in vitro plant regeneration from immature zygotic embryos of pearl millet [*Pennisetum glaucum* (L.) R. Br.] and *Sorghum bicolor* (L.) Moench. Plant Cell Rep 20:416–421

Chapter 18

Somatic Embryogenesis and Plant Regeneration in the Culture of *Arabidopsis thaliana* (L.) Heynh. Immature Zygotic Embryos

Malgorzata D. Gaj

Abstract

Immature zygotic embryos (IZEs) of *Arabidopsis thaliana* (L.) Heynh., a model species for plant genomics, provide efficient explants for a simple, rapid, and effective system for inducing somatic embryogenesis (SE) under in vitro culture. The process of SE can be induced directly from explant tissue, or indirectly through a callus stage, and the mode of morphogenesis depends on the developmental stage of the IZEs that are used. Auxin treatment, preferably with 2,4-D, results in the formation of embryogenic callus tissue in cultures derived from IZEs less advanced in development, i.e., at globular and torpedo stages, while IZE at the late cotyledonary stage rapidly produces somatic embryos, mostly via a direct pathway. In the best SE-responsive genotypes, including the commonly used Col-0 ecotype, up to 90% of the late cotyledonary-stage zygotic embryos undergo rapid and efficient SE. The subculture of somatic embryos onto auxin-free medium results in their conversion into plantlets with an average frequency of 80%. Such a high frequency of somatic embryos developing rapidly from explant tissue, followed by efficient regeneration of fertile plants with a low level of somaclonal variation, is the recommended system for wide application in studies on mechanisms governing plant totipotency; and especially for identifying genetic factors controlling embryogenic transition of somatic plant cells. In this chapter, the induction, development, and maturation of somatic embryos leading to subsequent regeneration of *Arabidopsis* plantlets in culture of IZEs are presented.

Key words: Auxin treatment, Conversion rate, Immature zygotic embryo culture, Plant regeneration, Somaclonal variation, Somatic embryogenesis

1. Introduction

Molecular mechanisms governing plant cell plasticity have become of central interest for modern developmental biology. The process of somatic embryogenesis (SE) illustrates a unique phenomenon of plant cell developmental plasticity, and

Trevor A. Thorpe and Edward C. Yeung (eds.), *Plant Embryo Culture: Methods and Protocols*,
Methods in Molecular Biology, vol. 710, DOI 10.1007/978-1-61737-988-8_18,

thus provides an attractive model system for studies on the genetic and physiological factors controlling the fate of plant cells. During SE, the already differentiated somatic cells undergo numerous changes related to erasing of the existing developmental program, followed by induction of a new embryonic pathway of development. Understanding the key factors promoting vegetative-to-embryogenic transition, and especially the identification of genes involved in the induction of competence for embryogenesis and subsequent embryo development, present a challenge for modern molecular biology. Moreover, SE shares considerable similarity with zygotic embryogenesis (ZE), and thus provides an attractive experimental model system for studying molecular and cellular mechanisms determining in vivo plant embryogensis (1, 2).

Since 1950s, the carrot SE system (3, 4) has been widely used in studies of plant embryogenesis, due to its simplicity and reproducibility in the production of somatic embryos from undifferentiated callus tissue in large quantities (1). However, the discrepancy between the easy and routinely used SE system in carrot, and our limited knowledge of the structure and function of the carrot genome, presents obstacles to further progress in the molecular analysis of plant embryogenesis in this species. Thus, numerous efforts were taken to establish a method for SE induction in *Arabidopsis thaliana* (L.) Heynh, a model species in structural and functional plant genomics, including studies on developmental processes (5).

In the last decade, an efficient system of SE in *Arabidopsis* was established, based on the use of immature zygotic embryos (IZEs) as explants. In general, two developmentally different pathways lead to somatic embryo formation in vitro: (1) a rapid and efficient process of direct embryogenesis and (2) a much less efficient, slower pathway of indirect embryogenesis, which is preceded by cell dedifferentiation and callus formation. In *Arabidopsis* cultures, both direct and indirect developmental pathways can be induced, and a strictly defined developmental stage of IZE explants is critical for the particular pathway of SE (6). The utilization of IZEs in the globular to torpedo stages results in the induction of embryogenic callus cultures (7–9), while the culture of IZEs presenting a more advanced, namely, the late cotyledonary (LC), stage enables efficient production of somatic embryos via a direct pathway (6). Histological studies on SE induction in LC explants indicated that protodermal and subprotodermal explant cells are involved in the formation of embryogenic-like centers, while in the direct formation of somatic embryos protoderm-derived cells are involved (10). Analysis of genetic chimerism in SE-derived plants (11), and histological observations (10), indicated single and multicell origins of IZE-derived somatic embryos.

In contrast to IZE explants, the postembryonic tissue able to produce embryogenic cultures in *Arabidopsis* was only found in certain mutants, i.e., primordia timing (*pt*), altered meristem program (*amp1* allelic to *pt*), and *clavata* (*clv* 1 and *clv3*), in which germinating seeds and seedlings were shown to form somatic embryos (9, 12). The *pt/amp1* mutant forms cotyledon-like leaves (13) and, as a consequence, embryogenic competence is prolonged until the seedling stage, and thus postembryonic cells in shoot apical meristems can respond in vitro and form somatic embryos (12). In contrast to *pt* and *clv* mutants, in which cells retain competency for SE at the seedling stage, the *lec* mutants exhibited strongly impaired embryonic potential of IZE-derived cultures (14). This phenotype of the *lec* mutants results from their enhanced maturation process and seedling characteristics during embryonic development, leading to leaf-like cotyledons, "leafy cotyledons" (15).

In addition to an appropriate explant type, to induce efficient SE, a specific hormonal treatment of cultured tissue is also a common prerequisite (16). In dicots, the induction of SE usually requires a single hormonal treatment, and in most species, including *Arabidopsis*, an auxin, preferably 2,4-D, is recommended (17). Somatic embryo induction is followed by transfer onto hormone-free medium for embryo maturation and conversion into plantlets. Within a procedure leading to production of plantlets via SE, several steps can be distinguished, including (1) induction and maintenance of embryogenic tissue, (2) development and maturation of somatic embryos, (3) their germination and conversion, and (4) regeneration of SE-derived plantlets.

The present protocol describes a rapid and efficient system enabling the production of *Arabidopsis* somatic embryos from cultures of IZEs. Somatic embryos are frequently formed through direct somatic embryogenesis (DSE), and the protocol can be used for a wide range of *A. thaliana* ecotypes, including Col-0 which is the most commonly used in molecular studies.

2. Materials

1. E5 induction medium: stock salt formulations and organic additives used in the media are presented in Table 1. The E5 medium containing the basic B5 Gamborg salts and organic components (18) supplemented with 5 μM 2,4-D, 20 g/L sucrose and 8.0 g/L Oxoid agar are used for SE induction.
2. MS20 medium for embryo development and conversion: Murashige and Skoog (MS) basic medium (19) supplemented

Table 1
Composition of basal MS and B5 media

Component	MS medium (mg/L)	B5 medium (mg/L)
Macronutrients		
NH_4NO_3	1,650	–
KNO_3	1,900	3,000
$MgSO_4 \cdot 7H_2O$	370	500
KH_2PO_4	170	–
$NaH_2PO_4 \cdot 2H_20$	–	150
$(NH_4)_2SO_4$	–	134
$CaCl_2$	440	150
Micronutrients		
KI	0.83	0.75
H_3BO_3	6.2	3.0
$MnSO_4 \cdot 4H_2O$	22.3	13.2
$ZnSO_4 \cdot 7H_2O$	8.6	2.0
$CuSO_4 \cdot 5H_2O$	0.025	0.025
$Na_2MoO_4 \cdot 2H_2O$	0.25	0.25
$CoCl_2 \cdot 6H_2O$	0.025	0.025
$FeSO_4 \cdot 7H_2O$	27.8	27.8
$Na_2EDTA \cdot 2H_2O$	37.3	37.3
Organics		
Myo-inositol	100.0	100.0
Nicotinic acid	0.5	1.0
Pyridoxine-HCl	0.5	1.0
Thiamine-HCl	0.1	10.0
Glycine	2.0	–

From refs. (18, 19)

with 20 g/L sucrose and solidified with 3.5 g/L Phytagel. The media pH is adjusted to 5.8 before adding the solidifying agent. The medium is placed in plastic Petri dishes (35 mm) for continual embryo development and in larger Petri dishes (60 mm), jars, or test tubes for embryo conversion.

3. The seeds are of *A. thaliana* (L.) Heynh., ecotype Col-0 (see Note 1).
4. Sterilizing solution: 2% solution of sodium hypochlorite with three drops of Tween 20/100 mL as surfactant.
5. Culture plates and containers: Petri dish (35 × 10 mm; SIGMA, No. C 6296); glass jars (SIGMA, No. V 0633) with Magenta B-caps (SIGMA, No. B 8648); glass tubes (DURAN® 25 × 200 mm).

3. Methods

3.1. Cultivation of Arabidopsis Plants Delivering Explants

1. Place the seeds at 4°C for 2–4 days to overcome dormancy, and allow the seeds to germinate synchronously.
2. Fill the pots with top soil mixed with vermiculite (1:1) (see Note 2).
3. Sprinkle the seeds onto the surface of the soil (five seeds per pot of 10 cm in diameter) (see Note 3).
4. Place the pots at the temperature of 20–22°C under a 16-h photoperiod with a white light of 100 μmol/m^2/s intensity.
5. Keep the soil slightly damp, but not saturated (see Note 4).

3.2. Explant Preparation and Culture Initiation

1. Collect the siliques with immature seeds from 7- to 8-week-old plants (see Note 5).
2. Sterilize the siliques for 20 min in the sterilizing solution, and then rinse three times in sterile water. From this step onwards, all procedures are done under sterile conditions.
3. Open the sterile siliques with very fine needles under a dissecting microscope.
4. Select the IZEs at an advanced late stage (LS) of development; the appropriate embryos are >600 μm in length (at 11–13 DAP) and display fully developed, green and bent cotyledons (see Note 6).
5. Place ten of the isolated LS immature embryos in a Petri dish (35 × 10 mm) with E5 solid medium, Table 1 (see Note 7), containing 5 μM 2,4-D (see Note 8) and seal with Parafilm. Keep cultures at 21–23°C under a 16-h photoperiod of 40 μmol/m^2/s of white, fluorescent light.

3.3. Embryo Induction and Development

1. Maintain the explants for 15 days on E5 induction medium (see Notes 9 and 10).
2. Transfer the explants exhibiting embryogenic response (see Note 11) onto Petri dishes (60 × 10 mm) with solid MS20 medium (Table 1).

3.4. Conversion of Somatic Embryos into Plants

1. After 10 days, dissect the somatic embryos from the primary explant tissue with the use of fine needles (see Notes 12–15).
2. Place the separated embryos in 60 × 10 mm Petri dishes, or glass jars with Magenta B-caps on MS20 agar medium (see Note 16) for their further growth and conversion into plants (see Note 17). Keep the culture for 10 days at 20–22°C under a 16-h photoperiod with 40 μmol/m^2/s light intensity.

3. Transfer the developing plantlets into glass tubes (DURAN®) with MS20 medium (see Note 18), and grow the regenerants at 18–20°C (see Note 19) under a 16-h photoperiod with a white light of 100 µmol/m²/s intensity.
4. Harvest the seeds (see Notes 20 and 21).

4. Notes

1. The protocol can be used for different genotypes, but with different efficiencies. The frequency of explants forming somatic embryos (SE efficiency) ranges from 80–90% for the most responsive genotypes (e.g., Col-0, RLD) to 60% in those which are less embryogenic (e.g., Wassilewskija and Landsberg *erecta*) (6, 20).
2. To eliminate pests, the soil can be presoaked with distilled water and then sterilized in an autoclave for at least 20 min.
3. The seeds require light for germination and should not be covered with soil.
4. Insects, and frequently aphids, can cause substantial damage to *Arabidopsis* plants, and to minimize the risk of aphids invasion the plants, and especially the underside of the rosette leaves, the stems, and the base of the inflorescence should be regularly inspected. Washing plants with a mild detergent is recommended, but to eliminate the insects entirely the use of systemic pesticides is required. The special care of the plants is very important as the explants isolated from the aphid-damaged plants show drastically reduced capacity for SE.
5. First inflorescences with flower buds are usually seen after 5–6 weeks, and an additional 2 weeks is needed for silique development.
6. The efficiency of SE increases with the age of the IZE used as the explant. In cultures of Col-0, 73–90% of zygotic embryos at the advanced late stage (fully developed, green and bent cotyledons) undergo embryogenic transition, while only 22–29% of younger, heart to torpedo-shaped, zygotic embryos produce somatic embryos (6).
7. When a liquid medium is used to culture IZE explants at the "early bent cotyledon" stage, callus is produced during a 21-day auxin treatment. In this system, the somatic embryos are formed indirectly from green embryogenic clusters developing in liquid cultures, following their transfer onto solid auxin-free medium (9).

8. Development of somatic embryos is also observed in the presence of NAA (10–20 μM), but the process is less efficient and is accompanied with strong callus tissue and hairy root structure formation (21).
9. Distinct morphological changes of the explant accompany the induction of somatic embryogenesis. During the first week, the straightening and expansion of previously bent cotyledons, and swelling of the cotyledonary node, is observed. By the second week (8–10 day) of the culture, the first somatic embryos appear on the adaxial side of cotyledons, in the area proximal to the cotyledonary node. Somatic embryos are produced asynchronously, and by the end of week 2 the cotyledon-part of the IZE is covered with somatic embryos in various stages of development. Indirect SE development starts later on, at the third week of the culture, from callus tissue formed mostly from the abaxial side of cotyledons (10).
10. To stimulate efficient embryogenic response, and to minimize the risk of other morphogenic responses, an optimal time of 15 days for 2,4-D treatment is recommended. A shorter time of auxin treatment (8 days) promotes frequent shoot and leaf organogenesis (22), while a prolonged 30 days of exposure to auxin stimulates secondary callus formation.
11. The explants not responding in SE produce callus which upon subculture onto fresh E5 medium develops somatic embryos with frequency below 20% (Col-0).
12. In practice, somatic embryos, contrary to other regenerative structures, can be easily detached from explant tissue, as they are not connected with explants through vascular tissue (10). However, it should be noted that not all embryo-like structures represent true somatic embryos, since partial embryos lacking a properly formed root pole can be produced (23). Analysis of the root pole in regenerative structures, with the use of a reporter auxin-activated GUS gene, indicated that the frequency of complete somatic embryos depends on length of 2,4-D treatment. The highest number (around 65%) of embryos developing root meristems was observed in culture induced for 15 days on E5 medium (Gaj and Manka, unpublished).
13. Variation in somatic embryo morphology is commonly observed. Single typical embryos of normal morphology are usually in minority, while the more frequent clusters of single trumpet-shaped embryos, or multiple embryos with fused cotyledons, are observed. The somatic embryos are usually larger (about 2 mm) than their zygotic counterparts (6, 9, 10).

14. The number of somatic embryos produced per explant (SE productivity) can range from 1 to 25, but the majority (80%) of embryogenic explants produce 2–10 somatic embryos with average number of 8.0 ± 4.8 per responding explant (6).
15. Alternatively, whole primary explants can be transferred onto MS20 medium. In such a case, the development of all types of regenerative structures into plants is observed, and bushes of plantlets/shoots are developed from each responding explant. The plants can be isolated and cultured separately, which leads to high multiplication of plant number obtained from one primary explant. Following this procedure, up to 40 plants can be derived from one explant (6).
16. Alternatively, the primary somatic embryo can be used as an explant to induce secondary SE (SSE (20)), and to establish callus embryogenic culture (24). The primary somatic embryos placed on E5 medium produce callus in which secondary somatic embryos are formed. These embryos develop into plants on hormone-free medium. The culture of SSE in the presence of 9 μM 2,4-D is recommended to establish and maintain embryogenic callus culture which upon monthly subculture retain their SE potential at least for 1 year (24).
17. Despite frequent developmental malformation of somatic embryos (see Note 13), high plant regeneration can be obtained, and no distinct differences in ability for conversion into plants were observed for normal and multiple/malformed somatic embryos. On the average, 66.2% of somatic embryos produced shoots with roots (6).
18. To promote rooting and better plant development, MS20 medium can be supplemented with 3 mL/L of 6% KH_2PO_4.
19. The temperature below 20°C is recommended to provide vigorous growth of regenerants.
20. Over 70% of the regenerants set seeds (6), and to maximize the number of seeds harvested per plant nonsterile conditions for the growth of rooted plants are recommended. Before planting into pots, plants (2–5 cm in height), with well-developed roots, should be selected, and the roots washed out gently under tap water to remove all agar clumps, then follow steps 2–5. Acclimatization conditions (shading, high humidity) are important to maximize survival of the transplanted regenerants.
21. The seed-derived progeny of regenerants display high genetic uniformity and fidelity, as confirmed with the use of cytogenetic analysis and a test on embryo-lethal and chlorophyll mutations (6, 11).

References

1. Zimmerman L (1993) Somatic embryogenesis: a model for early development in higher plants. Plant Cell 5:1411–1423
2. Berleth T (1998) Experimental approaches to *Arabidopsis* embryogenesis. Plant Physiol Biochem 36:69–82
3. Reinert J (1958) Morphogenese und ihre Kontrolle an Gewebekulturen aus Carotten. Naturwissenschaften 45:344–345
4. Steward FC, Mapes MO, Mears K (1958) Growth and organized development of cultured cells. I. Growth and division of freely suspended cells. Am J Bot 45:693–703
5. Somerville C, Koornneef M (2002) A fortunate choice: the history of *Arabidopsis* as a model plant. Nat Rev 3:883–889
6. Gaj MD (2001) Direct somatic embryogenesis as a rapid and efficient system for in vitro regeneration of *Arabidopsis thaliana* (L.) Heynh. Plant Cell Tissue Organ Cult 64:39–46
7. Wu Y, Haberland G, Zhou C, Koop HU (1992) Somatic embryogenesis, formation of morphogenic callus and normal development in zygotic embryos of *Arabidopsis thaliana* in vitro. Protoplasma 169:89–96
8. Luo Y, Koop HS (1997) Somatic embryogenesis in cultured immature zygotic embryos and leaf protoplasts of *Arabidopsis thaliana* ecotypes. Planta 202:387–396
9. Mordhorst AP, Voerman KJ, Hartog MV, Meijer EA, Van Went J, Koornneef M, deVries SC (1998) Somatic embryogenesis in *Arabidopsis thaliana* is facilitated by mutation in genes repressing meristematic cell divisions. Genetics 149:549–563
10. Kurczyńska EU, Gaj MD, Ujczak A, Mazur E (2007) Histological analysis of direct somatic embryogenesis in *Arabidopsis thaliana* (L.) Heynh. Planta 226:619–628
11. Gaj MD (2001) Somatic embryogenesis in in vitro cultures of *Arabidopsis thaliana* (L.). Heynh. (in Polish with English summary). Monograph of Silesian University, No. 1970. Silesian University, Katowice
12. von Recklinghausen IR, Iwanowska A, Kieft H, Mordhorst AP, Schel JHN, von Lammeren AAM (2000) Structure and development of somatic embryos formed in *Arabidopsis pt* mutant callus cultures derived from seedlings. Protoplasma 211:217–224
13. Conway LJ, Poethig RS (1997) Mutations of *Arabidopsis thaliana* that transform leaves into cotyledons. Proc Natl Acad Sci (USA) 94:10209–10214
14. Gaj MD, Zhang S, Harada JJ, Lemaux PG (2005) Leafy cotyledon genes are essential for induction of somatic embryogenesis of *Arabidopsis*. Planta 222:977–988
15. Harada JJ, Stone SL, Kwong RW, Lee H, Kwong LW, Pelletier J (2003) Leafy cotyledon genes and the control of embryo development. In: Vasil IK (ed) Plant biotechnology 2002 and beyond. Kluwer Academic Publishers, Dordrecht, pp 263–268
16. Jimenez VM (2005) Involvement of plant hormones and plant growth regulators on in vitro somatic embryogenesis. Plant Growth Regul 47:91–110
17. Gaj MD (2004) Factors influencing somatic embryogenesis induction and plant regeneration with particular reference to *Arabidopsis thaliana* (L.) Heynh. Plant Growth Regul 43:27–47
18. Gamborg OL, Miller RA, Ojima K (1968) Nutrient requirement of suspension cultures of soybean root cells. Exp Cell Res 50:151–158
19. Murashige T, Skoog F (1962) A revised medium for rapid growth and bioassays with tobacco tissue cultures. Physiol Plant 15:437–497
20. Gaj MD, Czubin M (2004) Efficiency of somatic embryogenesis in *Arabidopsis thaliana* (L.) Heynh. under various in vitro culture conditions. Biotechnology 64:221–235
21. Gaj MD, Trojanowska A, Ujczak A, Mędrek M, Kozioł A, Garbaciak B (2006) Hormone-response mutants of *Arabidopsis thaliana* (L.) Heynh. impaired in somatic embryogenesis. Plant Growth Regul 49:189–197
22. Raghavan V (2006) Can carrot and *Arabidopsis* serve as model systems to study the molecular biology of somatic embryogenesis? Curr Sci 90:1336–1343
23. Bassuner BM, Lam R, Lukowitz W, Yeung EC (2007) Auxin and root initiation in somatic embryos of *Arabidopsis*. Plant Cell Rep 26:1–11
24. Pillon E, Terzi M, Baldan B, Mariani P, Schiavo FL (1996) A protocol for obtaining embryogenic cell lines from *Arabidopsis*. Plant J 9:573–577

Chapter 19

Pine Somatic Embryogenesis Using Zygotic Embryos as Explants

Gerald S. Pullman and Kylie Bucalo

Abstract

Somatic embryogenesis (SE) has the potential to be the lowest-cost method to rapidly produce large numbers of high-value somatic seedlings with desired characteristics for plantation forestry. At least 24 of the 115–120 known *Pinus* species can undergo SE. Initiation for most species works best with immature megagametophytes as starting material, although a few pines can initiate SE cultures from isolated mature seed embryos. Successful initiation depends heavily on explant type, embryo developmental stage, and medium salt base. Most first reports of initiation used 2,4-D and BAP or a combination of cytokinins. More recent reports have optimized initiation for many *Pinus* spp., but still use mostly the combinations of auxin and cytokinins. Initiation can be stimulated with medium supplements including abscisic acid (ABA), brassinosteroids, ethylene inhibitors, gibberellin inhibitors, organic acids, putrescine, specific sugar types (maltose, galactose, d-*chiro*-inositol, and d-xylose), triacontanol, vitamins (B_{12}, biotin, vitamin E, and folic acid), or manipulation of environmental factors including pH, water potential, cone cold storage, gelling agent concentration, and liquid medium. Embryo development and maturation usually occur best on medium containing ABA along with water potential reduction (with sugars and polyethylene glycol) or water availability reduction (with raised gelling agent increasing gel-strength). Activated carbon and maltose may also improve embryo maturation. The main issues holding SE technology back are related to the high cost of producing a somatic seedling, incurred from low initiation percentages for recalcitrant species, culture loss, and decline after initiation and poor embryo maturation resulting in no or poor germination. Although vast progress has been made in pine SE technology over the past 24 years, fundamental studies on seed and embryo physiology, biochemistry, and gene expression are still needed to help improve the technology to a point where large-scale commercialization is economically viable for a broad range of pine species.

Key words: Conifer, Embryogenesis, *Pinus*, Somatic embryogenesis

1. Introduction

Approximately 115–120 species of *Pinus* are recognized depending on the authority. Most originate in the Northern Hemisphere where *Pinus* is the largest genus of conifers (1). Many species of

Trevor A. Thorpe and Edward C. Yeung (eds.), *Plant Embryo Culture: Methods and Protocols*, Methods in Molecular Biology, vol. 710, DOI 10.1007/978-1-61737-988-8_19,

Pinus are valued for their timber, wood, and recreational value. Large numbers of seedlings from several species are planted annually. For example, loblolly pine (*Pinus taeda* L., LP.) is the major species planted across the southern USA with 1–1.5 billion trees planted annually (2).

Forest productivity can be increased by tree plantations with large numbers of elite, high-value trees, and efforts are ongoing to propagate genetically superior conifer trees through clonal propagation by somatic embryogenesis (SE). SE can capture the benefits of breeding or genetic engineering programs by multiplying trees with improved wood quantity, quality, and uniformity. SE technology can also be highly useful to study embryo development, differentiation, and to help preserve endangered species.

Conifer SE proceeds through four steps: initiation, multiplication, maturation, and germination. A fifth step of cryogenic storage may be added when storage of embryogenic cultures is desired. SE has been demonstrated for many coniferous species including pines. At least 27 *Pinus* species are reported to go through SE (Table 1). Several comprehensive reviews on conifer SE and recent advances are available (3–12). Due to the commercial interest in SE technology, much additional information is available in published patent applications and issued patents. To illustrate the approaches used with SE in *Pinus*, *P. taeda* L. will be used as a model throughout this chapter.

1.1. Embryogenic Tissue Initiation and Maintenance

Somatic embryos can be generated from a variety of tissues including: immature megagametophytes or excised zygotic embryos (*Pinus strobus* (13, 14), *P. pinaster* (15–17), *P. sylvestris* (16, 18, 19), *P. caribaea* (20), *P. taeda* (21, 22), *P. monticola* (23)); excised full-term embryos (*P. geradiana* (24), *P. kesiya* (25), *P. koraiensis* (26), *P. lambertiana* (27), *P. massoniana* (28), *P. nigra* (29), *P. strobus* (30), *P. taeda* (31), *P. wallichiana* (32)); and recently, vegetative shoot apices from aged trees (*P. kesiya* (33), *P. sylvestris* (34), *P. patula* (35), and *P. roxburghii* (36)).

The most success has occurred with immature megagametophytes. Fertilized megagametophytes from surface-sterilized seeds are excised and placed on medium to permit the extrusion of embryogenic tissue from the micropylar end. Use of the whole megagametophyte avoids the laborious dissection process of embryo excision. Numerous somatic embryos often form in the extruded material which can then be subcultured to a multiplication medium (37–39). When using this method, particular attention needs to be paid to the stage of embryo development, rather than time of year, to optimize initiation (see Note 1).

A number of tools can improve embryogenic tissue initiation from coniferous tissue. These include traditional factors such as choice and optimization of salt mixtures, auxins, and cytokinins

Table 1
First and recent reports of somatic embryogenesis in *Pinus* spp.

Species	Explant	Induction media	Induction hormones	Initiation frequency (first/recent)	Response	Reference: first report	Reference: recent report
Pinus armandii Franch. Var. amamiana (Koidz.) Hatusima	MG-PC	Modified 1/2 EM	10 μM 2,4-D, 5 μM BAP	1.5%	SE, PL	(103)	–
P. banksiana Lamb.	MG, EE	1/2 Litvay; DCR	10 μM 2,4-D, 5 μM BAP	0.4%	SE, PL, CR	(104)	–
P. brutia TEN	MG-PC	Supplemented DCR	13.6 μM 2,4-D, 2.2 μM BAP	11.6%	SE, PL	(105)	–
P. bungeana Zucc. ex Endl.	EE-PC	DCR_1	10 mg/L 2,4-D, 4 mg/L BAP	84.4%	SE	(106)	–
P. caribaea Morelet	MG-PEP	LPG	10 μM 2,4-D, 5 μM BAP	5%	SE, PL, CR	(20)	(107, 108)
P. densiflora	MG-PC	Modified DCR; modified LP	10.0 μM 2,4-D, 5 μM BAP	2.9, 1.0/5%	SE, PL SE, PL	(109, 110)	(111, 112)
P. elliottii Engelm	EE-PEP	WPMG; MNCI	20 μM 2,4-D, 5 μM BAP; 20 μM 2,4-D, 2.5 μM BAP, 2.5 μM kinetin	2–6/9%	SE, PL	(113)	(114, 115)
P. geradiana Wall	EE-M	1/2 MSG	9.0 μM 2,4-D	81.2%	SE, PL	(24)	–
P. heldreichii	MG-PC	Gresshoff and Doy	2 mg/L 2,4-D, 0.5 mg/L BAP	6.7%	SE	(116)	–

(continued)

Table 1 (continued)

Species	Explant	Induction media	Induction hormones	Initiation frequency (first/recent)	Response	Reference: first report	Reference: recent report
P. kesiya Royle ex. Gord.	EE-M; VA; EE-PC	Modified 1/2 MS; 1/2 DCR	22.6 µM 2,4-D, 26.9 µM NAA, 8.87 µM BAP	ND/86/0–46%	SE, PL	(25)	(33, 117)
P. koraiensis Sieb et Zucc.	EE-M	Supplemented Litvay	10 µM 2,4-D, 5 µM BAP	14.7% (3 weeks)	SE	(26)	–
P. lambertiana Dougl.	EE-M	Modified DCR	3–500 mg/L 2,4-D	4–5%	SE, PL	(27)	(118)
P. massoniana Lamb.	EE-M	DCR	10 mg/L 2,4-D, 4 mg/L kinetin, 4 mg/L BAP	17–45%	SE, PL	(28)	–
P. monticola Dougl.	MG-PEP to PC	Modified Litvay	2.25 µM 2,4-D, 2.25 µM BAP	0.8–6.7%	SE, PL, CR	(23)	–
P. nigra Arn	MG-PC	DCR	2 mg/L 2,4-D, 0.5 mg/L BAP	2/7–9%	SE, PL	(119)	(49, 120)
P. palustris	MG-PC	Modified MSG; DCR	3.0 mg/L 2,4-D, 0.5 mg/L BAP	3.5%	SE	(46)	–
P. patula Schiede et Deppe	MG-PEP; VA	Supplemented DCR1	3.0 mg/L 2,4-D, 0.5 mg/L BAP	2.6–8.5%	SE, PL, CR	(121)	(35, 85, 122, 123)
P. pinaster Ait.	MG-PEP, EE-PEP; EE-PC	H medium; modified Litvay	2.2 mg/L 2,4-D, 1.1 mg/L BAP; 9 µM 2,4-D, 4.4 µM BAP	5–19/93%	SE, PL, CR	(15)	(16, 71)
P. radiata D. Don	MG-PC	Modified EM; modified SH	ND/±1–2 mg/L 2,4-D	ND/40%	SE, PL, CR	(124)	(53, 125–127)

P. rigida × *P. taeda*	MG-PEP	Modified P6	13.5 µM 2,4-D, 4.4 µM BAP	1.1%	SE, PL	(128)	–
P. roxburghii Sarg.	MG, EE-PC, SN	DCR	10.0 µM 2,4-D, 5 µM BAP	9.6/46–65%	SE, PL, CR	(129)	(44, 130)
P. serotina	MG, EE	Modified MS; modified DCR1	2–5 mg/L 2,4-D or NAA, 0–1 mg/L BAP	12%	SE	(131)	–
P. strobus L.	MG-PC	Modified DCR	2 mg/L 2,4-D, 1 mg/L BAP	54/2.6–23%	SE, PL, CR	(13)	(30, 132, 133)
P. sylvestris L.	MG-PEP; VA	MSG; modified Litvay	9.0 µM 2,4-D, 4.4 µM BAP	5/22%	SE, PL	(18)	(16, 34, 133, 134)
P. taeda L.	MG-PC; MG-PEP	Modified 1/2 MS	11.0 mg/L 2,4-D, 4.5 mg/L BAP, 4.3 mg/L kinetin	9–10/20–33%	SE, PL, CR	(94)	(51, 135, 136)
P. thunbergii	MG-PC	Modified DCR; modified LP	10.0 µM 2,4-D, 5 µM BAP	8.0; 3.0/2%	SE, PL SE, PL	(109, 110)	(137)
P. wallichiana A. B. Jacks	EE-M	MSG	9.0 µM 2,4-D, 2.0 µM 24-epibrassinolide	61–92%	SE, PL	(32)	–

MG megagametophyte, *EE* excised embryo, *PC* precotyledonary embryos, *PEP* poly embryony phase, *C* cotyledonary, *M* mature, *VA* vegetative apices from mature trees, *SN* secondary needles from mature trees, *BAP* 6-benzylaminopurine, 6-benzyladenine, N6-benzyladenine, benzyladenine, BA, 6-BAP, *SE* somatic embryos, *PL* plants, *CR* success in cryopreservation

Note: *ND* no data presented, only the most successful medium is listed in induction medium column for each species

(4, 9, 22); media supplementation with abscisic acid (ABA) (40–42), brassinosteroids (32, 43) (see Note 2), triacontanol (35, 44), putrescine (24), ethylene inhibitors (45), specific sugar types (22, 46–51), vitamins B_{12}, biotin, vitamin E, and folic acid (32, 38, 39); organic acids (39); and gibberellin inhibitors (52). Cold storage of cones prior to seed preparation for explanting can significantly increase initiation results for *P. radiata* (53) and *P. taeda* (Pullman, unpublished). Control of environmental factors, including water potential (54, 55), pH (38, 56, 57), adsorption of medium components by activated carbon (AC), gelling agent content liquid media, and study of embryo gene expression patterns, has led to further improvements (22, 45, 58–65, 138). Using a combination of media and environmental agents has permitted great increases in embryogenic tissue initiation.

After initiation, somatic embryos are moved to a capture medium that usually contains reduced hormones. Further maintenance occurs bimonthly or weekly on gelled or liquid medium, again with auxins and cytokinins, to continue cleavage polyembryony and somatic embryo multiplication. Liquid media have the advantage of faster growth rates, decreasing variation, simplifying preparation of cells for cryostorage, ease of visualizing somatic embryos, and automation of cell suspension transfer (see Note 3).

1.2. Somatic Embryo Maturation

Maturation improvements have been achieved in many pines and other coniferous species (10). Most improvements occur through stage-specific medium optimization that involves, but is not limited to, manipulation of basal salts, ABA, choice or combination of sugars such as sucrose, maltose, and trehalose, addition of water potential mediators such as polyethylene glycol (PEG), gelling agents, carbohydrates, or other osmoticum, tissue partial desiccation, and addition of AC to control hormone pulse and/or remove unwanted compounds. Most maturation media for *Pinus* sp. contain 60–480 μM ABA (66). For *P. taeda*, 19.6–37.8 μM ABA works well for initial and subsequent medium transfers (67).

Pullman and Buchanan (139) analyzed stage-specific female gametophyte (FG) and embryo tissues of *P. taeda* for 14 key metals. A tenfold increase in cotyledonary embryos occurred along with improved gene expression patterns that were more similar to zygotic embryos when maturation medium was modified to better match boron, calcium, potassium, and iron levels present in similar stages of natural embryos (45).

Pullman and Gupta (68) combined AC and high concentrations of ABA (to compensate for ABA adsorption by AC) in gelled or liquid medium. Variations included AC-coated embryos plated on a maturation medium containing ABA, or plating embryos on filter paper coated with AC placed on ABA-containing medium. This combination worked extremely well for *Picea abies*, *Pseudotsuga menziesii*, and *P. taeda* resulting in improved embryo quantity and quality (68–70). Later, Lelu-Walter et al. (71) found these combinations also worked well for *P. pinaster*.

Osmotic conditions appear to control embryo development in many plant species (72). Water relation parameters have been partially investigated for zygotic and somatic embryos of *P. resinosa*, *P. taeda*, and *P. strobus* (54, 55, 73–75). These investigations show that seed tissue water potential values are much greater (in mmol/kg) than that measured in a typical plant tissue culture medium. So, it was no surprise that early researchers found ABA alone was often not enough to prevent premature conifer somatic embryo germination and medium water potential also needed to be altered through the addition of osmoticants (4).

There are two main methods used today to control osmotic conditions during embryo maturation in *Pinus*. The first method adds sucrose, maltose, sugar alcohols, PEG, or combinations of these osmoticants to the medium to lower water potential. The second method adds extra gelling agent to the maturation medium to increase gel-strength and reduce medium water availability (14). Combinations of these methods are also used (17, 23, 76). Mild desiccation of maintenance tissue causing 50% water loss is reported to stimulate maturation in *P. kesiya* and *P. patula* (33, 35).

Additional factors are beneficial for somatic embryo development in other coniferous genera and may play a role in *Pinus*. Early stage embryo development appears to occur best in a reducing environment while late-stage development requires shifting to a more oxidizing environment (77). Both embryo development and germination can be enhanced with the addition of redox chemicals to the medium ((10), Pullman, unpublished). Arabinogalactan proteins are glycosylated polypeptides consisting of up to 90% carbohydrate and are capable of stimulating SE when added to a weakly embryogenic cell line (78–80). Endogenous levels of ethylene are low in developing seeds but may be high in culture (81). Control of ethylene in vitro through biosynthesis or ethylene action inhibitors improved embryo morphology and postembryonic performance for *Picea glauca* and may also stimulate SE in some pines (82–84).

1.3. Somatic Embryo Germination

Somatic embryos are usually germinated on medium devoid of plant growth hormones and with reduced levels of salts, sugars, and nutrients. AC is often present to facilitate the removal of residual plant growth hormones. Unfortunately, germination of *Pinus* embryos depends heavily on genotype and is often difficult to achieve (17, 19, 61, 85). The eventual goal of many SE systems is to reduce costs and labor through delivery of embryos to soil through production of artificial seeds (86, 87).

Several treatments increase the germination of pine somatic embryos including: embryo drying or desiccation prior to germination (88), stratification (89), application of red light to pregermination and germinating embryos (90), direct or indirect exposure to the ectomycorrhizal fungus *Psolithus tinctorius* (91), and supplementation of medium with redox chemicals (92). Merkle et al. were able to increase conversion, root growth and germination in *P. taeda*, *P. elliottii*, and *P. palustris*, and hybrids of *P. elliottii* × *P. palustris* with red light treatments provided by light-emitting diodes (90). When 0.1 mM ascorbic acid (an antioxidant) was added to the germination medium, white spruce somatic embryos increased germination and germinants had enlarged apical regions causing more leaf primordia and larger shoots (92).

1.4. Embryogenic Tissue Cryostorage

Cryogenic storage maintains long-term cell viability at ultra low temperatures in liquid nitrogen allowing long storage periods that prevent culture decline often observed during repeated culture maintenance. Clones can be placed in cryostorage shortly after initiation and later revived for production after evaluation of field performance. Most cryopreservation protocols are modifications of Kartha et al. (93) developed for white spruce. Table 1 lists *Pinus* species that have shown successful cryostorage.

1.5. Toward Large-Scale SE Technology Commercialization: Challenges Ahead

The first pine somatic embryo was reported from *P. lambertiana* in 1986 and 1 year later success occurred with *P. taeda* (27, 94). Despite progress over the past 23 years, SE technology has yet to make a significant contribution to the billions of pine seedlings that are planted around the world. While at least one company produces several million somatic loblolly pine seedlings annually, the economic feasibility of this propagation system currently restricts its use to a fraction of the desired genetic material. Factors currently limiting commercialization of SE for LP include: (1) low initiation of recalcitrant high-value seed sources; (2) inability to maintain culture growth for many genotypes once initiation has occurred; (3) decline of cultures resulting in loss of plant regeneration potential; and (4) low quality of embryos produced, resulting in slow initial growth and low germination percentages. While these factors may be overcome for individual genotypes, they raise the cost of genotypes that can be produced.

1.6. Approaches to Improve SE Technology

Since the FG tissue that normally surrounds and feeds the natural embryo is not present during SE in vitro, the addition of compounds normally provided by the FG may be necessary to maximize somatic embryo growth. Duplication of hormonal, nutritional, and physical environments found in vivo, study of embryo gene expression patterns, and understanding how medium changes over time, such as AC adsorption and pH change, have advanced the protocol development in *P. taeda*.

Analyses of *P. taeda* seed tissues were conducted to determine the levels of ABA (67, 95), organic acid (96), and vitamins and sugars (97). A somatic embryo bioassay was used to evaluate growth-promotion of the individual compound added to medium at concentrations found in the zygotic tissue. Compounds that increased early stage embryo growth were then tested for effects on initiation. When tested for initiation, media supplementation with ABA, α-ketoglutaric acid, pyruvic acid, succinic acid, folic acid, biotin, vitamins E and B_{12}, d-xylose, and d-*chiro*-inositol caused statistically significant increases in initiation. A promising approach was recently shown by Carman et al. (98) where corrosion cavity fluids in Douglas fir were analyzed. Cyclitols, sucrose equivalents, erythrose, and arabinose were many-fold higher in corrosion cavity fluid than in whole seed tissues and provided a model for SE protocol improvement.

2. Materials

1. Seed (collected at specific developmental stages).
2. Media for *P. taeda*: initiation (2212, 2305), capture and maintenance (1250, 1133), embryo development (1562), germination (397), and cryopreservation (2007). Components are shown in Table 2.
3. Sterilizing solutions: 10% liquinox with 0.2% Tween 20; 20% H_2O_2.
4. Chemical reagents: reagent alcohol (70%), DMSO, and liquid nitrogen.
5. Consumable supplies: scalpel blades (sterile), pipettes (10, 50 mL), vacuum filters (0.2 μm, 250 mL), syringe filter (0.2 μm, 13 mm) Costar #3526 Well Culture Cluster Plates, Cryostorage vials (13.5 × 48.3 mm, 2.0 mL), and parafilm.
6. Equipment: CryoMed Freezing rate controller and chamber and CryoPlusII Liquid Nitrogen Storage Unit.

3. Methods

SE offers the advantage of rapid embryo multiplication in small spaces. The disadvantage is that, while fairly efficient for many plants and some steps of the process, the process currently does not produce a vigorous full-term loblolly pine somatic embryo. Understanding and improving SE is a major research effort at IPST. The SE process used in our laboratory is briefly described in Table 3 (45, 61, 99).

Table 2
***Pinus taeda* initiation, maintenance, maturation, and germination medium components**

Components (mg/L)	Medium number 2212	2305	1250	1133	1562	397	2007
NH_4NO_3	200.0	200.0	603.8	603.8	200	206.3	603.8
KNO_3	909.9	909.9	909.9	909.9	454.9	1,170	909.9
KH_2PO_4	136.1	136.1	136.1	136.1	136.1	85	136.1
$Ca(NO_3)_2 \cdot 4H_2O$	236.2	236.2	236.2	236.2	59.1	0	236.2
$CaCl_2 \cdot 2H_2O$	–	–	0	0	0	220	0
$MgSO_4 \cdot 7H_2O$	246.5	246.5	246.5	246.5	246.5	185.5	246.5
$Mg(NO_3)_2 \cdot 6H_2O$	256.5	256.5	256.5	256.5	256.5	0	256.5
$MgCl_2 \cdot 6H_2O$	101.7	101.7	101.7	101.7	101.7	0	101.7
KI	4.15	4.15	4.15	4.15	4.15	0.415	4.15
H_3BO_3	15.5	15.5	15.5	15.5	7.75	3.1	15.5
$MnSO_4 \cdot H_2O$	10.5	10.5	10.5	10.5	10.5	8.45	10.5
$ZnSO_4 \cdot 7H_2O$	14.668	14.668	14.4	14.4	14.4	4.3	14.4
$Na_2MoO_4 \cdot 2H_2O$	0.125	0.125	0.125	0.125	0.125	0.125	0.125
$CuSO_4 \cdot 5H_2O$	0.1725	0.1725	0.125	0.125	0.125	0.25	0.125
$CoCl_2 \cdot 6H_2O$	0.125	0.125	0.125	0.125	0.125	0.0125	0.125
$AgNO_3$	3.398	3.398	0	0	0	0	0
$FeSO_4 \cdot 7H_2O$	13.9	13.9	6.95	6.95	41.7	13.93	6.95
Na_2EDTA	18.65	18.65	9.33	9.33	55.95	18.65	9.33

Maltose	15,000	15,000	0	0	20,000	0	0
Sucrose	0	0	30,000	30,000	0	20,000	30,000
Acros PEG 8000	0	0	0	0	130,000	0	0
D-Xylose	100	100	0	0	0	0	0
Myo-inositol	20,000	20,000	1,000	1,000	100	100	1,000
Casamino acids	500	500	500	500	500	0	500
L-Glutamine[a]	450	450	450	450	450	0	450
Thiamine · HCl	1.0	1.0	1.0	1.0	1.0	1.0	1.0
Pyridoxine · HCl	0.5	0.5	0.5	0.5	0.5	0.5	0.5
Nicotinic acid	0.5	0.5	0.5	0.5	0.5	0.5	0.5
Glycine	2.0	2.0	2.0	2.0	2.0	2.0	2.0
MES	250	250	250	0	0	0	0
Biotin	0.05	0.05	0.05	0	0	0	0
Folic acid	0.5	0.5	0.5	0	0	0	0
Vitamin B_{12}[a]	0.1	0.1	0	0	0	0	0
Vitamin E[a]	0.1	0.1	0	0	0	0	0
α-Ketoglutaric acid	100	100	0	0	0	0	0
NAA	2.0	0.3	0	0	0	0	0
2,4-D	0	0	1.1	1.1	0	0	1.1
BAP	0.63	0.63	0.45	0.45	0	0	0.45
Kinetin	0.61	0.61	0.43	0.43	0	0	0.43
ABA[a]	0	9.0	1.3	1.3	5.2	0	1.3

(continued)

Table 2
(continued)

Components (mg/L)	Medium number 2212	2305	1250	1133	1562	397	2007
Sorbitol	0	0	0	0	0	0	0.4 M
Activated carbon	50	50	0	0	0	2,500	0
Brassinolide[a]	0.1 μM	0.1 μM	0	0	0	0	0
Gelrite	2,000	0	2,500	0	2,500	0	0
TC agar	0	0	0	0	0	8,000	0
pH	5.7	5.7	5.7	5.7	5.7	5.7	5.7

[a]Filter-sterilized stock solution added to medium after autoclaving

Table 3
Somatic embryogenesis protocol: major steps, growth regulators, sugars, and osmoticants

Step	Initiation	Multiplication (maintenance)	Maturation 1	Maturation 2	Maturation 3
Time	5–10 weeks	Weekly	4 weeks	4 weeks	4 weeks
Process	Induction of cleavage polyembryony	Multiplication, cleavage polyembryony	Development	Development	Development
Embryo stage	1–2	1–2.5	1–6	1–8	1–9.1
Hormones or inhibitors	Auxin/cytokinin, $AgNO_3$, ABA, brassinolide,	Auxin/cytokinin ± ABA	ABA 5 ppm	ABA 5–10 ppm	ABA 5–10 ppm
Sugar	2% maltose, d-xylose (100 mg/L)	Sucrose or maltose, 3%	Maltose 2%	Maltose 2%	Maltose 2%
Osmolality (mM/kg)	225	160	250	250	250
Extra vitamins	Folic acid, biotin, vitamins B_{12}, E	–	–	–	–
Other supplements	Activated carbon (50 mg/L), MES pH buffer, organic acids	–	–	–	–
Osmoticant to lower medium osmotic potential	2% myo-inositol	None	PEG 13%	PEG 13%	PEG 13%

3.1. Initiation of *P. taeda* Embryogenic Tissue

1. High-value seeds from breeding programs are used to initiate a culture. Cross-pollinated cones are collected in early to mid-July from clonal seed orchards, shipped on ice, and received within 24–48 h. Cones are stored at 4–5°C for 1–5 weeks (see Notes 4 and 5). Cones containing seeds with embryos mostly at stages 2–4 (Fig. 2 (100)) are used for initiation experiments as described by Pullman et al. (43).
2. Seeds are removed from cones, mixed in 10% liquinox with 0.2% Tween 20 for 10 min, rinsed in flowing tap water for 30 min, agitated aseptically in 20% H_2O_2 for 10 min, and rinsed five times for 5 min with sterile deionized water.
3. Aseptic dissection: seeds are opened, the integument and nucellus removed, and the whole megagametophyte is placed on 2 mL of medium 2212 (Table 2) in wells of Costar #3526 Well Culture Cluster Plates wrapped in Parafilm and incubated at 23–25°C in the dark.
4. After 14 days, 0.25 mL of medium 2305 (Table 2) is added (64). The liquid overlay contains fresh medium, ABA, greatly reduced NAA and functions to refresh medium contents and pH, and expose extruding tissues to supplemental materials such as ABA. Our best medium now further contains 100 mg/L d-xylose added to both the gelled and liquid medium (51).
5. Percent extrusion and initiation are evaluated after 9–10 weeks. In LP, about 50–60% of extrusions will progress to form somatic embryos.

The initiation sequence for *P. taeda* is shown in Fig. 1 (43) and described in more detail by Becwar and Pullman (37). *Pinus taeda* initiation occurs in three steps: extrusion at 1–4 weeks when one or more usually subordinate zygotic embryos expand out of the megagametophyte micropylar end; proliferating cells and

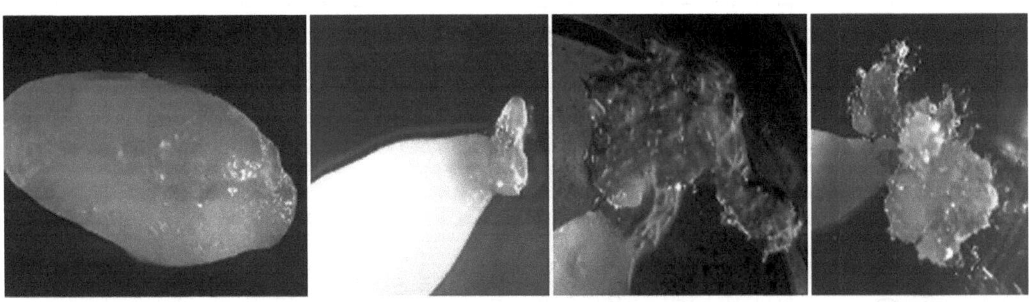

Fig. 1. Typical sequence of embryogenic tissue initiation in loblolly pine. Reproduced from Pullman et al. (43) with permission from Springer.

somatic embryos appear in extruded tissue at 5–7 weeks; and resulting embryogenic tissue multiplies to form a mass. These phases can be evaluated as percent extrusion, percent initiation (somatic embryos visible through a dissecting microscope), and percent of cultures achieving a target mass or size. Extrusion and initiation are routinely evaluated 9–10 weeks after placement of megagametophytes on medium. Care should be taken to distinguish between extrusion and initiation.

3.2. Maintenance and Multiplication of *P. taeda* Somatic Embryos

1. After 7–9 weeks, initiations are transferred to capture medium 1250 (Table 2) and re-cultured every 2–2.5 weeks until target masses of 200 mg/culture are reached. About 50% of the new initiations can be maintained on capture medium. The remaining 50% do not grow even though embryogenic tissue formed during initiation or initially grow slowly and then stop growth within several weeks to several months.
2. Cultures are maintained on gelled or liquid medium. Embryogenic cell suspensions are established by adding about 1 g of 10- to 14-day-old semisolid-grown tissue to 9 mL of liquid medium 1133 (Table 2) in a 250 mL Erlenmeyer flask rotating at 120 rpm. After 5–7 days, each flask is swirled vigorously to facilitate breakup of tissue, and 10 mL of medium is added. After another 7 days, contents of the flask are poured into sterile centrifuge tubes and settled for 20 min. Old liquid medium is removed and settled volumes are measured.
3. Cells are resuspended in medium at a density of 1 mL settled cells per 9 mL medium, rotated at 90–100 rpm, and maintained on a weekly transfer schedule at the same ratio of cells to medium. Cells replicate 2–6 times weekly, rapidly producing large numbers of somatic embryos.

3.3. Maturation of *P. taeda* Somatic Embryos

1. Replicate aliquots of 0.5–1 mL of suspension-grown cells are spread onto support material on 20 mL of maturation medium 1562 contained in 100 × 15 mm Petri plates ((45, 61), see Note 6).
2. Two additional transfers of support and tissue, each 4 weeks apart, occur on fresh medium for a total of 3 months on maturation medium.
3. Typically 10–100 cotyledonary embryos form and can be counted and categorized by stage. Using the embryo staging system of Pullman and Webb (100), embryos develop to a maximum of stage 9.1 compared with zygotic embryos that complete development with an additional 7–10 weeks of growth (stages 9.7–9.10) (45, 61). Some genotypes produce more embryos or advance development slightly if ABA is

increased to 10 mg/L during the second and third month on maturation medium or if a fourth month of maturation is added (67).

3.4. Germination of *P. taeda* Somatic Embryos

The last step is germination and acclimation to achieve growth in vivo and to produce somatic seedlings ready for planting in the field.

1. After 2.5–4 months on maturation medium, somatic embryos are selected that exhibit normal embryo shape.
2. Ten embryos are placed horizontally on 20 mL of germination medium 397 (101) contained in 100×20 mm Petri plates.
3. Plates are incubated for 7 days in the dark and then placed under fluorescent lights (16 h of 7 μmol photons/m^2/s light daily).
4. Embryos are scored at the end of 6 and 12 weeks for the presence of roots and shoots. An embryo is considered to have germinated when it contains both a root and a shoot. Our most advanced somatic embryos resemble stages 7–8 zygotic embryos in germination performance (61). Conversion, acclimation, and field testing procedures used in the past are described in Pullman et al. (61).

3.5. Cryogenic Storage

We have modified the successful cryopreservation procedure in Pullman et al. (61). Protocols have been formulated and tested for plant material originating from either a solid media or liquid cell suspension.

3.5.1. Embryogenic Tissue Grown on Gelled Medium

1. LP cultures growing on solid maintenance media (1250) are transferred every 2 weeks until total mass is at least 1 g. When processed for cryostorage, 1 g of tissue will produce three vials for storage.
2. Measured tissue is added to a sterile Erlenmeyer flask with liquid media 2007 (Table 2) at a ratio of 1 g tissue:1.5 mL medium.
3. Flasks are left on a shaker overnight.
4. After agitation, flasks are placed on ice in a laminar flow hood.
5. A total of 150 μL of filter-sterilized DMSO (cryoprotectant) is added per gram of tissue in five equal aliquots over 30 min, swirling after each DMSO addition to prevent premature freezing of cells, and promote mixing.
6. Aliquots of 1 mL are dispensed into sterile Nalgene cryogenic vials on ice.
7. Vials are transferred to a freezing chamber regulated by a controlled rate freezing program that cools cells to −35°C at a rate of 0.33°C/min.

8. Vials are then transferred to a storage container and s ubmerged in liquid nitrogen for long-term storage at −196°C.
9. To retrieve cultures, vials are removed from liquid N_2, immersed in warm-sterilized deionized water at 37°C for ~2 min, and removed when half of the vials look thawed.
10. After thawing, vials are wiped with 70% alcohol to remove potential contamination.
11. Vials are opened, flamed to sterilize the vial lip, and contents poured onto sterile nylon fabric (see Note 5) in a Petri dish (100×15 mm) on 20 mL semisolid medium 1133 minus ABA+2.5 g/L Gelrite.
12. After 1 h, the nylon overlain with cells is moved to a plate of fresh medium.
13. Eighteen hours later, cells are again transferred to fresh media. Plates are kept in the dark.

3.5.2. Embryogenic Tissue Grown in Liquid Medium

1. LP cultures in liquid maintenance media 1133 are transferred every 7 days until the settled cell volume reaches at least 30 mL. It is best if cells are collected at day 6–7 of the transfer cycle.
2. Settled cells are added to a sterile flask with liquid media 2007 at a ratio of 1:4 parts cells:medium. When 30 mL cells:120 mL medium is used, approximately 50 cryogenic vials are generated.
3. Flasks are left on a shaker overnight.
4. After agitation, flasks are removed from the shaker, placed on ice in a laminar flow hood, and contents are poured into a sterile glass graduated volumetric cylinder.
5. After 20 min of settling, a sterile pipette is used to remove liquid until 97 mL of mixture remains in the cylinder.
6. The mixture is placed back into the flask and put on ice.
7. Filter-sterilized DMSO (3.5 mL) is added per 30 mL of cells in three aliquots (1.2, 1.2, and 1.1 mL) over 30 min, swirling the flask after each DMSO addition.
8. 1 mL of prepared cells is dispensed into each vial on ice.
9. Vials are transferred to a freezing chamber and cooled to −35°C at 0.33°C/min.
10. Vials are then placed in a storage box and submerged in liquid N_2 in a storage chamber for long-term storage.
11. To retrieve cultures, the same procedure is used as indicated above. After 1–3 weeks, visible embryogenic tissue colonies begin to form. Resulting tissue can be grown on medium 1250 or liquid medium 1133.

4. Notes

1. Use of a staging system is critical for the initiation process to help understand variation in zygotic embryo development due to mother tree location, time of year, and genotype. A clear staging system becomes important again to help monitor somatic embryo development. We use the embryo staging system of Pullman and Webb (100) to evaluate morphological development in zygotic and somatic embryos (Fig. 2). Stages 1–8 are based on changes in embryo morphology while stage 9 zygotic embryos are categorized by the week they are collected; e.g., 9.1 (stage 9, week 1), 9.2 (stage 9, week 2). This system is based on the literature concerning embryology within the pine family and is detailed to enable close scrutiny of embryo development. We use the system to classify both zygotic and somatic embryos to compare early, mid-, and late-stage development.
2. Brassinosteroids of reliable quality are difficult to purchase. An excellent company, CIDtech Research Inc (Mississauga, Ontario, Canada), provided high-quality brassinosteroids for many years but disappeared in the early 2000s. High-quality brassinosteroids can be purchased from Sigma-Aldrich (St. Louis, MO) or Duchefa (Haarlem, The Netherlands).
3. It is important to keep clumps of embryogenic tissue small to increase surface area where new tissue grows most rapidly. Old brown and dying embryogenic tissue in the center of larger clumps should be removed along with nonembryogenic tissue forming callus (often hard and/or green). This selection process is important to maintain embryogenic tissue as the culture ages.

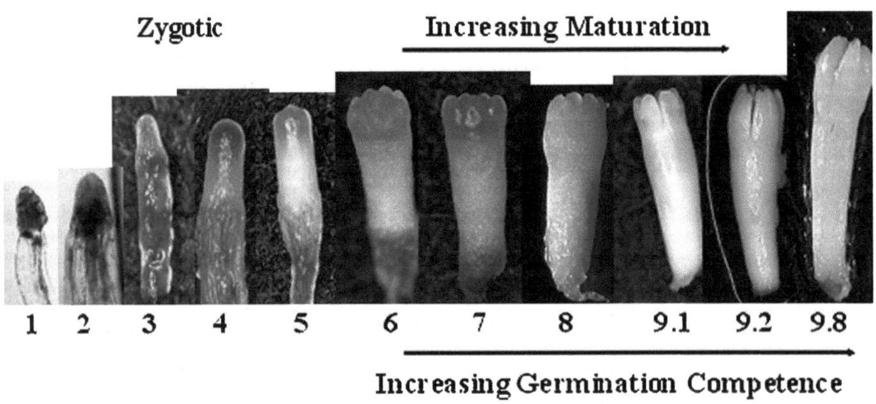

Fig. 2. Stages of zygotic embryos in loblolly pine. Stages 2–4 are optimal for loblolly pine embryogenic tissue initiation. Reproduced from Cairney and Pullman (65) with permission from Wiley-Blackwell.

4. In some cases, 5–6 weeks of cone storage at 4–5°C can significantly increase initiation (Pullman, unpublished). In several *Abies* species employing a cold treatment increased initiation (9). *Pinus taeda* cones collected when zygotic embryos are at stages 1–2 can be stored in the cold for up to 7–9 weeks. After 6–7 weeks, a browning, indicating microbial colonization, will be noted on the ovuliferous scale that begins on the cone surface and progresses toward the seed with time. When the brown coloration reaches the seed, contamination rates after sterilization will rapidly rise.
5. When late precotyledonary zygotic embryos (stages 4–5 (100)) are desired for initiation, a time- and labor-saving tetrazolium chloride embryo staining technique may be used to evaluate seed health and screen out seed sources with high proportions of dead embryos (102). Zygotic embryos earlier that stage 4 do not stain well.
6. Acceptable low-cost support materials for the maturation and cryostorage retrieval steps have been difficult to find. Early in our program, we used black filter paper (Ahlstrom Filtration, no. 8613-0425) placed on maturation or cryostorage medium (61). In 1999/2000, the filter paper manufacturing process changed, and the product dyes caused pH decreases that were detrimental to embryo growth and survival. Later, black cotton (100%) fabric (Beechwood Country Class Solid 6785 (61)) was substituted for filter paper, but embryo growth was still not optimal. Most recently, a low-cost nylon mesh, similar to that used for high-quality tea bags, has been found to work very well. Monofilament and spun nylon fabrics can be purchased from Decotex Inc., Pawling, New York. We currently use fabric 06400JP-72 with 162 × 162 fibers/in. and openings about 60 μm in size. We selected this fabric based on cost, tissue growth, and ease of cutting on a paper cutter. When last purchased (January 2007), the cost was $4.86/square yard.

References

1. Gernandt DS, Geada López G, Ortiz García S, Liston A (2005) Phylogeny and classification of *Pinus*. Taxon 54:29–42
2. Shultz RP (1999) Loblolly – the pine for the twenty-first century. New For 17:71–88
3. Tautorus TE, Fowke LC, Dunstan DI (1991) Somatic embryogenesis in conifers. Can J Bot 69:1873–1899
4. Attree SM, Fowke LC (1993) Embryogeny of gymnosperms: advances in synthetic seed technology of conifers. Plant Cell Tissue Organ Cult 35:1–35
5. Gupta PK, Grob JA (1995) Somatic embryogenesis in conifers. In: Jain S, Gupta P, Newton R (eds) Somatic embryogenesis in woody plants, vol 1. Kluwer Academic Publishers, Dordrecht, pp 81–98
6. von Aderkas P, Bonga JM (2000) Influencing micropropagation and somatic embryogenesis in mature trees by manipulation of phase change, stress and culture environment. Tree Physiol 20:921–928
7. von Arnold S, Sabala I, Bozhkov P, Dyachok J, Filonova L (2002) Developmental pathways of somatic embryogenesis. Plant Cell Tissue Organ Cult 69:233–249
8. Stasolla C, Kong L, Yeung EC, Thorpe TA (2002) Maturation of somatic embryos in

conifers: morphogenesis, physiology, biochemistry and molecular biology. In Vitro Cell Dev Biol Plant 38:93–105
9. Zoglauer K, Behrendt U, Rahmat A, Ross H, Taryono (2003) Somatic embryogenesis – the gate to biotechnology in conifers. In: Laimer M, Rücker W (eds) Plant tissue culture, 100 years since Gottlieb Haberlandt. Springer Wien, New York, pp 175–202
10. Stasolla C, Yeung EC (2003) Recent advances in conifer somatic embryogenesis: improving somatic embryo quality. Plant Cell Tissue Organ Cult 74:15–35
11. Jimenez VM (2005) Involvement of plant hormones and plant growth regulators on in vitro somatic embryogenesis. Plant Growth Regul 47:91–100
12. Haggman H, Vuosku J, Sarjala T, Jokela A, Niemi K (2005) Somatic embryogenesis of pine species: from functional genomics to plantation forestry. In: Mujib A, Samaj J (eds) Plant cell monograph, vol 2. Springer, Berlin, pp 119–140
13. Finer JJ, Kriebel HB, Becwar MR (1989) Initiation of embryogenic callus and suspension cultures of eastern white pine (*Pinus strobus* L.). Plant Cell Rep 8:203–206
14. Klimaszewska K, Smith DR (1997) Maturation of somatic embryos of *Pinus strobus* is promoted by a high concentration of gellan gum. Physiol Plant 100:949–957
15. Bercetche J, Paques M (1995) Somatic embryogenesis in maritime pine (*Pinus pinaster*). In: Jain S, Gupta P, Newton R (eds) Somatic embryogenesis in woody plants, vol 3. Kluwer Academic Publishers, Dordrecht, pp 221–242
16. Lelu MA, Bastien C, Drugeault A, Gouez ML, Klimaszewska K (1999) Somatic embryogenesis and plantlet development in *Pinus sylvestris* and *Pinus pinaster* on medium with and without growth regulators. Physiol Plant 105:719–728
17. Miguel C, Goncalves S, Tereso S, Marum L, Maroco J, Oliveria MM (2004) Somatic embryogenesis from 20 open-pollinated families of Portuguese plus trees of maritime pine. Plant Cell Tissue Organ Cult 76:121–130
18. Keinonen-Mettala K, Jalonen P, Eurola P, von Arnold S, von Weissenberg K (1996) Somatic embryogenesis of *Pinus sylvestris*. Scand J For Res 11:242–250
19. Haggman H, Jokela A, Krajnakova J, Kauppi A, Niemi K, Aronen T (1999) Somatic embryogenesis of Scots pine: cold treatment and characteristics of explants affecting induction. J Exp Bot 50:1769–1778
20. Laine E, David A (1990) Somatic embryogenesis in immature embryos and protoplasts of *Pinus caribaea*. Plant Sci 69:215–224
21. Becwar MR, Nagmani R, Wann SR (1990) Initiation of embryogenic cultures and somatic embryo development in loblolly pine (*Pinus taeda*). Can J For Res 20:810–817
22. Pullman GS, Johnson S (2002) Somatic embryogenesis in loblolly pine (*Pinus taeda* L.): improving culture initiation rates. Ann For Sci 59:663–668
23. Percy RE, Klimaszewska K, Cyr DR (2000) Evaluation of somatic embryogenesis for clonal propagation of western white pine. Can J For Res 30:1867–1876
24. Malabadi RB, Nataraja K (2007) Putrescine influences somatic embryogenesis and plant regeneration in *Pinus geradiana* Wall. Am J Plant Physiol 2:107–114
25. Malabadi RB, Choudhury H, Tandon P (2002) Plant regeneration via somatic embryogenesis in *Pinus kesiya* (Royle ex. Grod.). Appl Biol Res 4:1–10
26. Bozhkov PV, Ahn IS, Park YG (1997) Two alternative pathways of somatic embryo origin from polyembryonic mature stored seeds of *Pinus koraiensis* Sieb et Zucc. Can J Bot 75:509–512
27. Gupta PK, Durzan DJ (1986) Somatic polyembryogenesis from callus of mature sugar pine embryos. Biotechnology 4:643–645
28. Huang J-Q, Wei Z-M, Xu Z-H (1995) Somatic embryogenesis and plantlet regeneration from callus of mature zygotic embryos of masson pine. Acta Bot Sin 37:289–294
29. Radojevic L, Alvarez C, Fraga MF, Rodriguez R (1999) Somatic embryogenic tissue establishment from mature *Pinus nigra* Arn. Spp. Salzmannii embryos. In Vitro Cell Dev Biol Plant 35:206–209
30. Garin E, Isabel N, Plourde A (1998) Screening of large numbers of seed families of *Pinus strobus* L. for somatic embryogenesis from immature and mature zygotic embryos. Plant Cell Rep 18:37–43
31. Tang W, Guo Z, Ouyang F (2001) Plant regeneration from embryogenic cultures initiated from mature loblolly pine zygotic embryos. In Vitro Cell Dev Biol Plant 37:558–563
32. Malabadi RB, Nataraja K (2007) 24-Epibrassinolide induces somatic embryogenesis in *Pinus wallichiana* A. B. Jacks. J Plant Sci 2:171–178
33. Malabadi RB, Choudhury H, Tandon P (2004) Initiation, maintenance and maturation of somatic embryos from thin apical dome sections in *Pinus kesiya* (Royle ex.

Gord) promoted by partial desiccation and gellan gum. Sci Hort 102:449–459

34. Aronen T, Pehkonen T, Malabadi R, Ryynanen L (2008) Somatic embryogenesis of Scots pine – advances in pine tissue culture at Metla. In: Proceedings of the Nordic meeting on vegetative propagation of conifers for enhancing landscaping and tree breeding, Punkaharju, Finland, pp 68–71, 10–11 Sept 2008
35. Malabadi RB, van Staden J (2005) Somatic embryogenesis from vegetative shoot apices of mature trees of *Pinus patula*. Tree Physiol 25:11–16
36. Malabadi RB, Nataraja K (2006) Cryopreservation and plant regeneration via somatic embryogenesis using shoot apical domes of mature *Pinus roxburghii* Sarg. trees. In Vitro Cell Biol Plant 42:152–159
37. Becwar MR, Pullman GS (1995) Somatic embryogenesis in loblolly pine (*Pinus taeda* L.). In: Mohan JS, Gupta PK, Newton RJ (eds) Somatic embryogenesis in woody plants, Vol 3 – Gymnosperms. Kluwer Academic Publishers, Dordrecht, pp 287–301
38. Pullman GS, Johnson S, Van Tassel S, Zhang Y (2005) Somatic embryogenesis in loblolly pine (*Pinus taeda* L.) and Douglas fir (*Pseudotsuga menziesii*): Improving culture initiation with MES pH buffer, biotin, and folic acid. Plant Cell Tissue Organ Cult 80:91–103
39. Pullman GS, Chopra R, Chase KM (2006) Loblolly pine (*Pinus taeda* L.) somatic embryogenesis: improvements in embryogenic tissue initiation by supplementation of medium with organic acids, Vitamins B_{12} and E. Plant Sci 170:648–658
40. Handley L III (1997) Method for regeneration of coniferous plants by somatic embryogenesis in culture media containing abscisic acid. US Patent 5,677,185, 14 Oct 1997
41. Handley L III (1999) Method for regeneration of coniferous plants by somatic embryogenesis in culture media containing abscisic acid. US Patent 5,856,191, 5 Jan 1999
42. Pullman GS, Namjoshi K, Zhang Y (2003) Somatic embryogenesis in loblolly pine (*Pinus taeda* L.): improving culture initiation with abscisic acid, silver nitrate, and cytokinin adjustments. Plant Cell Rep 22:85–95
43. Pullman GS, Zhang Y, Phan B (2003) Brassinolide improves embryogenic tissue initiation in conifers and rice. Plant Cell Rep 22:96–104
44. Malabadi RB, Nataraja K (2007) Plant regeneration via somatic embryogenesis using secondary needles of mature trees of *Pinus roxburghii* Sarg. Int J Bot 3:40–47
45. Pullman GS, Montello P, Cairney J, Xu N, Feng X (2003) Loblolly pine (*Pinus taeda* L.) somatic embryogenesis: maturation improvements by metal analyses of zygotic and somatic embryos. Plant Sci 164:955–969
46. Nagmani R, Diner AM, Sharma GC (1993) Somatic embryogenesis in longleaf pine (*Pinus palustris*). Can J For Res 23: 873–876
47. Denchev P, Attree SM, Kong L, Tsai CJ, Radley RA, Lobatcheva II (2004) Method for reproducing conifers by somatic embryogenesis using galactose containing compounds as a carbon and energy source. US Patent 20040203150A1, 14 Oct 2004 (patent pending)
48. Gupta PK, Holmstrom DG, Budworth D (2004) Embryogenic culture initiation of Douglas-fir by maltose. US Patent 20040237130, 25 Nov 2004 (patent pending)
49. Salajova T, Salaj J (2005) Somatic embryogenesis in *Pinus nigra*: embryogenic tissue initiation, maturation and regeneration ability of established cell lines. Biol Plant 49:333–339
50. Steiner N, Vieira FDN, Maldonado S, Guerra MP (2005) Effect of carbon source on morphology and histodifferentiation of *Araucaria angustifolia* embryogenic cultures. Braz Arch Biol Technol 48:895–903
51. Pullman GS, Chase KM, Skryabina A, Bucalo K (2008) Conifer embryogenic tissue initiation: improvements by supplementation of medium with d-*chiro*-inositol and d-xylose. Tree Physiol 29:147–156
52. Pullman GS, Mein J, Johnson S, Zhang Y (2005) Gibberellin inhibitors improve embryogenic tissue initiation in conifers. Plant Cell Rep 23:596–605
53. Chandler SF, Young R (1995) Somatic embryogenesis in *Pinus radiata* Don. In: Jain S, Gupta P, Newton R (eds) Somatic embryogenesis in woody plants, vol 3. Kluwer Academic Publishers, Dordrecht, pp 243–255
54. Pullman GS (1997) Osmotic measurements of whole ovules during loblolly pine embryo development. In: TAPPI biological sciences symposium, San Francisco, CA, pp 41–48, 19–23 Oct 1997
55. Pullman GS, Johnson S (2009) Osmotic measurements in whole megagametophytes and embryos of loblolly pine (*Pinus taeda* L.) during embryo and seed development. Tree Physiology 29:819–827
56. Van Winkle SC, Pullman GS (2003) The combined impact of pH and activated carbon

on the elemental composition of plant tissue culture media. Plant Cell Rep 22:303–311

57. Pullman GS, Johnson S (2009) Loblolly pine (*Pinus taeda* L.) female gametophyte and embryo pH changes during embryo and seed development. Tree Physiology 29:829–836
58. Xu N, Johns B, Pullman GS, Cairney J (1997) Rapid and reliable differential display from minute amounts of tissue: mass cloning and characterization of differentially expressed genes from loblolly pine embryos. Plant Mol Biol Rep 15:377–391
59. Cairney J, Xu N, MacKay J, Pullman G (2000) Transcript profiling: a tool to assess the development of conifer embryos. In Vitro Cell Dev Biol Plant 36:155–162
60. Pullman GS, Peter G (2002) Methods of initiating embryogenic cultures in plants. US Patent 6,492,174B1, 8 Oct 2002
61. Pullman GS, Johnson S, Peter G, Cairney J, Xu N (2003) Improving loblolly pine somatic embryo maturation: comparison of somatic and zygotic embryo morphology, germination, and gene expression. Plant Cell Rep 21:747–758
62. Van Winkle S, Johnson S, Pullman GS (2003) The impact of Gelrite and activated carbon on the elemental composition of plant tissue culture media. Plant Cell Rep 21: 1175–1182
63. Pullman GS, Peter G (2006) Methods for increasing conifer somatic embryo initiation, capture, and multiplication. US Patent 20060051868, 9 March 2006
64. Pullman GS, Skryabina A (2007) Liquid medium and liquid overlays improve embryogenic tissue initiation in conifers. Plant Cell Rep 26:873–887
65. Cairney J, Pullman GS (2007) The cellular and molecular biology of conifer embryogenesis. New Phytol 176:511–536
66. Klimaszewska K, Trontin JF, Becwar MR, Devillard C, Park YS, Lelu-Walter MA (2007) Recent progress in somatic embryogenesis of four *Pinus* spp. Tree For Sci Biotechnol 1:11–25
67. Vales T, Fang X, Ge L, Xu N, Cairney J, Pullman GS, Peter GF (2007) Improved somatic embryo maturation in loblolly pine by monitoring ABA-responsive gene expression. Plant Cell Rep 26:133–143
68. Pullman GS, Gupta PK (1991) Method for reproducing coniferous plants by somatic embryogenesis using adsorbent materials in the development stage. US Patent 5034326, 23 July 1991
69. Gupta PK, Pullman GS (1991) Method for reproducing coniferous plants by somatic embryogenesis using abscisic acid and osmotic potential variation. US Patent 5036007, 30 July 1991
70. Pullman GS, Gupta PK, Timmis R, Carpenter C, Kreitinger M, Welty E (2005) Improved Norway spruce somatic embryo development through the use of abscisic acid combined with activated carbon. Plant Cell Rep 24:271–279
71. Lelu-Walter MA, Bernier-Cardou M, Klimaszewska K (2006) Simplified and improved somatic embryogenesis for clonal propagation of *Pinus pinaster* (Ait.). Plant Cell Rep 25:767–776
72. Bradford KJ (1994) Water stress and the water relations of seed development: a critical review. Crop Sci 34:1–11
73. Gates JC, Greenwood MS (1991) The physical and chemical environment of the developing embryo of *Pinus resinosa*. Am J Bot 78:1002–1009
74. Dumont-BeBoux N, Mazari A, Livingston NJ, von Aderkas P, Becwar MR, Percy RE, Pond SE (1996) Water relations parameters and tissue development in somatic and zygotic embryos of three pinaceous conifers. Am J Bot 83:992–996
75. Klimaszewska K, Bernier-Cardou M, Cyr DR, Sutton BCS (2000) Influence of gelling agents on culture medium gel strength, water availability, tissue water potential, and maturation response in embryogenic cultures of *Pinus strobus* L. In Vitro Cell Dev Biol Plant 36:279–286
76. Klimaszewska KK, Park YS, Overton C, MacEacheron I, Bonga JM (2001) Optimized somatic embryogenesis in *Pinus strobus* L. In Vitro Cell Dev Biol Plant 37:392–399
77. Belmonte MF, Macey J, Yeung EC, Stasolla C (2005) The effect of osmoticum on ascorbate and glutathione metabolism during white spruce (*Picea glauca*) somatic embryo development. Plant Physiol Biochem 43:337–346
78. Kreuger M, van Holst GJ (1993) Arabinogalactan proteins are essential in somatic embryogenesis of *Dacus carota* L. Planta 189:243–248
79. Showalter AM (2001) Arabinogalactan-proteins: structure, expression and function. Cell Mol Life Sci 58:1399–1417
80. von Arnold S, Bozhkov P, Clapham D, Dyachok J, Filonova L, Högberg KA, Ingouff M, Wiweger M (2005) Propagation of Norway spruce via somatic embryogenesis. Plant Cell Tissue Organ Cult 81:323–329
81. Kong L, Attree SM, Evans DE, Binarova P, Yeung EC, Fowke LC (1999) Somatic embryogenesis in white spruce: studies of

embryo development and cell biology. In: Mohan JS, Gupta PK, Newton RJ (eds) Somatic embryogenesis in woody plants, vol 4. Kluwer Academic Publishers, Dordrecht, pp 1–28

82. Kong L, Yeung EC (1994) Effects of ethylene and ethylene inhibitors on white spruce somatic embryo maturation. Plant Sci 104:71–80
83. Kong L, Yeung EC (1995) Effects of silver nitrate and poly ethylene glycol on white spruce (Picea glauca) somatic embryo development: enhancing cotyledonary embryo formation and endogenous ABA content. Physiol Plant 93:298–304
84. El Meskaoui A, Desjardins Y, Tremblay FM (2000) Kinetics of ethylene biosynthesis and its effects during maturation of white spruce somatic embryos. Physiol Plant 109:333–342
85. Jones NB, van Staden J (1995) Plantlet production from somatic embryos of *Pinus patula*. J Plant Physiol 145:519–525
86. Timmis R (1998) Bioprocessing for tree production in the forest industry: conifer somatic embryogenesis. Biotechnol Prog 14:156–166
87. Carlson WC, Hartle JE (1995) Manufactured seed of woody plants. In: Jain S, Gupta P, Newton R (eds) Somatic embryogenesis in woody plants, vol 3. Kluwer Academic Publishers, Dordrecht, pp 253–263
88. Hay EI, Charest PJ (1999) Somatic embryo germination and desiccation tolerance in conifers. In: Mohan JS, Gupta PK, Newton RJ (eds) Somatic embryogenesis in woody plants, vol 4. Kluwer Academic Publishers, Dordrecht, pp 61–69
89. Gupta PK, Holmstrom DG, Larson B, Zucati J (2005) Development and stratification of pine somatic embryos using a liquid system. US Patent 20050026281, 3 Feb 2005
90. Merkle SA, Montello PM, Xia X, Upchurch BL, Smith DR (2006) Light quality treatments enhance somatic seedling production in three southern pine species. Tree Physiol 26:187–194
91. Niemi K, Haggman H (2002) *Pisolithus tinctorius* promotes germination and forms mycorrhizal structures in Scots pine somatic embryos in vitro. Mycorrhiza 12:263–267
92. Stasolla C, Yeung EC (1999) Ascorbic acid improves conversion of white spruce somatic embryos. In Vitro Cell Dev Biol Plant 35:316–319
93. Kartha KK, Fowke LC, Leung NL, Caswell KL, Hakman I (1988) Induction of somatic embryos and plantlets from cryopreserved cell cultures of white spruce (*Picea abies*). J Plant Physiol 132:529–539
94. Gupta PK, Durzan DJ (1987) Biotechnology of somatic polyembryogenesis and plantlet regeneration in loblolly pine. Biotechnology 5:147–151
95. Kapik RH, Dinus RJ, Dean JF (1995) Abscisic acid and zygotic embryogenesis in *Pinus taeda*. Tree Physiol 15:405–409
96. Pullman GS, Buchanan M (2006) Identification and quantitative analysis of stage-specific organic acids in loblolly pine (*Pinus taeda* L.) zygotic embryo and female gametophyte. Plant Sci 170:634–647
97. Pullman GS, Buchanan M (2008) Identification and quantitative analysis of stage-specific carbohydrates in loblolly pine (*Pinus taeda*) zygotic embryo and female gametophyte tissues. Tree Physiol 28:985–996
98. Carman JG, Reese G, Fuller RJ, Ghermay J, Timmis R (2005) Nutrient and hormone levels in Douglas-fir corrosion cavities, megagametophytes, and embryos during embryony. Can J For Res 35:2447–2456
99. Pullman GS, Cairney J, Peter G (1998) Clonal forestry and genetic engineering: where we stand, future prospects, and potential impacts on mill operations. TAPPI J 81:57–64
100. Pullman GS, Webb DT (1994) An embryo staging system for comparison of zygotic and somatic embryo development. In: TAPPI R&D division biological sciences symposium, Minneapolis, Minnesota, pp 31–34, 3–6 Oct 1994
101. Murashige T, Skoog F (1962) A revised medium for rapid growth and bioassays with tobacco tissue cultures. Physiol Plant 15:473–497
102. Pullman GS, Johnson S, Bucalo K (2009) Douglas fir embryogenic tissue initiation. Plant Cell Tissue Organ Cult 96:75–84
103. Maruyama E, Hosoi Y, Ishii K (2007) Somatic embryogenesis and plant regeneration in yakutanegoyou, *Pinus armandii* Franch. var. *amamiana* (Koidz.) Hatusima, an endemic and endangered species in Japan. In Vitro Cell Dev Biol Plant 43:28–34
104. Park YS, Bonga JM, Cameron SI, Barrett JD, Forbes K, DeVerno LL, Klimaszewska K (1999) Somatic embryogenesis in Jack pine (*Pinus banksiana* Lamb). In: Jain S, Gupta P, Newton R (eds) Somatic embryogenesis in woody plants, vol 4. Kluwer Academic Publishers, Dordrecht, pp 491–504
105. Yildrim T, Kaya Z, Isik K (2006) Induction of embryogenic tissue and maturation of

somatic embryos in *Pinus brutia* TEN. Plant Cell Tissue Organ Cult 87:67–76

106. Zang C-X, Li Q, Kong L (2007) Induction, development and maturation of somatic embryos in Bunge's pine (*Pinus bungeana* Zucc. Ex Endl.). Plant Cell Tissue Organ Cult 91:273–280
107. Laine E, David A (1992) Recovery of plants from cryopreserved embryogenic cell suspensions of *Pinus caribaea*. Plant Cell Rep 11:295–298
108. David A, Laine E, David H (1995) Somatic embryogenesis in *Pinus caribaea*. In: Jain S, Gupta P, Newton R (eds) Somatic embryogenesis in woody plants, vol 3. Kluwer Academic Publishers, Dordrecht, pp 145–181
109. Taniguchi T (2001) Plant regeneration from somatic embryos in *Pinus thunbergii* (Japanese black pine) and *Pinus densiflora* (Japanese red pine). In: Morohoshi N, Komamine A (eds) Molecular breeding of woody plants. Elsevier, Amsterdam, pp 319–324
110. Ishii K, Maruyama K, Hosoi Y (2001) Somatic embryogenesis of Japanese conifers. In: Morohoshi N, Komamine A (eds) Molecular breeding of woody plants. Elsevier science, Amsterdam, pp 297–304
111. Maruyama E, Hosoi Y, Ishii K (2005) Propagation of Japanese red pine (*Pinus densiflora* Zieb.et.Zucc.) via somatic embryogenesis. Prop Ornam Plants 5:199–204
112. Shoji M, Sato H, Nakagawa R, Funada R, Kubo T, Ogita S (2006) Influence of osmotic pressure on somatic embryo maturation in *Pinus densiflora*. J For Res 11:449–453
113. Jain M, Dong N, Newton RJ (1989) Somatic embryogenesis in slash pine (*Pinus elliottii*) from immature embryos cultured in vitro. Plant Sci 65:233–241
114. Newton RJ, Marek-Swize KE, Magallanes-Cedeno ME, Dong N, Sen S, Jain SM (1995) Somatic embryogenesis in slash pine (*Pinus elliottii* Engelm.). In: Jain S, Gupta P, Newton R (eds) Somatic embryogenesis of woody plants, vol 3. Kluwer Academic Publishers, Dordrecht, pp 183–195
115. Newton RJ, Tang W, Jain SM (2005) Slash pine (*Pinus elliottii* Engelm.). In: Jain S, Gupta P (eds) Protocol for somatic embryogenesis in woody plants. Springer, Dordrecht, pp 1–10
116. Stojicic D, Uzelac B, Janosevic D, Culafic L, Budimir S (2007) Induction of somatic embryogenesis in *Pinus heldreichii* culture. Arch Biol Sci Belgrade 59:199–202
117. Choudhury H, Kumaria S, Tamon P (2008) Induction and maturation of somatic embryos from intact megagametophyte explants in Khasi pine (*Pinus kesiya* Royle ex. Gord.). Curr Sci 95:1433–1438
118. Gupta PK (1995) Somatic embryogenesis in sugar pine (*Pinus lambertiana*). In: Jain S, Gupta P, Newton R (eds) Somatic embryogenesis in woody plants, vol 3. Kluwer Academic Publishers, Dordrecht, pp 197–205
119. Salajova T, Salaj J (1992) Somatic embryogenesis in European black pine (*Pinus nigra* Arn.). Biol Plant 34:213–218
120. Salaj T, Moravcikova J, Salaj J (2006) Somatic embryogenesis in *Pinus nigra* Arn: some physiological, structural and molecular aspects. In: Mujib A, Samaj J (eds) Plant cell monograph, vol 2. Springer, Berlin, pp 141–156
121. Jones NB, van Staden J, Bayley AD (1993) Somatic embryogenesis in *Pinus patula*. J Plant Physiol 142:366–372
122. Ford CS, Jones NB, van Staden J (2000) Cryopreservation and plant regeneration from somatic embryos of *Pinus patula*. Plant Cell Rep 19:610–615
123. Ford CS, Fisher LJ, Jones NB, Nigro SA, Makunga NP, van Staden J (2005) Somatic embryogenesis in *Pinus patula*. In: Jain S, Gupta P (eds) Protocol for somatic embryogenesis in woody plants. Springer, Dordrecht, pp 121–139
124. Smith DR, Singh AP, Wilton L (1985) Zygotic embryos of Pinus radiata in vivo and in vitro, in Proceedings of the international workshop on conifer tissue culture. In: Abstracts of the 3rd meeting International conifer tissue culture work group, Rotorua, pp 12–16
125. Aquea F, Poupin MJ, Matus JT, Gebauer M, Medina C, Arce-Johnson P (2008) Synthetic seed production from somatic embryos of *Pinus radiata*. Biotechnol Lett 30:1847–1852
126. Hargreaves C, Smith DR (1992) Cryopreservation of *Pinus radiata* embryogenic tissue. Int Plant Propag Soc Combin Proc 42:327–333
127. Hargreaves CL, Grace LJ, Holden DG (2002) Nurse culture for efficient recovery of cryopreserved *Pinus radiata* D. Don embryogenic cell lines. Plant Cell Rep 21:40–45
128. Kim YW, Moon HK (2007) Regeneration of plant by somatic embryogenesis in *Pinus rigida* x *P. taeda*. In Vitro Cell Dev Biol Plant 43:335–342

129. Arya S, Kalia RK, Arya ID (2000) Induction of somatic embryogenesis in *Pinus roxburghii* Sarg. Plant Cell Rep 19:775–780
130. Mathur G, Alkutkar VA, Nadgauda RS (2003) Cryopreservation of embryogenic culture of *Pinus roxburghii*. Biol Plant 46:205–210
131. Becwar MR, Wann SR, Johnson MA, Verhagen SA, Feirer RP, Nagmani R (1988) Development and characterization of in vitro embryogenic systems in conifers. In: Ahuja MR (ed) Somatic cell genetics of woody plant. Kluwer Academic Publishers, Dordrecht, pp 1–18
132. Garin E, Bernier-Cardou M, Isabel N, Klimaszewska K, Plourde A (2000) Effect of sugars, amino acids, and culture technique on maturation of somatic embryos of *Pinus strobus* on medium with two gellan gum concentrations. Plant Cell Tissue Organ Cult 62:27–37
133. Haggman H, Ryynanen LA, Aronen TS, Krajnakova J (1998) Cryopreservation of embryogenic cultures of Scots pine. Plant Cell Tissue Organ Cult 54:45–53
134. Niemi K, Sarjala T, Chen X, Haggman H (2007) Spermidine and the ectomycorrhizal fungus *Psolithus tinctorius* synergistically induce maturation of Scots pine embryogenic cultures. J Plant Physiol 164: 629–635
135. Becwar M, Clark J, Chowdhury K, Leshaun L, Nehra N, Victor J (2008) A liquid-based method for producing plant somatic embryos from proliferative cells of conifers. PCT International Application, WO 2008030423 A2 20080313 CAN 148:302902 AN, 13 March 2008
136. Becwar MR, Krueger SA (2004) Recovering cryopreserved conifer embryogenic cultures. US Patent 6,682,931, 27 Jan 2004
137. Maruyama E, Hosoi Y, Ishii K (2005) Somatic embryo production and plant regeneration of Japanese black pine (*Pinus thunbergii*). J For Res 10:403–407
138. Becwar M, Chesick E, Handley L III, Rutter M (1995) Method for regeneration of coniferous plants by somatic embryogenesis. US Patent 5,413,930, 9 May 1995
139. Pullman GS, Buchanan M (2003) Loblolly pine (Pinus taeda L). Stage-specific elemental analyses of zygotic embryo and female gametophyte tissue. Plant Sci 164:943–954

Chapter 20

Micropropagation of *Phalaenopsis* Orchids via Protocorms and Protocorm-Like Bodies

Kee Yoeup Paek, Eun Joo Hahn, and So Young Park

Abstract

Phalaenopsis orchids have high economic value in the floriculture industry. Hybridization or cross-pollination in the breeding program have proven to be very reliable techniques for the production of a wide range of successful cultivars with attractive combinations of spray length, bud number, flower color and type, fragrance, seasonality, and compactness. In vitro propagation makes it possible to clonally mass propagate hybrids of commercial value and conserved species. However, in vitro culture technologies are still a challenge because of the slow growth of plantlets, low multiplication rate, poor rooting, and somaclonal variation. Although seed-raised plants can be used for conservation and breeding for the selection of superior features, genetic characteristics including seasonality, inflorescence, flower color, and type are not uniform. In this regard, micropropagation through protocorm-like bodies obtained from germinating embryos and somatic tissues is an important strategy in obtaining genetically stable plants and the improvement of quality. However, not all genotypes of *Phalaenopsis* respond to the same protocol under the same culture conditions and often result in the development of undesirable characteristics. In this chapter, plantlet production in *Phalaenopsis* orchids via the culture of protocorms from seeds and protocorm-like bodies from leaf sections and root tips are detailed.

Key words: Asymbiotic germination, Leaf segment, Micropropagation, Protocorm, Protocorm-Like body, *Phalaenopsis*, Somaclonal variation

1. Introduction

For centuries, orchids have held a fascination for people; there is an air of mystery surrounding them such as sizes, shapes, colors, and fragrances. With an estimated 750 different genera with at least 25,000–30,000 species worldwide, it is probably the largest family (Orchidaceae) in the plant kingdom, larger even than the sunflower family (Asteraceae). The number of bi- and pluri-generic hybrids registered is fast approaching over 100,000 due

Trevor A. Thorpe and Edward C. Yeung (eds.), *Plant Embryo Culture: Methods and Protocols*, Methods in Molecular Biology, vol. 710, DOI 10.1007/978-1-61737-988-8_20,

to the high degree of compatibility among genera and species, which has increased public awareness and stimulated an exciting industry of hybridization.

Many of these hybrids are important commercial plants used as cut flowers and potted plants to satisfy a strong demand, both locally and internationally. Over the years, it has evolved from a hobbyist's market into a highly commercial market and the potential remains high for further growth. Large-scale cultivation of orchid cut flowers and potted orchids is now the trend of the horticultural industry. Mass cultivation became possible with a breakthrough in orchid seed germination. This laid the foundation for intensive breeding and selection of new commercial orchid hybrids. The discovery and the development of asymbiotic germination of orchid seeds by Knudson (1, 2) have also paved the way for the development of plant tissue culture techniques for micropropagation of orchids. The availability of asymbiotic seed germination and clonal propagation in vitro has also made commercial-scale orchid cultivation possible and created an economically feasible market (3).

Since orchid seeds are minute and dust-like, a single fruit or capsule has the potential to produce up to a million seeds, depending on the species. However, orchid seeds have little food storage and rely on a specific type of fungus from orchid mycorrhizae for their germination and development. Orchid seeds can be germinated in vitro using relatively simple sugar-containing nutrient media, since the survival rate in the wild habitat is very low. Generally, orchid seed germination in vitro is not a difficult or complex procedure, but it does require the acquisition of certain skills and knowledge. Many of the terrestrial native orchid species, especially those from temperate latitudes, do not germinate well on media used for tropical orchids but can be germinated in the presence of a symbiotic fungus or aseptically on more complex media using either mature seeds or ovules gathered from green pods. For example, there exists two distinctly different organogenic pathways between terrestrial and subtropical or tropical *Cymbidium* in vitro. Organogenesis of *Cymbidium* from an asymbiotic seed or shoot-tip culture can be attained by way of protocorms (the small spherical tuber-like bodies formed by germinating orchid seeds) or protocorm-like bodies (PLBs: the structures that resemble protocorms but are formed by tissue explants and/or callus in vitro); but in temperate *Cymbidium*, organogenesis is via the rhizome which is recalcitrant to regeneration when compared to the protocorm system (4–6). Micropropagation of orchids via protocorm systems can be widely used for the purpose of a breeding program and for the conservation of endangered native ones.

Orchid seeds differ from the majority of flowering plants by having minute embryos. Their apical meristems and cotyledons are not usually present at the time of seed dispersal and a variety of embryo developmental patterns, especially in the suspensor morphology, can be found (7–9). Since orchid seeds germinate only by fungal infection, a practical seed germination method utilizing fungi has been utilized for a long time. After Knudson (1, 2) discovered that orchid seeds could germinate on a relatively simple mineral and sugar-containing medium, asymbiotic procedures gained wide acceptance instead of symbiotic methods (10, 11). This approach also made it possible to produce magnificent orchid hybrids, especially colored *Phalaenopsis* hybrids which early growers might not have imagined.

The clonal propagation of highly heterozygous varieties of orchid became possible on a large scale following the work of Morel (12, 13), who showed that the in vitro culture of shoot tips of *Cymbidium* and other orchid genera may lead to the formation of structures like seedling protocorms. These structures sometimes produced a great number of PLBs and could be multiplied by cuttings. By repeating this process, large tissue stocks of any one clone could be obtained within a relatively short time (14, 15). If the culture is left undisturbed, many plantlets with normal shoots and roots can be regenerated from the PLBs. Somatic tissue cultures such as shoot meristem, leaf (16), root tip (17), and flower stalk (18) in orchids usually result in either direct formation of PLBs, or in the formation of callus from which protocorms are regenerated (19). As orchid PLBs represent an early embryonic stage, protocorm formation in orchids is considered either direct or indirect embryogenesis (20). Thus, the callus formed from orchid seeds or explants is considered to be embryogenic callus from which protocorms are regenerated. Plants produced via tissue culture of vegetative orchid tissues, especially meristems, are often sold as mericlones. A mericlone could therefore be defined as an orchid plant originating from vegetatively propagated PLBs derived from somatic tissue of a single mother plant. In this chapter, plantlet production in *Phalaenopsis* orchids via the culture of protocorms by asymbiotic seeds is detailed. Since PLBs produced from somatic tissues have a similar developmental pattern as protocorms, the protocol of plantlet formation from PLBs produced from leaf segments and root tips is also detailed (Fig. 1). This serves to illustrate the importance of PLBs production in in vitro orchid propagation.

1.1. Practical Considerations

In vitro cultures of orchid seeds and somatic tissues including shoot and root tip, flower stalk, and leaf segment can be complicated depending on genus, species and genotypes, so it requires certain equipment, skill, experience and knowledge. The general outline of these skills as well as media and apparatus presented in this chapter are mostly from commercial tissue culture labs and references (see Note 1). Media used for orchid tissue culture and seed germination may reflect both a special requirement for each species and the preference of the investigators who carried out the

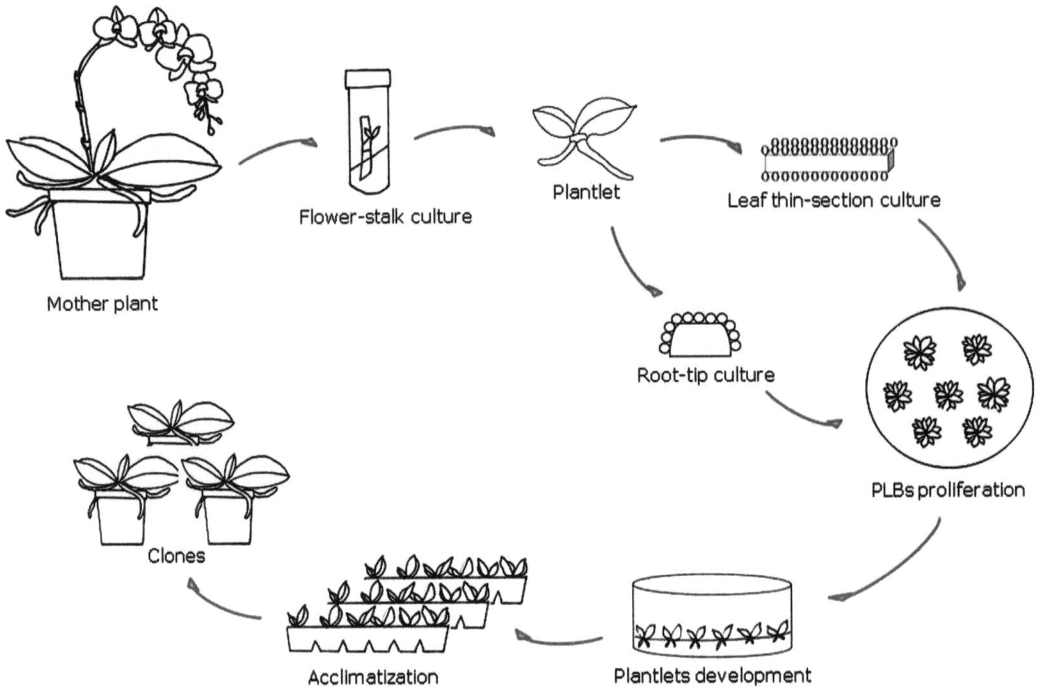

Fig. 1. Vegetative propagation of elite *Phalaenopsis* via PLB multiplication.

initial research (8). The simple media used for the germination such as Vacin and Went (VW) (21), Knudson C (2), or Hyponex (22) are generally employed for mericlone culture. More complex media such as Murashige and Skoog (MS) (23) have been described and may be required in some genera. Because there are no vitamins or amino acids in simple media, many laboratories often add undefined natural complex addenda as supplementary substances. Coconut water and homogenate of green bananas and potatoes are used in many cases. Low concentrations of auxin and cytokinin are sometimes added to the media at the initial stage of protocorm proliferation, but they are unnecessary at the later stage for the development of shoot and root from protocorms. The effects of auxins in a certain species and genotype may be different from those in other orchids. As with auxins, the choice of cytokinins and their concentration are based on species and genotypes. Sucrose is added to the media for initial culture of explants, but occasionally protocorm formation is improved in its absence. Transfer of a culture to a medium lacking sugar is essential in some genera to promote greening and plantlet differentiation such as in *Phalaenopsis* (24). Activated charcoal improves the growth of *Paphipedilum* (25) and *Phalaenopsis* (11) and led to the development of charcoal supplemented media, which gained wide acceptance in a short time. One possible explanation of the effect of charcoal on orchid seedlings or tissue culture-derived plantlets can be the improvement of aeration. The second possibility is that the charcoal absorbs ethylene (11) and phenolic inhibitors that inhibit growth and development. However, charcoal also absorbs plant growth regulators and other compounds such as vitamins in the medium and, therefore, should be used with caution. When preparing media, it is important to follow instructions carefully and strictly as given in recipes and to measure and weigh all compounds accurately (see Note 1).

Stock solutions save work because only one weighing is necessary to prepare enough concentrate of 10, 100, or even 1,000 L. Methods recommended for preparation and usage can vary considerably from one laboratory to another. Individual stock solutions should be prepared for each macroelement, vitamin, amino acid, and plant growth regulator, while all microelements should be combined into one stock solution. Since stock solutions containing nitrogen such as NO_3^-, NH_4^+, and urea tend to become contaminated with time, they must be kept in a refrigerator before use. Plant growth regulators, vitamins, and amino acids may not be stable for a prolonged period, so that it is best to prepare only small volumes of stock solutions. Ascorbic acid and glutamine, for example, are known to be broken down in a solution, even at low temperatures. Stock solutions of inositol, sugar or agar should not be made. The pH of the medium is mostly adjusted to 5.5–5.8 using NaOH, KOH, or HCl after all the components in the medium have been mixed prior to autoclaving. Selection of a relatively high pH before autoclaving is usually thought to be necessary in agar media to ensure gelation (see Note 1).

Culture media, tools, working space, and tissues must be sterilized to avoid contamination in cultures. A number of methods are used to ensure sterility by using autoclaves, filtering, microwave ovens, open flame, solvent (ethanol), and liquids (hypochlorite solution). Culture media are usually autoclaved at 121°C under a 1.05 kg/cm^2 (103.4 KPa). To become sterile, a solution is required to reach 121°C and be kept at this temperature for 15 min. The time taken for the sterilization in an autoclave increases according to the volume of the liquid in a vessel. A large number of autoclave models and sizes are available. The standard conditions for sterilization can be obtained automatically in autoclaves or in pressure cookers. Some media components are destroyed by elevated temperatures so they cannot be heat sterilized. Solutions containing these substances may be sterilized by passing them through very fine sterilizing filters (Millipore filter) which permit the passage of liquids but not particles larger than 0.22 or 0.45 μm. Disposable filter membranes are made for various pieces of apparatus and both small and large volumes of liquid can be treated with them. Heating can be used to sterilize tools and the surface of culture vessels for cultures. A natural gas burner such as Bunsen burner is the best because it produces a clean, nonsmoking, high temperature flame. If it is not available, an alcohol flame can also be used, but it may not be hot enough. Another possibility is to dip the tools in alcohol and ignite the liquid with an alcohol flame to sterilize their surfaces. Surface sterilization can be carried out with several different germicidal reagents. Clearly, the best sterilization method is cheap and non-toxic to both plants and people, and effective on a wide range of plant materials. The most commonly used materials are the hypochlorite ion and simple alcohols. As the germicide should make the best possible contact with plant materials, it is advisable to add a few drops of wetting agent to the aqueous sterilization solution. Liquid sterilants containing 4.75 or 5.25% (w/v) NaOCl (5% available chlorine) such as household bleaches (Clorox, Javex, and Purex) are simple and excellent sterilants for sterilization of work area, tools, and tissues. If used to sterilize seeds, capsules, and tissues, these bleaches should be diluted to avoid damage of culture materials. A saturated solution of calcium hypochlorite can be used to surface-sterilize tissues and seeds. This solution is prepared by dissolving 10 g calcium hypochlorite in 140 mL water, stirring vigorously, and allowing the solution to stand for 3–5 min. This is repeated until the precipitate has settled and then filtered. It should be used within 12 h (26). While calcium hypochlorite may be less effective than NaOCl in removing contaminants, it is equal in activity and less liable to induce tissue browning or injury (27), possibly due to the high concentration of calcium ions in the solution (see Note 1).

Containers used for plant tissue culture need to be translucent so that cultures may be illuminated and inspected easily. The culture vessel should become larger as the culture stage proceeds, e.g., from test tubes to Erlenmeyer flasks and to large jars. The container size and type can directly influence the multiplication rate, growth, and quality. The optimum size to use depends on the types of plant material being multiplied and has to be determined by culture stage. Glass test tubes are commonly used for the initiation of cultures because there is a danger that infection from contaminated explants can easily spread when several explants are placed together in a single vessel. The use of disposable culture vessels considerably reduces the cost of washing, but they are too expensive for most commercial tissue culture labs. PVC pots and jars manufactured for the food industry are particularly useful in terms of repeated autoclaving and cost reduction. Polypropylene and polycarbonate tubes, jars and boxes may prove to be cheaper alternatives, but they often become cloudy after repeated autoclaving, which is a disadvantage due to the reduction of light transmission (see Note 1).

2. Materials

1. Seed germination medium (see Table 1): Hyponex (N:P:K = 6.5:6:19, 2 g/L) medium supplemented with 2% (w/v) sucrose and 0.2% (w/v) Gelrite or 0.8% (w/v) Bacto agar, and adjusted to pH 5.5.
2. First transplantation (1st TP) medium (see Table 2): same as the Hyponex seed germination medium supplemented with 0.03% (w/v) activated charcoal.
3. Second transplantation (2nd TP) medium (see Table 2): same as the 1st TP medium supplemented with 30–100 g/L of green banana or unsprouted potato homogenate.

Table 1
Composition of modified MS, VW, and Knudson media

Component	MS (mg/L)	VW (mg/L)	Knudson (mg/L)
Macronutrients			
NH_4NO_3	825		
$(NH_4)_2SO_4$		500	500
$Ca_3(PO_4)_2$		200	
$Ca(NO_3)_2 \cdot 4H_2O$			1,000
$CaCl_2 \cdot H_2O$	220		
$MgSO_4 \cdot 7H_2O$	185	250	250
KNO_3	950	525	
KH_2PO_4	85	250	250
Micronutrients			
Na_2EDTA	18.65		
$FeSO_4 \cdot 7H_2O$	13.9		25
$Fe_2(C_4H_4O_6)_3 \cdot 2H_2O$		28	
H_3BO_3	3.1		
$CoCl_2 \cdot 6H_2O$	0.0125		
$CuSO_4 \cdot 5H_2O$	0.0125		
$MnSO_4 \cdot 4H_2O$	11.15	7.5	
KI	0.415		
$Na_2MoO_4 \cdot 2H_2O$	0.125		
$ZnSO_4 \cdot 4H_2O$	4.3		
Organics			
Glycine	2		
Myo-inositol	100		100
Nicotinic acid	0.5		1
Pyridoxine	0.5		
Thiamine HCl	0.1		1
Adenine sulfate	10		

From refs. (9–11)*EDTA* ethylenediamine tetraaceticacid

Table 2
Composition of modified Hyponex media

Component	Seed germination	Protocorm multiplication (PM)	First trans-planting (1st TP)	Second trans-planting (2nd TP)
Hyponex (g/L)				
N:P:K=6.5:6:19	3.0	1.0	1.0	1.0
N:P:K=20:20:20		1.0	1.0	1.0
Adenine sulfate (mg/L)	5.0			
Peptone (g/L)		2.0	2.0	3.0
Coconut water (%)	20	10	10	
Potato or banana homogenate (g/L)		30–100		30–100
Activated charcoal (%)	0.05	0.05	0.05	0.05

From refs. (17, 22, 28, 33)

4. Potting mix for seedlings: sphagnum moss or a mixture of coarse vermiculite, perlite, and peat moss (1:1:1), or small particles of bark can be used as potting media.
5. Protocorm multiplication (PM) medium (see Table 2): Hyponex (N:P:K=6.5:6:19, 1 g/L), Hyponex (N:P:K=20:20:20, 1 g/L), 2 g/L peptone, coconut water 10% (v/v), unsprouted potato homogenate 30 g/L, activated charcoal 0.05%, and 0.8% agar, and adjusted to pH 5.5.
6. Flower stalk culture medium: VW or Hyponex medium (see Tables 1 and 2), 2% (w/v) sucrose, 20% (v/v) coconut water, and 1% Bacto agar.
7. Leaf thin-section culture medium: half-strength MS medium (23) supplemented with 2.0 mg/L TDZ or 10.0 mg/L BA, 10 mg/L adenine sulfate, 2% (w/v) sucrose, 20% (v/v) coconut water, and 0.23% Gelrite. Adjust the pH of the medium to 5.7 before adding Gelrite.
8. Root tip culture medium: half-strength MS medium, 1 mg/L TDZ, 20% (v/v) coconut water, 10 mg/L adenine sulfate, and 0.23% (w/v) Gelrite.
9. Coconut water: liquid endosperm of coconut is drained from ripe nuts, filtered through two layers of cheese cloth or coffee filter paper and either used immediately or frozen in deep freezer (−70°C).

10. Mature green banana homogenate: peel off coat and homogenize 30–100 g green banana with 200 mL water in blender for 30 s at high speed. The liquid extract is mixed immediately with other medium components.
11. Unsprouted potato homogenate: prepare in the same manner as banana homogenate. Potatoes that have not had long storage give the best results. Homogenate obtained from sprouted potato can cause medium browning after culture.
12. Activated charcoal: finely divided activated charcoal has frequently been added to media (0.01–0.05% w/v) at different stages of tissue culture. Just as in the case of agar, there has been the idea that various brands of activated charcoal are in and themselves far better than others for use in orchid tissue culture. It is recommended that vegetable charcoal be used since this has a much higher percentage (95–99%) of active charcoal (Sigma) than charcoal obtained from other sources.
13. Sterilization solutions: 70% ethanol, 1.5% sodium hypochlorite solution with two drops of Tween 20/100 mL for seed sterilization, 3% sodium hypochlorite solution for vegetative explants.

3. Methods

3.1. Asymbiotic Seed Germination and Seedling Establishment

1. Collect seeds from a dehisced (mature) capsule. Place seeds into 125 mL Erlenmeyer flasks containing sodium hypochlorite solution with Tween 20 for 15 min by agitating on an orbital shaker at a speed of 50–60 rpm. The hypochlorite solution is conveniently removed by filtration with a Buchner funnel. Wash seeds three times with sterile water.
2. Undehisced capsules are first washed with soap and water, dipped in 70% ethanol for 10 s and then soaked in 3% sodium hypochlorite solution for 20 min. Wash three times with sterile distilled water and sterilize the capsule surface by open flame for a second before opening the capsule on a clean bench (see Note 2). To remove immature seeds, cut green capsules with a sterile knife, scalpel, razor blade, or spatula. Scrape out seeds with a sterile spatula.
3. Sow seeds on the seed germination medium in wide-mouth 150-mL Erlenmeyer flasks containing 50 mL medium. Distribute the seeds evenly over the surface of the medium using a spatula and a small amount of sterile water, which normally condenses on the agar. Mature seeds can be placed on the medium in the same manner as immature seeds.

4. Place flasks in a culture room at 20–25°C under 16–24 h photoperiod at 40 μmol/m²/s. Swollen embryos with several rhizoid breaks from testa start to form protocorms after 10–14 days in culture and the culture will appear crowded.
5. Collect protocorms which have one developed leaf (about 5 mm) and transfer them onto the 1st TP medium (Table 2). Be careful not to damage clumps of protocorms during separation of the proliferated protocorm clumps and transfer. Damaged protocorms or seedlings are one of the sources of growth-inhibiting phenolic compounds which turn the medium brown (see Note 3).
6. After 2–3 months of the first transfer, transplant seedlings with two more leaves and roots to the 2nd TP medium (Table 2). Addition of 5% (v/v) coconut water into the medium is favorable for the growth of seedlings. Required culture period of second transfer is about 3 months.
7. Acclimatization conditions are important to maximize the survival rate and to stimulate the vigorous growth of seedlings. After 6 months of in vitro culture, seedlings may develop 4–5 leaves and 3–4 roots. Wash away the medium from the roots in tap water. The need to remove agar is due to the agar trapping sucrose and other organic compounds in roots causing disease and infection. It is probably unnecessary to remove the agar if the culture medium does not contain sugar. The seedlings are put into a potting mix and hardened in a greenhouse setting. The most common method of hardening small batches of seedlings is to place them in benches as planted in pots or plugs covered with clear plastic lids or with plastic domes made of thin film. Seedlings kept in a greenhouse after removal from flasks must be shaded up to 90% from direct light for 7–10 weeks. Direct light can be gradually increased up to 70% for photosynthesis of the seedlings (19, 21, 22). The most suitable temperature range for acclimatization of seedlings is 20–25°C under high humidity.

3.1.1. In Vitro Multiplication of Protocorms

1. Percentages of seed germination and protocorm formation are dependent on genotype, seed maturity, media composition, and culture environment. Select well-growing greenish globular protocorms for proliferation.
2. Remove small shoots (one leaf) developed on top of protocorms and transfer the protocorms to fresh PM medium (Table 2) at 4-week intervals. For more rapid proliferation, protocorms can be divided into 2–4 pieces longitudinally. One piece of protocorm produces 10–20 protocorms after 4 weeks in culture.
3. Discard small yellowish protocorms and divide the protocorm clumps into one protocorm before subculture. Subculture in

time to prevent old and/or deteriorated protocorms which give rise to abnormal shoots and the retarding of growth both in vitro and ex vitro. Do not subculture for more than 1 year to reduce the number of off-type plantlets. Production of less than 30,000 plantlets per capsule (Standard *Phalaenopsis*) is advisable for commercial purposes.

4. Place 20 pieces of protocorms in one disposable Petri-dish (10 cm in diameter) containing 25 mL of PM medium. Maintain the cultures at 25°C for 4 weeks with a 16 h photoperiod under 30 μmol/m²/s.
5. Once the seedlings develop the first leaf, they can be transplanted onto the 1st TP medium as for asymbiotic seed germination and subsequently to soil.

3.2. The Induction of Protocorm-Like Bodies from Leaf Thin Sections and Root Tip Explants

3.2.1. PLB Formation from Leaf Thin-Section Culture

1. The flower stalks which have 3–5 open flowers with buds in their node are good material to induce adventitious shoots (Fig. 1).
2. Wipe-trimmed stalks are cut into sections with one lateral bud in the center of a 3–4 cm length of stalk. Immerse nodal sections into 3% sodium hypochlorite solution containing one drop of Tween 20 for 10 min and then rinse three times with sterilized water. Remove bleached end of sections before placing flower stalk section into medium with its base submerged (28).
3. Place trimmed flower stalk section on VW medium (Table 1) with 2% (w/v) sucrose, 20% (v/v) coconut water, and 1% (w/v) agar. Addition of 3.0 mg/L BAP or 1.0 mg/L TDZ in medium stimulates shoot development. Place the cultures in the culture room at 26–28°C under 16 h photoperiod at 30 μmol/m²/s.
4. Shoots with two or three leaves generally appear within 1–2 months depending on the species and genotypes.
5. Leaves from flower stalk culture are a suitable source of material for making thin leaf segments (28). Five to seven 1 mm segments are cut transversely using a surgical blade from the proximal (basal) portion of the youngest leaf from each plantlet (see Note 3).
6. Soak thin-sectioned segments into half-strength MS liquid medium for 2 h.
7. Forty sections are placed cut side down onto Petri-dishes (10 cm in diameter) containing 30 mL of leaf thin-section culture medium (see Note 4).
8. Cultures are incubated for 1 week at 27°C in the dark, transferred to a tissue culture room at 25°C under a 16 h photoperiod at a 20 μmol/m²/s and maintained for 6 weeks.

3.2.2. PLB Formation from Root Tip Culture

1. Root tips (less than 0.5 cm long) dissected from in vitro plantlets derived from flower stalk culture (see Subheading 3.2) can be used. They are placed on root tip culture medium with cut side down. TDZ is found to be more effective cytokinin in the induction of PLBs from root tip than BAP or zeatin.
2. About 20 root tips are cultured in a plastic Petri-dish (10 cm diameter) containing 25 mL of medium and culture conditions are the same as in leaf thin-section culture.
3. Two to six PLBs can be obtained from a root tip after 8 weeks of culture.

3.2.3. Subculture and Multiplication of PLBs

1. PLBs developed from the leaf segments or root tip are proliferated using the same protocol as the protocorms from asymbiotic seed germination (see Subheading 3.1.1, Note 4).
2. For subculture and further proliferation of PLBs, use the upper portion of PLBs as explants for proliferation to reduce the occurrence of variation. Discard the lower portion of PLBs (see Note 5). Sucrose-free VW medium with 20% coconut water or the PM medium can be used for multiplication of PLBs.
3. Discard all abnormal PLBs during subculture. Generally, 4-week intervals are advisable (see Note 6).
4. Cultures in liquid medium respond better when agitated on a horizontal gyratory (100 rpm) or a vertical-wheel type (2–3 rpm) shaker.
5. Distinguish normal and off-type PLBs to reduce the frequency of somaclonal variation after transplanting plantlets to the greenhouse. Off-type PLBs are categorized into two types: One is translucent and turgid, and another is small and branched. Both types are difficult to develop into plantlets, and the latter has the characteristics of differentiating new PLBs from the surface (29).
6. Collect normal PLBs having one leaf (about 5 mm) after separation of PLB clumps and transfer them onto the 1st TP medium (Table 2). For the PLB-derived shoot development and acclimatization, the protocol for protocorm seedling can be used (see Subheading 3.1).

4. Notes

1. For a general outline of techniques including preparations of media and stock solutions, procedures, equipment, and facilities, see refs. (10, 26, 30, 31).

2. The size of capsule varies from 5 cm long in native species to 10 cm long in large flower cultivars. For seed maturation of a pollinated flower, it takes 5–7 months depending on genotypes.
3. In thin leaf section culture, wounding caused by the excision process plays an important role in PLB production (28, 32). Polyphenolics are released to the medium immediately from the cut side of explants, which subsequently oxidize and cause browning of the medium. Finally, the explants die. Therefore, one or two times of subculture of explants (at 1–2 week intervals) to fresh media at an early stage of culture are desirable to increase survival and PLB formation (28).
4. Formation of PLBs and callus-like bodies is more stimulated by ethylene released from thin leaf sections compared with thick leaf segment (over 5 mm) culture. Changes of ethylene concentration in the culture vessel during the culture period are closely related to the percentage of PLB forming explants (19, 33).
5. Endoreduplication is variable according to tissue types, ages, and parts in one tissue. Although the proliferation rate is high when PLBs are used as explants, this often results in somaclonal variation, especially tetraploidy which tends toward high levels of endoreduplication. Shoot apical meristem in the upper part of PLBs having low degree of endoreduplication is more suitable for stable in vitro culture compared with that in lower part of PLBs (29, 34).
6. Days required for initial PLB formation depend on genotypes. Some genotypes show relatively high percentages (up to 70%) of off-type plantlets. It is very difficult to distinguish the PLBs which eventually develop into off-type plantlets. Also the occurrence of off-types does not exactly match with subculture time and culture period but is mainly related to the genotypic traits of clones (34–36).

References

1. Knudson L (1922) Non-symbiotic germination of orchid seeds. Bot Gaz 73:1–25
2. Knudson L (1946) A new nutrient solution for the germination of orchid seed. Am Orchid Soc Bull 14:214–217
3. Hew CS, Yong WH (2004) The physiology of tropical orchids in relation to the industry. World Scientific Publishing, Singapore
4. Paek KY, Kozai T (1998) Micropropagation of temperate *Cymbidium* via rhizome culture. Horttechnology 8:283–288
5. Paek KY, Yeung EC (1991) The effect of 1-naphthalenacetic acid and N-benzyladenine on the growth of *Cymbidium forrestii* rhizomes in vitro. Plant Cell Tissue Organ Cult 24:65–71
6. Paek KY, Murthy HN (2002) Temperate oriental *Cymbidium* species. In: Kull T, Arditti J (eds) Orchid biology: reviews and perspective, VIII. Kluwer Academic Publishers, Dordrecht, pp 235–286
7. Lee YI, Yeung EC, Chung MC (2007) Embryo development of orchids. In: Chen WH, Chen HH (eds) Orchid biotechnology. World Scientific Publishing, Singapore, pp 23–44
8. Yam TW, Yeung EC, Ye XL, Zee SY, Arditti J (2002) Orchid embryos. In: Kull T, Arditti J

(eds) Orchid biology: reviews and perspective, VIII. Kluwer Academic Publishers, Dordrecht, pp 288–377
9. Yeung EC, Law SK (1997) Ovule and megagametophyte development in orchid. In: Arditti J, Pridgeon AM (eds) Orchid biology: reviews and perspective, VII. Kluwer Academic Publishers, Dordrecht, pp 31–73
10. Arditti J, Clements MA, Fast G, Hadley G, Nishimura G, Ernst R (1982) Orchid seed germination and seedling culture – a manual. In: Arditti J (ed) Orchid biology – reviews and perspectives, II. Cornell University Press, Ithaca, pp 245–370
11. Ernst R (1975) Studies in asymbiotic culture of orchids. Am Orchid Soc Bull 44:12–18
12. Morel G (1960) Producing virus-free cymbidiums. Am Orchid Soc Bull 29:495–497
13. Morel G (1964) Clonal propagation of orchids by meristem culture. Cymbidium Soc News 20:3–11
14. Yam TW, Arditti J (2009) History of orchid propagation: a mirror of the history of biotechnology. Plant Biotechnol Rep 3:1–56
15. Arditti J (1992) Fundamentals of orchid biology. Wiley, New York
16. Tanaka M (1977) Clonal propagation of *Phalaenopsis* by leaf tissue culture. Am Orchid Soc Bull 46:733–737
17. Park SY, Murthy HN, Paek KY (2003) Protocorm-like body induction and subsequent plant regeneration from root tip culture of *Doritaenopsis*. Plant Sci 164:919–923
18. Tanaka M (1978) Factors affecting the growth ofin vitro cultured lateral buds from *Phalaenopsis* flower stalks. Sci Hort 8:169–178
19. Ichihashi S (1997) Research on micropropagation of *Cymbidium*, nobile-type *Dendrobium*, and *Phalaenopsis*. In: Arditti J, Pridgeon AM (eds) Orchid biology: reviews and perspective, VII. Kluwer Academic Publishers, Dordrecht, pp 285–310
20. George EF (1996) Plant propagation by tissue culture. Exegetics Limited, Edington
21. Vacin E, Went FW (1949) Some pH changes in nutrient solutions. Bot Gaz 110:605–613
22. Kano K (1965) Studies on the media for orchid seed germination. Mem Fac Agric Kagawa Univ 20:1–68
23. Murashige T, Skoog F (1962) A revised medium for rapid growth and bioassays with tobacco tissue cultures. Physiol Plant 15:473–497
24. Intuwwong O, Sagawa Y (1974) Clonal propagation of *Phalaenopsis* by shoot tip culture. Am Orchid Soc Bull 43:893–895
25. Ernst R (1974) The use of activated charcoal in asymbiotic seedling culture of *Phaphiopedilum*. Am Orchid Soc Bull 43:35–38
26. Arditti J, Ernst R (1993) Micropropagation of orchids. Wiley, New York
27. Sweet HC, Bolton WE (1979) The surface decontamination of seeds to produce axenic seedling. Am J Bot 66:692–698
28. Park SY, Murthy HN, Paek KY (2002) Rapid propagation of *Phalaenopsis* from floral stalk derived leaves. In Vitro Cell Dev Biol Plant 38:168–172
29. Park SY, Paek KY (1999) Occurrence of abnormal plants and their morphological characteristics as affected by clones and culture periods in clonally propagated *Phalaenopsis* by leaf culture. J Kor Soc Hort Sci 40:731–734
30. Brown DCW, Thorpe TA (1981) Requirements for a tissue culture facility. In: Thorpe TA (ed) Plant tissue culture – methods and application in agriculture. Academic Press, New York, pp 1–20
31. Arditti J (1977) Clonal propagation of orchids by means of tissue culture – a manual. In: Arditti J (ed) Orchid biology – reviews and perspectives, I. Cornell University Press, Ithaca, pp 203–294
32. Chang WC (2007) In vitro morphogenesis and micropropagation of orchid. In: Chen WH, Chen HH (eds) Orchid biotech nology. World Scientific Publishing, Singapore, pp 45–64
33. Park SY, Yeung EC, Chakrabarty D, Paek KY (2002) An efficient direct induction of protocorm-like bodies from leaf subepidermal cells of *Doritaenopsis* hybrid using thin-section culture. Plant Cell Rep 21:46–51
34. Vajrabhaya T (1977) Variations in clonal propagation. In: Arditti J (ed) Orchid biology – reviews and perspectives, I. Cornell University Press, Ithaca, pp 177–202
35. Park SY, Murthy HN, Chakrabarthy D, Paek KY (2009) Detection of epigenetic variation in tissue-culture-derived plants of *Doritaenopsis* by methylation-sensitive amplification polymorphism (MSAP) analysis. In Vitro Cell Dev Biol Plant 45:104–108
36. Jeon MH, Ali MB, Hahn EJ, Paek KY (2005) Effects of photon flux density on the morphology photosynthesis and growth of a CAM orchid, *Doritaenopsis* during post-micropropagation acclimatization. Plant Growth Regul 45:139–147

Part V

Transformation Using Zygotic Embryos as Explants

Chapter 21

Genetic Transformation Protocols Using Zygotic Embryos as Explants: An Overview

Muhammad Tahir, Ejaz A. Waraich, and Claudio Stasolla

Abstract

Genetic transformation of plants is an innovative research tool which has practical significance for the development of new and improved genotypes or cultivars. However, stable introduction of genes of interest into nuclear genomes depends on several factors such as the choice of target tissue, the method of DNA delivery in the target tissue, and the appropriate method to select the transformed plants. Mature or immature zygotic embryos have been a popular choice as explant or target tissue for genetic transformation in both angiosperms and gymnosperms. As a result, considerable protocols have emerged in the literature which have been optimized for various plant species in terms of transformation methods and selection procedures for transformed plants. This article summarizes the recent advances in plant transformation using zygotic embryos as explants.

Key words: Agrobacterium, Biolistic transformation, Embryo transformation, Plant cell tissue and embryo culture, Plant transformation protocols, Zygotic embryogenesis

1. Introduction

The concept and practice of "genetic modification (GM) of plants" is not new. Spontaneous mutations followed by the natural selection has been the principal force behind the GM of plants for millions of years as reflected in the evolution of plant species which we see today (1). The slow natural process of GM has been accelerated by human interventions in selecting plants with desirable traits. This rapid increase in GM of plants is specially witnessed in agriculturally and/or economically important plant species (1) which has been tailored according to human needs and preferences, and this practice is persistent for at least 10,000 years even without an understanding about the nature of genetic

Trevor A. Thorpe and Edward C. Yeung (eds.), *Plant Embryo Culture: Methods and Protocols*,
Methods in Molecular Biology, vol. 710, DOI 10.1007/978-1-61737-988-8_21,

material (2). The discovery of Mendel's laws of inheritance in early 1900s was followed by dramatic developments in understanding the nature of genetic material (DNA). GM of plants, especially agricultural crops, has become a key tool employed by plant breeders (searchable database: http://www.agbios.com/dbase.php).

The breakthrough in GM of plants was triggered by the landmark discovery of Frederick Griffith in 1928 when he demonstrated the "transforming principal" (3, 4), whereas the DNA could be transferred between bacterial cells causing the cells to behave differently in causing the disease pneumonia. Later on, classic research work during 1940s conducted by Colin Macleod and Maclyn MacCarthy proved Griffith's transforming principal and termed the GM by foreign DNA as transformation (3). With better understanding about the structure and replication of genetic material (DNA), and codevelopment of tissue culture or biotechnological techniques, the concept of GM evolved from GM at the massive level to the modification of only a specific or few genes without alteration in the rest of the genome.

GM technology now also referred as transgenic technology rapidly developed and expanded in the past decade, although GM or genetically modified plants/crops started becoming a commercial reality in 1990s. The boom in the development and use of transgenic crops is a result of cumulative advancements in the fields of molecular genetics, transformation techniques, plant cell and tissue culture, regeneration of transformed cells/tissues, and developmental processes underlying the plant growth (5). No wonder that the year 2007 witnessed the production of GM crops on 114.3 million hectares in 23 countries of the world (1). Samples of successful plant transformation in angiosperms and gymnosperms are presented in Tables 1 and 2.

The actual procedure of producing the transgenic plants (6) involves the introduction of foreign DNA into a plant tissue and then regenerating the plants containing the introduced DNA. The success of the procedure is directly proportional to the successful transformation as measured by the change in the phenotype of an organism by the insertion of foreign DNA to its genome. Essentially, the requirements for production of transgenic plants (4, 6, 7) can be categorized into (1) a suitable target tissue and/or cells which have the ability to regenerate plants, (2) an effective method to introduce DNA into the regenerative cells/tissue, and (3) an appropriate procedure to select the transformed plants in sufficient numbers.

For production of transgenic plants, the first most important prerequisite is the selection of suitable target tissue. Different plant tissues including leaf tissue, immature cotyledons, roots, stems, shoot apices, embryogenic suspension cells, somatic and zygotic embryos, and whole seedlings can be used as a target tissue (5, 8–10). Immature zygotic embryo has been the widely used explant source to develop embryogenic callus lines, cell suspensions, and protoplasts for transformation of cereal crops

Table 1
Transformation in angiosperms using zygotic embryos as explants

Species	Method	Explant	References
Maize	*Agrobacterium*	Immature	(15, 44, 45)
Maize	*Agrobacterium*	Mature	(11, 46)
Maize	Biolostic	Immature	(14, 15, 17, 47–55)
Wheat	*Agrobacterium*	Immature	(18, 56)
Wheat	Biolistic	Immature	(57)
Wheat	Biolistic	Mature	(27)
Rice	*Agrobacterium*	Mature	(58)
Rice	*Agrobacterium*	Mature	(59)
Rice	Biolistic	Immature	(60)
Rice	Biolistic	Immature	(9)
Oat	Biolistic	Immature	(61–64)
Oat	Biolistic	Mature	(65)
Barley	*Agrobacterium*	Immature	(66, 67)
Barley	Biolistic	Immature	(19, 20, 68–73)
Barley	Biolistic	Mature	(74, 75)
Sorghum	*Agrobacterium*	Immature	(76)
Sorghum	Biolistic	Immature	(77, 78)
Millet	Biolistic	Immature	(79, 80)
Cotton	*Agrobacterium*	Mature	(34)
Arabidopsis	*Agrobacterium*	Mature	(33)
Peanut	*Agrobacterium*	Mature	(32, 81)
Brassica	Biolistic	Mature	(30)
Avocado	*Agrobacterium*	Mature	(29)
Papaya	Biolistic	Immature	(25, 26)
Pearl millet	Biolistic	Immature	(17)
Soybean	*Agrobacterium*	Immature	(22)
Soybean	*Agrobacterium*	Mature	(11)
Soybean	Biolistic	Immature	(23)
Sunflower	Other	Immature	(24)
Cowpea	*Agrobacterium*	Mature	(11)
Black henbane	*Agrobacterium*	Mature	(28)
Thorn apple	*Agrobacterium*	Immature	(82)
White lead tree	*Agrobacterium*	Immature	(10)

Table 2
Transformation in gymnosperms using zygotic embryos as explants

Species	Method	Explant	References
Hinoki cypress	Other	Immature	(35)
White spruce	*Agrobacterium*	Mature	(11, 83)
Pinus	*Agrobacterium*	Mature	(36)
Loblolly pine	*Agrobacterium*	Mature	(37)
Larix gmelinii	Biolistic	Mature	(38)
Terminalia chebula	*Agrobacterium*	Mature	(39)
Chir pine	Biolistic	Mature	(40)
Christmas tree species	*Agrobacterium*	Mature	(41)
Picea abies	Biolistic	Mature	(42)

including maize, wheat, rice, oat, barley, sorghum, and millet (8). There are several other reports (see Tables 1 and 2) in which zygotic embryos have been used as a explants for the production of transgenic plants.

The second prerequisite is an effective method to introduce DNA into the regenerative cells/tissue. There are different methods by which DNA can be introduced into the target tissues. DNA can be delivered into cells via *Agrobacterium*-mediated transformation, microinjection, electroporation, and/or polyethylene glycol (PEG)-mediated protoplast transformation, pollen tube pathway, ultrasonication-mediated DNA transfer, and whiskers-mediated DNA transfer (5, 9–11). With the various gene transfer methods currently available, simple placement or transfer of DNA into a plant cell is no longer a limiting factor (6). However, both the mechanism for DNA transfer to a plant cell and targeting of the DNA to a complex tissue or organ competent for regeneration is still a major limitation (12). The third requirement for the production of transgenic plants is an appropriate procedure to select the transformed plants (reviewed in (7, 13)) in sufficient numbers. One of the best methods to select the transformed plants is the use of vector-assisted selectable markers. The selectable markers are of three types, i.e., kanamycin resistance, hygromycin resistance, and bleomycin resistance. In spite of selectable markers scorable makers (NPT II activity, opine production, β-glucuronidase (GUS) activity, chloramphenicol acyl transferase activity, luciferase activity) can also be used as reporter genes. These genes can be placed under the control of a specific promoter. If the necessary cell machinery is present, then the promoter will be activated, and RNA polymerase will make the mRNA of the reporter gene and it will be translated. To determine if the gene was activated, plant tissue is treated with the appropriate substrate and expression can be monitored. If expression of the reporter gene is detected then the expression pattern of the promoter can be determined.

This review is an effort to highlight the advancements in procedures which are being employed for transformation of zygotic embryos and to provide basic concepts related to the transformation of angiosperm and gymnosperm species.

2. Transformation of Angiosperms

2.1. Transformation of Angiosperms Using Immature Zygotic Embryos

The efficiency of transformation can be measured by studying the parameters such as the ability of the transformed embryos to produce secondary embryos, regeneration ability of the embryos and the transient gene expression. In maize, Aulinger et al. (14) showed that immature embryos can potentially be transformed by particle bombardment, since they responded positively to all the studied parameters, although with lower efficiencies than fully mature embryos. In particular, differences were found in the rate of secondary embryogenesis and the density of transformed cells. In another study (15), production of transgenic maize from bombarded zygotic embryo-derived type II callus was studied. These researchers observed the effect of gold particle size and callus morphology on transformation efficiency. They presented a routine and efficient protocol for year-round production of fertile transgenic maize plants. Type II callus derived from maize immature zygotic embryos were transformed using the PDS 1000/He biolistic gun and selected on bialaphos. In an effort to improve the transformation protocol, they also investigated the effects of gold particle size and callus morphology on transformation efficiency. The average transformation efficiency of preembryogenic, early embryogenic, and late embryogenic callus did not vary significantly. In an earlier study (16), efficient transformation of scutellar tissue of immature maize embryos was established by improving transformation conditions for the particle bombardment, such as the amount of gold particles used per bombardment, particle velocity, preculture time of the scutellum prior to bombardment, and osmotic treatment of the target tissue before and after bombardment. Fertile transgenic plants selected on Basta-containing medium were regenerated for three inbred lines and two hybrids. The transformation frequency ranged from 2 to 4% and a total of 29 transgenic plant lines were obtained and verified with Southern blot analysis. All of the transgenic plants were fertile and set seeds. The R1 progeny of single plants was analyzed and a Mendelian segregation of the transgenes was observed for all of the transformants. For one of the candidates, stable inheritance and stable expression of the transgenes were followed up to the R4 generation. Further improvements were reported by O'Kennedy et al. (17) in the elite white maize using the particle inflow gun followed by detailed analysis of a low-copy integration event. The culture of immature zygotic embryos of selected elite white maize lines on medium containing 2 mg/L 2,4-dichlorophenoxyacetic acid and 20 mM L-proline proved to be most successful explants for transformation.

Wheat, barley, and pearl millet are other monocot species where immature zygotic embryos have been successfully transformed. Wu et al. (18) developed a protocol for *Agrobacterium*-mediated transformation of bread and durum wheat using freshly isolated immature embryos. In barley, fertile transgenic plants were obtained by bombarding the embryonic axis of immature embryos (19). The transformed plant produced 98 fertile spikes where integration and inheritance of the transferred *nptII* gene was confirmed by Southern blot hybridization. Although present as several copies, the transferred gene was inherited as a single Mendelian locus into the T2 progeny. In another study (20), a large numbers of independently transformed fertile barley plants were obtained from immature zygotic embryos. A total of 91 independent bialaphos-resistant callus lines expressed functional phosphinothricin acetyltransferase and integration of the marker gene was confirmed by DNA hybridization in the 67 lines analyzed. Transmission of the transgenes to T1 progeny was demonstrated in the five families analyzed by DNA hybridization. O'Kennedy et al. (21) studied Pearl millet transformation system using the positive selectable marker gene phosphomannose isomerase. Proliferating immature zygotic embryos were used as target tissue for bombardment using a particle inflow gun. Different culture and selection strategies were assessed in order to obtain an optimized mannose selection protocol. Stable integration of the manA gene into the genome of pearl millet was confirmed by PCR and Southern blot analysis. Stable integration of the manA transgene into the genome of pearl millet was demonstrated in T1 and T2 progeny of two independent transformation events with no more than four to ten copies of the transgene.

Within dicotyledonous species, transformation of soybean was achieved (22) by *Agrobacterium tumefaciens* mediated transfer of genes in immature zygotic cotyledon explants. Sato et al. (23) studied stable transformation via particle bombardment in two different soybean regeneration systems. The first system was multiple shoot proliferation from shoot tips obtained from immature zygotic embryos of the cultivar Williams 82, and the second was somatic embryogenesis from a long-term proliferative suspension culture of the cultivar Fayette. Bombardment of shoot tips with tungsten particles, coated with precipitated DNA containing the gene for GUS, produced GUS-positive sectors in 30% of the regenerated shoots. However, none of the regenerants which developed into plants continued to produce GUS-positive tissue. Bombardment of embryogenic suspension cultures produced GUS-positive globular somatic embryos which proliferated into GUS-positive somatic embryos and plants. An average of 4 independent transgenic lines were generated per bombarded flask of an embryogenic suspension. Particle bombardment delivered particles into the first two-cell layers of either shoot tips or somatic embryos.

In sunflower, it was found (24) that immature embryos were more suitable for transformation than their mature counterparts. The highest level of transient GUS expression after 3 and 14 days was obtained with embryos (1.5 mm) precultured for 3 days in the presence of NAA and BAP. It was further reported that after 2 and 4 weeks of culture, following bombardment with plasmids harboring a doubled CaMV 35S and a stress-inducible promoter, GUS activity increased. In mature embryos GUS-expressing cells were mostly observed in the epidermal layer, while in immature embryos they were located between the epidermis and the fourth mesophyll layer. The performance of the two biolistic equipments was also comparable. Under any condition, GUS expression declined with increasing culture time.

Cai et al. (25) developed an efficient transformation protocol for *Carica papaya* L. and provided useful tips for improving the process. In this study, they obtained a total of 83 transgenic papaya lines expressing the nontranslatable coat protein gene of papaya ring spot virus (PRSV). The transformation efficiency was very high as 100% of the bombarded plates produced transgenic plants. In another study on papaya (26), stable transformation was achieved via microprojectile bombardment. Three types of embryogenic tissues, including immature zygotic embryos, freshly explanted hypocotyl sections, and somatic embryos derived from both, were bombarded with tungsten particles carrying chimeric NPTII and GUS genes. Upon transfer to 2,4-D-free medium containing 150 mg/L kanamycin sulfate, ten putative transgenic isolates produced somatic embryos and five regenerated leafy shoots. Leafy shoots were produced 6–9 months following bombardment. Tissues from 13 of these isolates were assayed for NPTII activity and 10 were positive. Six out of 15 isolates assayed for GUS expression were positive. Three isolates were positive for both NPTII and GUS.

2.2. Transformation of Angiosperms Using Mature Zygotic Embryos

Successful transformation of mature zygotic embryos has been reported in several angiosperm species. One such example (27) is in wheat where gene transfer into intact scutellum cells was achieved by electroporating zygotic embryos without any special pretreatment. The use of plasmids carrying either two chimeric anthocyanin regulatory genes or a chimeric *gusA* gene allowed clear identification of transformed cells in the scutellum. Moreover, it was observed that gene transfer by electroporation was tissue specific as scutellum cells were found to be much more susceptible to gene transfer than other cell types of the embryo. An efficient *Agrobacterium*-based transformation technology termed SAAT (sonication-assisted *Agrobacterium*-mediated transformation) was reported to be applicable to both monocots and dicots and was claimed to overcome the barriers of transformation and enhance DNA transfer (11). The SAAT involves subjecting the plant tissue to brief periods of ultrasound in the presence of *Agrobacterium*. Scanning electron and light microscopy revealed that SAAT treatment produced small and uniform fissures and channels throughout the tissue allowing the *Agrobacterium* easy access to internal plant tissues. Unlike other transformation methods, this system has the potential to transform meristematic tissue buried under several cell layers. It was also revealed that SAAT treatment was necessary to obtain stable transformation in soybean. Tu et al. (28) studied transformation of *Hyoscyamus Niger* by *A. tumefaciens*. They inoculated leaf, root, stem, petiole, hypocotyl, and zygotic embryo explants, as well as pollen embryoids, and redifferentiated tissues from pollen embryoid-derived plantlets with *Agrobacterium* harboring the binary vectors (pGS Gluc1) and then cultured on media containing kanamycin. They found that transient GUS activity and kanamycin-resistant callus formation were influenced by explant origin. Transgenic calluses were obtained at a frequency of up to 30% from all the explants tested. They confirmed that transformation by the ability of the cells to produce kanamycin-resistant callus, GUS histochemical and flurometric assays, polymerase chain reaction, and Southern blot analyses. Their results showed that embryos may be an alternative source for both efficient transformation and regeneration of transgenic plants in recalcitrant species. In avocado, transformed somatic embryos were regenerated by *A. tumefaciens* – mediated transfer of *uidA* (GUS) and *nptII* genes in embryogenic cultures (29). Embryogenic avocado cultures derived from zygotic embryos of "Thomas" and consisting of proembryonic masses were gently separated and cocultivated with disarmed, acetosyringone-activated *A. tumefaciens* strain A208, which contained the cointegrative vector pTiT37-ASE::pMON9749 (9749 ASE). Kanamycin-resistant embryogenic suspension cultures were selected in two steps: (1) initial selection in maintenance medium, consisting of MS basal medium, supplemented with 0.1 mg/L picloram and 50 mg/L kanamycin sulfate for 2–4 months and (2) subsequent selection in maintenance medium with 100 mg/L kanamycin sulfate for 2 months in order to eliminate chimeras. Somatic embryo maturation was initiated by subculture onto semisolid maturation medium followed by transfer to maturation medium with 100 mg/L kanamycin sulfate. Genetic transformation of embryogenic cultures and somatic embryos was confirmed by the X-gluc reaction, and integration of *nptII* and *uidA* into the avocado genome was confirmed by PCR and Southern hybridization, respectively.

A method that allows the transfer of genes into single cells of excised zygotic embryos by particle bombardment was demonstrated in *Brassica juncea* or Indian mustard (30). The fate of single, genetically marked cells was followed during in vitro embryogenesis. A simple and defined embryo culture medium has been designed on which zygotic *B. juncea* embryos, excised at the globular or at later stages, develop normally into mature, fully grown embryos. The embryos grow on the surface of solid medium without embedding and are freely accessible to microprojectile bombardment. Shooting at globular, transition, and early heart-shaped embryos using both a particle inflow gun and a microtargeting particle accelerator resulted in transient expression of genes encoding visible markers. For both particle-acceleration devices the shooting conditions have been optimized based on transient GUS expression. Bombarding embryos under optimal conditions had no deleterious effects on in vitro embryogenesis. Multicellular GUS-expressing sectors were obtained, showing that bombarded cells can survive and resume normal development. The examination of these sectors has provided new information about cell division patterns characterizing early *B. juncea* embryogenesis. To be able to follow the development of particular genetically marked sectors, the authors tried to identify reporter genes that, in contrast to the *uidA* gene (which encodes GUS), can be nondestructively assayed in embryonic cells. Preliminary data has shown that expression of the firefly luciferase gene (*Luc*) can be detected in bombarded embryos without affecting their viability. Rochange et al. (31), reported DNA delivery into *Eucalyptus globulus* zygotic embryos through a biolistic approach. They optimized the procedure in terms of biological and physical parameters of bombardment for two different particle guns using the transient expression of a reporter gene as a test. Six-day-old cultured embryos were the best target material, and osmotic treatment increased the expression rate. The conditions of bombardment (particle acceleration and quality of the particle:DNA mix) were optimized and up to 130 GUS expression events per embryo with a good distribution over the tissue were reported.

In another study (32), *Agrobacterium*-mediated transformation of peanut (*Arachis hypogaea* L.) embryo axes led to the development of transgenic plants. The zygotic embryo axes from mature peanut seed were cultured with *A. tumefaciens* (strain EHA101) harboring a binary vector that contained the genes for the scorable marker GUS and the selectable marker neomycin phosphotransferase II. It was observed that 9% of the germinated seedlings were GUS-positive. Polymerase chain reaction analysis confirmed that GUS-positive shoots and T1 progeny contained T-DNA. Molecular characterization of one primary transformant and its T1 and T2 progeny plants established that T-DNA was integrated into the host genome.

An efficient procedure for *Agrobacterium*-mediated transformation of zygotic embryos derived from different *Arabidopsis thaliana* ecotypes has also been developed (33). This procedure yielded an average transformation rate of 76% for ecotype C24, and 15–20% for ecotypes Landsberg-erecta and Columbia. Light and electron microscopical studies showed that, during preculture, procambium cells of embryos became highly susceptible to *Agrobacterium* infection. Transformed cells developed callus and regenerated shoots within 4–5 weeks of culture. A total of 1,500 fertile transgenic plants were regenerated.

Leelavathi et al. (34) developed a simple and rapid *Agrobacterium*-mediated transformation protocol for cotton (*Gossypium hirsutum* L.). Embryogenic callus, derived from zygotic embryos, were cocultivated with *Agrobacterium* carrying the cry1Ia5 gene and then cultured under dehydration stress and antibiotic selection for 3–6 weeks to generate several transgenic embryos. An average of 75 globular embryo clusters were observed on selection plates and these embryos were cultured on multiplication medium followed by development of cotyledonary embryos on embryo maturation medium to obtain an average of 12 plants per Petri plate of cocultivated callus. About 83% of these plants have been confirmed to be transgenic by Southern blot analysis. An efficiency of ten kanamycin-resistant plants per Petri plate of cocultivated embryogenic callus was obtained. The simplicity of the procedure and the efficiency of the initial material allow transformation of any variety where a single regenerating embryogenic callus line can be obtained. In addition, multiple transformations can be performed either simultaneously or sequentially. The method is extremely simple, reliable, efficient, and much less laborious than any other existing method for cotton transformation.

3. Transformation of Gymnosperms

3.1. Transformation of Gymnosperms Using Immature Zygotic Embryos

Compared to angiosperms, there are very few reports available on the genetic transformation of coniferous species. In one study (35), a plant regeneration system from immature zygotic embryos of Hinoki cypress (*Chamaecyparis obtusa*) via somatic embryogenesis was established. According to this investigation, embryogenic tissues derived from immature zygotic embryos were successfully induced on three kinds of Smith media from mega-gametophyte explants containing precotyledonary embryos of *C. obtusa* plus-trees. The addition of 30 g/L maltose to the medium had a positive effect on embryo maturation, but sucrose was ineffective. The mature somatic embryos germinated at a germination frequency of approximately 60%, and the presence of activated charcoal was effective in stimulating plantlet growth. The plantlets acclimatized successfully in a greenhouse.

3.2. Transformation of Gymnosperms Using Mature Zygotic Embryos

Many labs have been interested in developing protocols for transformation of gymnosperms using mature zygotic embryos. A study by Charity et al. (36) reported *Agrobacterium*-mediated transformation of *Pinus radiata* cotyledons. This experiment resulted in up to 55% of cotyledons transiently expressing the reporter gene *uid*A. The authors developed a transformation protocol for both cotyledons and the apical meristematic dome, which is the portion of the embryo remaining after cotyledons were detached, and from which the apical shoot and axillary shoots regenerate. Molecular analysis of putatively transformed shoots regenerated either adventitiously from cotyledons or via axillary shoots from apical domes, indicated the presence of *uid*A and *npt*II genes in some of these shoots. Biochemical analysis of putatively transformed shoots using *npt*II ELISA indicated that they contained the *npt*II enzyme. However, Southern hybridization indicated stable integration of *npt*II only in one shoot which was regenerated from an apical dome. Shoots regenerated from cotyledons appeared to exhibit chimeric expression and were not stably transformed. Based on a comparison of time for regeneration, technical difficulty, and molecular and biochemical analysis, apical domes may be more suitable as explants for transformation and subsequent regeneration of transclones than detached cotyledons. Tang et al. (37) studied the regeneration of transgenic loblolly pine from zygotic embryos transformed with *A. tumefaciens.* Embryos of 24 open-pollinated families were used as explants. The *A. tumefaciens* strain GV3101 harboring thc plasmid was used to transform mature zygotic embryos of seven families of loblolly pine. The frequency of transformation varied among families infected with *A. tumefaciens.* The highest frequency (100%) of transient GUS-expressing embryos was obtained from family 11 to 1,029 with over 300 blue spots per embryo.

A new protocol for stable genetic transformation of *Larix gmelinii* was reported (38) using particle bombardment of zygotic embryos. Thirty mature zygotic embryos precultured for 3 days on solid medium supplemented with benzyladenine were bombarded with plasmids pUC-GHG (GUS, HPT, and green fluorescent protein (GFP) genes) or pBI221-HPT (HPT and GUS genes). After a 2-month culture on selection medium, hygromycin-resistant callus appeared on the surfaces of the necrotic embryos. The frequencies of embryos with resistant callus were 18.4 and 17.4% in the transformations with pUC-GHG and pBI221-HPTDNA, respectively. More than 20 adventitious shoots formed from each of the transgenic calluses. Of 17 elongated shoots selected for culturing on a rooting medium, five shoots rooted after 2 months. Expression of the GFP and GUS genes was detected in the resistant tissues by microscopic observations and by a histological GUS activity assay, respectively. PCR and Southern analysis confirmed the stable insertion of the introduced DNA into the genome. In a recent study (39) on *Terminalia chebula*, multiple explants such as cotyledon, hypocotyl, and excised mature zygotic embryos were transformed using *A. tumefaciens* strain C-58 followed by detection of tannin in transformed tissue. The transformed callus was subjected to nopaline assay using paper electrophoresis. The analysis indicated the transformed nature of the callus with the presence of nopaline and its absence in nontransformed control callus. Transformed callus grown on fresh MS basal medium showed more than twofold increase in the growth after 4 weeks of culture compared to normal control callus. Normally, no growth was observed in untransformed control callus. The transformed callus was analyzed for the presence of tannins using thin layer chromatography, which indicated the presence of tannic acid in the transformed callus. Genetic transformation of *T. chebula* and detection of tannin in transformed callus can be used to study the tannin biosynthetic pathway using biochemical and molecular approaches. In Chir Pine (*Pinus roxbughii Sarg.*), a particle inflow gun was used to transfer the plasmid pAHC25 containing the *bar* gene conferring resistance to glufosinate and the *gus*A reporter gene, each driven by the maize ubiquitin promoter, to mature (40). High levels of transient expression were obtained when embryos were cultured for 6 days on 10 μM benzyl adenine-containing medium and then exposed to high osmoticum (0.5 M sucrose) before and after bombardment. Tang and Newton (41) worked on transgenic Christmas trees regenerated from *A. tumefaciens* mediated transformation of zygotic embryos using the green fluorescence protein as a reporter. Mature zygotic embryos of recalcitrant Christmas tree species Fraser fir and Nordmann fir (*Abies nordmanniana* L.k.), and Virginia pine (*Pinus virginiana* Mill.) were used as explants for *A. tumefaciens* strain GV3850-mediated transformation using the GFP gene as a reporter. A high transformation frequency was obtained on TE medium containing 50 mg/L acetosyringone and using 500 mg/L timentin to eliminate bacteria. Transient gene expression was observed in all three Christmas tree species, but transgenic plants were only produced from Virginia pine. These results demonstrated that a stable transformation system has been established in Virginia pine and this system would provide an opportunity to transfer economically important genes into Christmas tree species. A particle inflow gun enabled efficient production of transgenic plantlets of *Picea abies* from embryogenic suspension cultures generated from mature zygotic embryos was reported (42). In transient assays, the *Zea* ubiquitin promoter was 12–16 times as active as the 35S promoter. Cells were maintained from 1 to 3 h before bombardment on proliferation medium supplemented with 0.25 M myo-inositol and, from day 8, supplemented with Basta as selective agent. Embryogenic colonies resistant to Basta appeared 2 months after bombardment. Of over 100 independent Basta-resistant sublimes tested, 65% expressed the cotransformed reporter gene, even when it was not linked to the selectable marker. Over 80% of the sublines retained their embryogenic potential. Of 11 transformants analyzed, 4 contained transgenes in low copy number (1–3), the rest contained transgenes in up to 15–20 copies. Over 200 Basta-resistant sublines from four cell lines have been established, of which 138 are confirmed as transformed. Plantlets have been regenerated and grown on in pots.

4. An Overview on Regeneration Methods Following Transformation

The previous sections deal with transformation procedures using *Agrobacterium* or biolistic approaches. However, success in transformation is often related to the ability of regenerating transformed cells into viable plants. Regeneration is achieved through either somatic embryogenesis, that is the ability of somatic cells to form bipolar structures, i.e., embryos or shoot organogenesis.

As reviewed by others (43) somatic embryogenesis can be subdivided into induction, maintenance, and development. All these steps must be well executed and rely on different levels of plant growth regulators. As a general rule, the induction phase, that is the formation of embryogenic tissue from the transformed cells is achieved under high levels of auxins and cytokinins. In maize (15), the type-2 (embryogenic) callus formation from the base of the scutellum of the transformed zygotic embryos is induced by high auxin levels. A similar requirement of auxin, in the form of 2,4-D was also described in the induction phase of cotton (34). High levels of auxin, in conjunction with cytokinins were necessary to induce embryogenic tissue from transformed Norway spruce embryos (42). Embryogenic tissue is usually easily recognizable from the nonembryogenic counterpart. Apparent differences are visible in coniferous species where embryogenic tissue is translucent and characterized by the presence of many protruding immature embryos. This is in contrast to the nonembryogenic tissue, which is usually dark and more compact (43).

Once generated, embryogenic tissue can be maintained and regularly subcultured on media supplemented with auxin and cytokinins (42) and embryo development can be induced by either removing plant growth regulators, as often observed for angiosperm species (15) or by adding abscisic acid (35, 42), which is a key requirement for many gymnosperm species. Embryo production is also favored by application of osmoticum agents, which reduce water availability in the tissue and induce proper growth and histodifferentiation. The most commonly used agent is PEG, which is a nonpermeating compound unable to cross the plasma membrane (43).

Regeneration of transformed cells via organogenesis appears to be the preferred method for many plant species. While the requirements for callus induction are similar to those described for somatic embryogenesis, with auxin and cytokinins as the main plant growth regulators, shoot formation requires levels of cytokinins which need to be optimized for each species. This requirement is unique to both angiosperms and gymnosperms (19, 37, 38).

5. Concluding Remarks

Plant transformation has become a common tool to introduce useful traits in crop species as well as to gain information about gene function. Over the past few years several protocols have been developed for both angiosperms and gymnosperms which utilize biological and nonbiological methods for gene integration. Despite great advancements in this area, there are still several aspects of plant transformation that need to be elucidated. One above all is the mechanism whereby the transgene is integrated in the host genome. This information is crucial for the design of optimized protocols and for the application of genetic transformation to a larger variety of plant species. Another limitation of plant transformation is represented by the ability to regenerate viable plants from the transformed cells. Despite the advancements in tissue culture practices which have occurred over the past decades, rapid and efficient regeneration procedures are still required for assisting the transformation of recalcitrant species or varieties.

References

1. James C (2008) Global status of commercialized biotech/GM crops: 2008. ISAAA Brief No. 39. ISAAA, Ithaca
2. Chassy BM, Parrot WA, Roush R (2005) Crop biotechnology and the future of food: a scientific assessment. In: CAST commentary, Ames, Oct 2005
3. Fink GR (2005) A transforming principle. Cell 120:153–154
4. Vasil IK (2007) A short history of plant biotechnology. Phytochem Rev 7:387–394
5. Sharma KK, Mathur PB, Thorpe TA (2005) Genetic transformation technology: status and problems. In Vitro Cell Dev Biol Plant 41:102–112
6. Birch RG (1997) Plant transformation: problems and strategies for practical application. Annu Rev Plant Physiol Plant Mol Biol 48:297–326
7. Vain P (2006) Thirty years of plant transformation. Plant Biotechnol J 5:221–229
8. Sticklen MB, Oraby HF (2005) Shoot apical meristem: a sustainable explant for genetic transformation of cereal crops. In Vitro Cell Dev Biol Plant 41:187–200
9. Jones HD, Sparks CA (2008) Stable transformation of plants. Methods Mol Biol 513:111–130

10. Jube S, Borthakur D, Jube S, Borthakur D (2009) *Agrobacterium*-mediated transformation protocol for the tree-legume *Leucaena leucocephala* using immature zygotic embryos. Plant Cell Tissue Organ Cult 96:325–333
11. Trick HN, Finer JJ (1997) SAAT: sonication-assisted *Agrobacterium*-mediated transformation. Transgenic Res 6:329–336
12. Herschbach C, Kopriva S (2002) Transgenic trees as tools in tree and plant physiology. Trees 16:250–261
13. Sundar IK, Sakthivel N (2008) Advances in selectable marker genes for plant transformation. J Plant Physiol 165:1698–1716
14. Aulinger IE, Peter SO, Schmid JE, Stamp P (2003) Gametic embryos of maize as a target for biolistic transformation: comparison to immature zygotic embryos. Plant Cell Rep 21:585–591
15. Frame BR, Zhang H, Cocciolone SM, Sidorenko LV, Dietrich CR, Pegg SE, Zhen S, Schnable PS, Wang K (2000) Production of transgenic maize from bombarded type II callus: effect of gold particle size and callus morphology on transformation efficiency. In Vitro Cell Dev Biol Plant 36:21–29
16. Brettschneider R, Lorz HBD (1997) Efficient transformation of scutellar tissue of immature maize embryos. Theor Appl Genet 94:737–748
17. O'Kennedy MM, Burger JT, Berger DK (2001) Transformation of elite white maize using the particle inflow gun and detailed analysis of a low-copy integration event. Plant Cell Rep 20:721–730
18. Wu H, Doherty A, Jones HD (2009) *Agrobacterium*-mediated transformation of bread and durum wheat using freshly isolated immature embryos. Methods Mol Biol 478:93–103
19. Ritala A, Aspegren K, Kurtén U, Salmenkallio-Marttila M, Mannonen L, Hannus R, Kauppinen V, Teeri TH, Enari TM (1994) Fertile transgenic barley by particle bombardment of immature embryos. Plant Mol Biol 24:317–325
20. Wan Y, Lemaux PG (1994) Generation of large numbers of independently transformed fertile barley plants. Plant Physiol 104:37–48
21. O'Kennedy MM, Burger JT, Botha FC (2004) Pearl millet transformation system using the positive selectable marker gene phosphomannose isomerase. Plant Cell Rep 22:684–690
22. Yan B, Serinivasa MS, Collins RGB, Dinkins RD (2000) *Agrobacterium tumefaciens* mediated transformation of soybean using immature zygotic cotyledon explants. Plant Cell Rep 19:1090–1097
23. Sato S, Newell C, Kolacz K, Tredo L, Finer L, Hinchee M (1993) Stable transformation via particle bombardment in two different soybean regeneration systems. Plant Cell Rep 12:408–413
24. Hunolda R, Burrus M, Bronnera R, Duretc J-P, Hahne G (1995) Transient gene expression in sunflower (*Helianthus annuus*) following microprojectile bombardment. Plant Sci 105:95–109
25. Cai W, Gonsalves C, Tennant P, Fermin G, Souza M, Sarindu N, Zhu F, Gonsalves D (1999) A protocol for efficient transformation and regeneration of *Carica papaya* L. In Vitro Cell Dev Biol Plant 35:61–69
26. Fitch MMM, Manshardt RM, Gonsalves D, Slightom JL, Santord JC (1990) Stable transformation of papaya via microprojectile bombardment. Plant Cell Rep 9:189–194
27. Kloti A, Iglesias VA, Wfinn J, Burkhardt PK, Datta SK, Potrykus I (1993) Gene transfer by electroporation into intact scutellum cells of wheat embryos. Plant Cell Rep 12:671–675
28. Tu S, Sangwan RS, Raghavan V, Verma DPS, Sangwan-Norreel BS (2005) Transformation of pollen embryo-derived explants by *Agrobacterium tumefaciens* in *Hyoscyamus niger*. Plant Cell Tissue Organ Cult 81:139–148
29. Cruz-Hernández A, Witjaksono RE, Litz M, Lim G (1998) *Agrobacterium tumefaciens* – mediated transformation of embryogenic avocado cultures and regeneration of somatic embryos. Plant Cell Rep 17:497–503
30. Kost B, Leduc N, Sautter C, Potrykus I, Neuhaus G (1996) Transient marker-gene expression during zygotic in-vitro embryogenesis of *Brassica juncea* (Indian mustard) following particle bombardment. Planta 198:211–220
31. Rochange E, Serrano L, Marque C, Teulitres Z, Boudet AM (1995) DNA delivery into *Eucalyptus globulus* zygotic embryos through biolistics: optimization of the biological and physical parameters of bombardment for two different particle guns. Plant Cell Rep 14:674–678
32. McKently AH, Moore GA, Doostdar H, Niedz RP (1995) *Agrobacterium-mediated* transformation of peanut (*Arachis hypogaea* L.) embryo axes and the development of transgenic plants. Plant Cell Rep 14:699–703
33. Sangwan RS, Bourgeois Y, Sangwan-Norreel BS (1991) Genetic transformation of *Arabidopsis thaliana* zygotic embryos and identification of critical parameters influencing transformation efficiency. Mol Gen Genet 230:475–485
34. Leelavathi S, Sunnichan VG, Kumria R, Vijaykanth GP, Bhatnagar RK, Reddy VS (2004)

A simple and rapid *Agrobacterium*-mediated transformation protocol for cotton (*Gossypium hirsutum* L.): embryogenic calli as a source to generate large numbers of transgenic plants. Plant Cell Rep 22:465–470

35. Taniguchi T, Kurita M, Itahana N, Kondo T (2004) Somatic embryogenesis and plant regeneration from immature zygotic embryos of Hinoki Cypress (*Chamaecyparis obtusa* Sieb. et Zucc.). Plant Cell Rep 23:26–31
36. Charity JA, Holland L, Donaldson SS, Grace L, Walter C (2002) *Agrobacterium*-mediated transformation of *Pinus radiata* organogenic tissue using vacuum-infiltration. Plant Cell Tissue Organ Cult 70:51–60
37. Tang W, Sederoff R, Whetten R (2001) Regeneration of transgenic loblolly pine from zygotic embryos transformed with *Agrobacterium tumefaciens.* Planta 213:981–989
38. Lin X, Zhang W, Takechi K, Takio S, Ono K, Takano H (2005) Stable genetic transformation of *Larix gmelinii* L. by particle bombardment of zygotic embryos. Plant Cell Rep 24:418–425
39. Shyamkumar B, Anjaneyulu C, Giri CC (2007) Genetic transformation of *Terminalia chebula* Retz. and detection of tannin in transformed tissue. Curr Sci 92:3–10
40. Parasharami VA, Naik VB, von Arnold S, Nadgauda RS, Clapham DH (2006) Stable transformation of mature zygotic embryos and regeneration of transgenic plants of chir pine (*Pinus roxbughii* Sarg.). Plant Cell Rep 24:708–714
41. Tang W, Newton RJ (2005) Transgenic Christmas trees regenerated from *Agro bacterium tumefaciens* mediated transformation of zygotic embryos using the green fluorescence protein as a reporter. Mol Breed 16:235–246
42. Clapham D, Demel P, Elfstrand M, Koop H-U, Sabala I, von Arnold S (2000) Gene transfer by particle bombardment to embryogenic cultures of *Picea abies* and the production of transgenic plantlets. Scand J For Res 15:151–160
43. Stasolla C, Kong L, Yeung EC, Thorpe TA (2002) Maturation of somatic embryos in conifers: morphogenesis, physiology, biochemistry, and molecular biology. In Vitro Cell Dev Biol Plant 38:93–105
44. Ishida Y, Saito H, Ohta S, Hiei Y, Komari T, Kumashiro T (1996) High efficiency transformation of maize (*Zea mays* L.) mediated by *Agrobacterium tumefaciens.* Biotechnology 14:745–750
45. Zhu H, Muthukrishnan S, Krishnaveni S, Wilde G, Jeoung JM, Liang GH (1998) Biolistic transformation of sorghum using a rice chitinase gene. J Genet Breed 52:243–252
46. Zhao ZY, Gu WN, Cai TS, Tagliani L, Hondred D, Bond D, Schroeder S, Rudert M, Pierce D (2001) High throughput genetic transformation mediated by *Agrobacterium tumefaciens* in maize. Mol Breed 8:323–333
47. Bohorova N, Frutos R, Royer M, Estanol P, Pacheco M, Rascon Q, McLean S, Hoisington D (2001) Novel synthetic *Bacillus thuringiensis* cry1B gene and the cry1B-cry1Ab translational fusion confer resistance to southwestern corn borer, sugarcane borer and fall armyworm in transgenic tropical maize. Theor Appl Genet 103:817–826
48. Brettschneider R, Becker D, Lorz H (1997) Efficient transformation of scutellar tissue of immature maize embryos. Theor Appl Genet 94:737–748
49. Geest AHM, Petolino JF (1998) Expression of a modified green fluorescent protein gene in transgenic maize plants and progeny. Plant Cell Rep 17:760–764
50. Howe AR, Gasser CS, Brown SM, Padgette SR, Hart J, Parker GB, Fromm ME, Armstrong CL (2002) Glyphosate as a selective agent for the production of fertile transgenic maize (*Zea mays* L.) plants. Mol Breed 10:153–164
51. Hueros G, Gomez E, Cheikh N, Edwards J, Weldon M, Salamini F, Thompson RD (1999) Identification of a promoter sequence from the BETL1 gene cluster able to confer transfer-cell-specific expression in transgenic maize. Plant Physiol 121:1143–1152
52. Rasco-Gaunt S, Liu D, Li CP, Doherty A, Hagemann K, Riley A, Thompson T, Brunkan C, Mitchell M, Lowe K (2003) Characterization of the expression of a novel constitutive maize promoter in transgenic wheat and maize. Plant Cell Rep 21:569–576
53. Sangtong V, Moran DL, Chikwamba R, Wang K, Woodman Clikeman W, Long MJ, Lee M, Scott MP (2002) Expression and inheritance of the wheat Glu-1DX5 gene in transgenic maize. Theor Appl Genet 105:937–945
54. Songstad DD, Armstrong CL, Petersen WL, Hairston B, Hinchee MAW (1996) Production of transgenic maize plants and progeny by bombardment of Hi-II immature embryos. In Vitro Cell Dev Biol Plant 32:179–183
55. Wright M, Dawson J, Dunder E, Suttie J, Reed J, Kramer C, Chang Y, Novitzky R, Wang H, Artim L (2001) Efficient biolistic transformation of maize (*Zea mays* L.) and wheat (*Triticum aestivum* L.) using the phosphomannose isomerase gene, pmi, as the

selectable marker. Plant Cell Rep 20:429–436

56. Khanna HK, Daggard GE (2003) *Agrobacterium tumefaciens*-mediated transformation of wheat using a superbinary vector and a polyamine-supplemented regeneration medium. Plant Cell Rep 21:429–436
57. Takumi S, Shimada T (1995) Effects of three promoters on stable integration of the herbicide resistance gene in wheat culture cells through particle bombardment. Bull RIAR Ishikawa Agr Coll 4:9–16
58. Ahmad A, Maqbool S, Riazuddin S, Sticklen MB (2002) Expression of synthetic Cry1Ab and Cry1Ac genes in Basmati rice (*Oryza sativa* L.) variety 370 via *Agrobacterium*-mediated transformation for the control of the European corn borer (*Ostrinia nubilalis*). Dev Biol 38:213–220
59. Pons MJ, Marfa V, Mele E, Messeguer J (2000) Regeneration and genetic transformation of Spanish rice cultivars using mature embryos. Euphytica 114:117–122
60. Baruah WJ, Harwood WA, Lonsdale DA, Harvey A, Hull R, Snape JW (1999) Luciferase as a reporter gene for transformation studies in rice (*Oryza sativa* L.). Plant Cell Rep 18:715–720
61. Hiruki C, Kakuta H, Hashidoko Y, Ge Z, Figueiredo G, Mizutani J (1993) Biolistic delivery of foreign DNA or genomic transcripts of plant virus full-length cDNA clones into monocotyledonous and dicotyledonous plant tissues. Proc Jpn Acad Ser B Phys Biol Sci 69:244–247
62. Perret SJ, Valentine J, Leggett JM, Morris P (2003) Integration, expression and inheritance of transgenes in hexaploid oat (*Avena sativa* L.). J Plant Physiol 160:931–934
63. Kuai B, Perret S, Wan SM, Dalton SJ, Bettany AJE, Morris P (2001) Transformation of oat and inheritance of bar gene expression. Plant Cell Tissue Organ Cult 66:79–88
64. Somers DA, Rines HW, Gu W, Kaeppler HF, Bushnell WR (1992) Fertile, transgenic oat plants. Biotechnology 10:1589–1594
65. Torbert KA, Rines HW, Somers DA (1998) Transformation of oat using mature embryo-derived tissue cultures. Crop Sci 38:226–231
66. Fang YD, Akula C, Altpeter F (2002) *Agrobacterium*-mediated barley (*Hordeum vulgare* L.) transformation using green fluorescent protein as a visual marker and sequence analysis of the T-DNA barley genomic DNA junctions. J Plant Physiol 159:1131–1138
67. Horvath H, Rostoks N, Brueggeman R, Steffenson B, von Wettstein D, Kleinhofs A (2003) Genetically engineered stem rust resistance in barley using the Rpg1 gene. Proc Natl Acad Sci USA 100:364–369
68. Bregitzer P, Tonks D (2003) Inheritance and expression of transgenes in barley. Crop Sci 43:4–12
69. Choi HW, Lemaux PG, Cho MJ (2003) Long-term stability of transgene expression driven by barley endosperm-specific hordein promoters in transgenic barley. Plant Cell Rep 21:1108–1120
70. Harwood WA, Ross SM, Bulley SM, Travella S, Busch B, Harden J, Snape JW (2002) Use of the firefly luciferase gene in a barley (*Hordeum vulgare*) transformation system. Plant Cell Rep 21:320–326
71. Koprek T, Hansch R, Nerlich A, Mendel RR, Schulze J (1996) Fertile transgenic barley of different cultivars obtained by adjustment of bombardment conditions to tissue response. Plant Sci 119:79–91
72. Manoharan M, Dahleen LS (2002) Genetic transformation of the commercial barley (*Hordeum vulgare* L.) cultivar Conlon by particle bombardment of callus. Plant Cell Rep 21:76–80
73. Zhang Y, Darlington H, Jones HD, Halford NG, Napier JA, Davey MR, Lazzeri PA, Shewry PR (2003) Expression of the gamma-zein protein of maize in seeds of transgenic barley: effects on grain composition and properties. Theor Appl Genet 106:1139–1146
74. Abumhadi N, Trifonova A, Takumi S, Nakamura C, Todorovska E, Getov L, Christov N, Atanassov A (2001) Development of the particle inflow gun and optimizing the particle bombardment method for efficient genetic transformation in mature embryos of cereals. Biotechnol Biotech 15:87–96
75. Gurel F, Gozu Kirmizi N (2000) Optimization of gene transfer into barley (*Hordeum vulgare* L.) mature embryos by tissue electroporation. Plant Cell Rep 19:787–791
76. Zhao ZY, Cai TS, Miller M, Wang N, Pang H, Rudert M, Schroeder S, Hondred D, Seltzer J, Pierce D (2000) *Agrobacterium*-mediated sorghum transformation. Plant Mol Biol 44:789–798
77. Jeoung JM, Krishnaveni S, Muthukrishnan S, Trick HN, Liang GH (2002) Optimization of sorghum transformation parameters using genes for green fluorescent protein and beta-glucuronidase as visual markers. Hereditas 137:20–28
78. Tadesse Y, Sagi L, Swennen R, Jacobs M (2003) Optimisation of transformation conditions and production of transgenic sorghum (*Sorghum bicolor*) via microparticle bombardment. Plant Cell Tissue Organ Cult 75:1–18

79. Goldman JJ, Hanna WW, Fleming G, Ozias-Akins P (2003) Fertile transgenic pearl millet plants recovered through microprojectile bombardment and phosphinothricin selection and apical meristem-, inflorescence-, and immature embryo-derived embryogenic tissues. Plant Cell Rep 21:999–1009
80. Lambe P, Dinant M, Deltour R (2000) Transgenic pearl millet (*Pennisetum glaucum*): I. Transgenic crops. Biotechnol Agric For 46:84–108
81. McKently AH, Moore GA, Doodstdar H, Niedz RP (1995) *Agrobacterium*-mediated transformation of peanut (*Arachis hypogaea* L.) embryo axes and the development of transgenic plants. Plant Cell Rep 14:699–703
82. Ducrocq C, Sangwan RS, Sangwan-Norreel BS (1994) Production of *Agrobacterium*-mediated transgenic fertile plants by direct somatic embryogenesis from immature zygotic embryos of *Datura innoxia*. Plant Mol Biol 25:995–1009
83. Clapham D, Damel P, Elfstrand M, Koop H-U, Sabala I, von Arnold S (2000) Gene transfer by particle bombardment to embryogenic cultures of *Picea abies* and the production of transgenic plantlets. Scand J For Res 15: 151–161

Chapter 22

Genetic Transformation Using Maize Immature Zygotic Embryos

Bronwyn Frame, Marcy Main, Rosemarie Schick, and Kan Wang

Abstract

Epidermal and subepidermal cells in the abaxial, basal region of the maize (*Zea mays* L.) immature zygotic embryo (IZE) scutellum can be induced by exogenous auxin to proliferate and undergo somatic embryogenesis. Successful genetic transformation of IZEs depends not only on optimizing transformation parameters for these totipotent cells, but also on achieving high embryogenic callus induction frequency (ECIF) in a population of targeted explants. In maize, ECIF is strongly influenced by genotype, the tissue culture media used, and the interaction of these two factors. Altering tissue culture media components to increase ECIF and/or transformation frequency (TF) has been one approach used to expand the range of maize genotypes amenable to genetic transformation using the IZE. This chapter outlines such an approach – an *Agrobacterium*-mediated transformation protocol is used for direct-targeting IZEs of the hybrid Hi Type II and inbred B104 lines. Two different media regimes are used for successful culture and transformation of two distinct genotypes.

Key words: *Agrobacterium tumefaciens*, B104, Callus induction frequency, Embryogenic callus, Genetic transformation, Hi II, Immature zygotic embryo, Maize

1. Introduction

Evidence that scutellar cells of the maize (*Zea mays* L.) immature zygotic embryo (IZE) can be induced to produce embryogenic callus in the presence of exogenous auxin was first reported over 30 years ago (1). Regeneration of fertile plants from this callus (2) demonstrated the totipotent nature of these epidermal and subepidermal meristematic cells found in the abaxial, basal region of the IZE scutellum (3). Both compact Type I and friable Type II embryogenic callus phenotypes (4) were observed to originate from these scutellar cells (3).

Trevor A. Thorpe and Edward C. Yeung (eds.), *Plant Embryo Culture: Methods and Protocols*,
Methods in Molecular Biology, vol. 710, DOI 10.1007/978-1-61737-988-8_22,

In the last two decades, numerous studies have shown that callus and cell suspensions derived from these totipotent cells are also transformation competent (5, 6). Of particular impact were reports demonstrating that the IZE scutellum can itself be directly targeted for genetic transformation using electroporation (7), the biolistic gun (8, 9) and *Agrobacterium*-mediated methods (10–12). Expected progeny segregation ratios for the inherited transgene provided evidence that, whether targeted embryos formed Type I (7, 10, 13) or Type II (9, 12, 14) callus, transformation occurred in a single cell in this meristematic region of the embryo scutellum.

A major benefit of direct-targeting the IZE for genetic transformation is a reduction in the in vitro culture period required to recover transgenic plants (8). This not only reduces the amount of labor required for routine maintenance of cell cultures, it also minimizes aberrant plant phenotypes caused by culture-induced somaclonal variation thereby favoring transgenic plant fertility (9). A practical drawback to using immature embryos for transformation is the dependency on year-round, high quality greenhouse space for growing embryo donor plants. Perhaps the greatest hurdle to using maize IZEs for transformation is achieving an adequate Type I or Type II embryogenic callus induction frequency (ECIF) in a targeted explant population (9, 13, 15, 16). While high ECIF does not guarantee success (13, 17) it is a necessary prerequisite for achieving a robust transformation protocol using the IZE. In maize, the frequency of embryogenic callus induction is genotype specific (18–20) and influenced by factors such as tissue culture media components (4, 19, 21, 22), embryo size (23), and environmental conditions of the embryo donor plants (23). Transformation and cocultivation parameters can themselves affect ECIF and need to be optimized while maintaining adequate ECIF after transgene delivery (9, 12, 15, 16, 24). Maize genotypes which exhibit high ECIF (~100% Type I or Type II callus phenotype) in culture such as the hybrid genotype Hi Type II or Hi II (25), and inbred lines A188 or H99 (20) have been successfully transformed using super-binary (10, 14, 16) or standard-binary (12, 26) *Agrobacterium* vectors to direct-target IZEs. Efforts to extend these routine transformation protocols to elite or diverse inbred lines have focused on breeding responsiveness into the genotype of choice (27), optimizing an array of transformation parameters (14, 28, 29), or altering culture media components to improve ECIF (13,17) or transformation frequency (TF) (12–14, 16, 26).

This chapter describes side by side protocols for using a standard-binary *Agrobacterium* vector and two media regimes to transform IZEs from two distinct maize genotypes: the Hi II hybrid line (25) and inbred line B104 (30).

2. Materials

2.1. Plant Materials

1. Hi II. F_1 seed of the hybrid Hi II line (25) is produced in the field (Ames, IA) each summer by pollinating Hi II parent A silks with Hi II parent B pollen (Hi II pA x Hi II pB, see Note 1). These two parent seed germplasms were obtained from the Maize Genetics Coop (https://maizecoop.cropsci.uiuc.edu/request/). F_2 IZEs used for all Hi II transformation experiments are produced from sib-pollinated F_1 plants grown year round in the ISU Plant Transformation Facility greenhouse in Ames, Iowa as described in our greenhouse protocol at: http://www.agron.iastate.edu/ptf/protocol/Greenhouse%20Protocol.pdf. Nine (in summer) to eleven (in winter) day-old ears are harvested when embryo size is between 1.2 and 1.8 mm. After harvest, maize ears (in their husks and inside their pollination bag) are stored in the refrigerator (4°C) in a loosely sealed dark plastic bag. Ears are stored for at least 1 and at most 4 days before being used for *Agrobacterium*-mediated genetic transformation experiments (see Notes 2 and 3).
2. B104. Greenhouse or field grown (see Note 4) embryo donor ears of maize inbred line B104 (30) are harvested 10 (from summer greenhouse) to 13 (from summer field) days after cross pollination when IZEs are 1.5–2 mm long (see Note 5). B104 seed can be obtained from the Iowa State University Committee for Agriculture Development (http://www.ag.iastate.edu/centers/cad/corn.html). Greenhouse care of B104 plants and storage of ears after harvest are identical to that described for Hi II (see Note 6).

2.2. Plasmids and *A. tumefaciens* Strains Used for Hi II and B104 Genetic Transformation

The cloning vector used routinely for Hi II and B104 *Agrobacterium*-mediated transformation of IZEs is pTF101.1 (31) – a derivative of the pPZP binary vector with a broad host range pVS1 origin of replication (32). pTF101.1 is an 11.6 kb standard binary vector in *A. tumefaciens* strain EHA101 (33) and contains a spectinomycin-resistant marker gene (*aadA*) for bacterial selection. In this vector, the herbicide resistant *bar* selectable marker gene (34) is driven by the cauliflower mosaic virus (CaMV) double 35S promoter (2 × P35S). The tobacco etch virus (TEV) translational enhancer (35) was inserted at the 5′ end of the *bar* gene and the soybean vegetative storage protein terminator (36) was cloned to its 3′ end. A multiple cloning site for introducing any gene of interest (GOI) into pTF101.1 between the right border region and the plant selectable marker gene carries unique restriction sites for *Bam*H I, *Eco*R I, *Hin*d III, *Sac* I, *Sma* I, and *Xba* I (see Note 7). The stock solutions and culture media for *A. tumefaciens* are as follows.

1. Spectinomycin sulfate (Sigma, St. Louis, MO, USA): 100 mg/mL stock in ddH_2O. Sterilize by filtration through a 0.2 μm membrane (Fisher Scientific Inc, Pittsburgh, PA, USA), aliquot (0.05 mL) and store at –20°C for up to 6 months (see Note 8).
2. Kanamycin sulfate (Sigma): 10 mg/mL stock in ddH_2O. Sterilize by filtration. Dispense in 0.25 mL aliquots in eppendorf tubes and store at –20°C for up to 6 months.
3. YEP medium (37): 5 g/L yeast extract, 10 g/L peptone, 5 g/L NaCl, pH 6.8. For solid medium, add 15 g/L Bacto agar. Appropriate antibiotics are added to autoclaved medium after it cools to 50°C. For the strain EHA101 containing pTF101.1, the final antibiotic concentrations are: 50 mg/L kanamycin (for maintaining of the disarmed Ti plasmid pEHA101), 100 mg/L spectinomycin (for maintaining the binary vector plasmid pTF101.1).

2.3. Culture Media for Maize Transformation

2.3.1. Stock Solutions for Transformation of Hi II and B104

1. N6 vitamin stock (38): 1.0 g glycine, 0.5 g thiamine HCl, 0.25 g pyridoxine HCl, and 0.25 g nicotinic acid are dissolved in 500 mL ddH_2O. This stock solution (1,000×) is filter sterilized, and stored at –20°C in 40 mL aliquots, which are thawed and used over a period of weeks as needed.
2. MS vitamin stock (39) (modified, see Note 9): 1.0 g glycine, 0.25 g thiamine HCl, 0.25 g pyridoxine HCl, and 0.025 g nicotinic acid are dissolved in 500 mL ddH_2O. This stock solution (1,000×) is filter sterilized, and stored at –20°C in 40 mL aliquots which are thawed and used over a period of weeks.
3. 2,4-D: 200 mg of powdered 2,4-dichlorophenoxyacetic acid (2,4-D) is dissolved in 5 mL of 1 N KOH on low heat and brought up to a final volume of 200 mL with ddH_2O. The stock solution (1 mg/mL) is stored at 4°C (see Note 10).
4. Dicamba: 0.0663 g of Dicamba (3,6,dichloro-*o*-anisic acid) is dissolved in 1 mL 1 N KOH on low heat and brought up to a final volume of 10 mL with ddH_2O. The stock solution (30 mM) is stored at 4°C.
5. Bialaphos: 100 mg of Bialaphos (Gold Biotechnology, Duchefa, St. Louis, USA) is dissolved in 100 mL of ddH_2O. The stock solution (1 mg/mL) is filter sterilized and stored at 4°C for up to 6 months.
6. Glufosinate: 100 mg of glufosinate ammonia is dissolved in 100 mL of ddH_2O. The stock solution (1 mg/mL) is filter sterilized and stored at 4°C for up to 6 months.
7. Acetosyringone (AS): 0.392 g of AS is dissolved in 10 mL of dimethyl sulfoxide (DMSO). This solution is diluted 1:1 with

ddH_2O and filter-sterilized. Aliquots (0.5 mL) of stock solution (100 mM) are stored at −20°C for up to 6 months (see Note 11).

8. Cysteine: 500 mg of L-cysteine (Sigma) is dissolved in 5 mL of ddH_2O. The stock solution (100 mg/mL) is filter sterilized and added the same day to autoclaved, cooled cocultivation medium for a final concentration of 300 mg/L. Any unused stock solution is discarded.
9. Silver Nitrate: 0.85 g of silver nitrate is dissolved in 100 mL of ddH_2O. The stock solution (50 mM) is filter sterilized and stored in a foil-wrapped duran at 4°C for up to 1 year.
10. Cefotaxime: 1.0 g of cefotaxime (Phytotechnology Laboratories, Overland Park, KS, USA) is dissolved in 5 mL ddH_2O. The stock solution (200 mg/mL) is filter sterilized, aliquoted (0.250 mL) and stored at −20°C for up to 1 month.
11. Vancomycin: 1.0 g of vancomycin hydrochloride (Phytotechnology Laboratories) is dissolved in 5 mL ddH_2O. The stock solution (200 mg/mL) is filter sterilized, aliquoted (0.250 mL), and stored at −20°C for up to 1 month.
12. Carbenicillin: 1.0 g of carbenicillin (Phytotechnology Laboratories) is dissolved in 10 mL ddH_2O. The stock solution (100 mg/mL) is filter sterilized, aliquoted (1.25 mL) and stored at −20°C for up to 3 months (see Note 12).

2.3.2. Media for Agrobacterium-Mediated Transformation of Hi II

Media 1–5 are after Zhao et al. (14) with the addition of cysteine (300 mg/L) to cocultivation medium and the use of cefotaxime and vancomycin instead of carbenicillin for counter-selection of *Agrobacterium* after cocultivation. Solid media (Media 2–5) use 100×25 mL Petri plates and are stored at room temperature.

1. Infection (liquid): 4 g/L N6 salts (38), 1 mL/L N6 vitamin stock, 1.5 mg/L 2,4-D, 0.7 g/L L-proline, 68.4 g/L sucrose, and 36 g/L glucose, pH 5.2. This medium is filter sterilized and stored at 4°C. AS (100 μM) is added prior to use.
2. Cocultivation (see Note 13): 4 g/L N6 salts, 1.5 mg/L 2, 4-D, 0.7 g/L L-proline, 30 g/L sucrose, and 3 g/L Gelrite (bioWorld PlantMedia, Dublin, OH, USA), pH 5.8. Filter sterilized N6 vitamin stock (1 mL/L), silver nitrate (5 μM), AS (100 μM), and L-cysteine (300 mg/L) are added after autoclaving.
3. Resting: 4 g/L N6 salts, 1.5 mg/L 2,4-D, 0.7 g/L L-proline, 30 g/L sucrose, 0.5 g/L 2-(4-morpholino)-ethanesulfonic acid (MES), and 8 g/L purified agar (Sigma), pH 5.8. Filter sterilized N6 vitamin stock (1 mL/L), cefotaxime (100 mg/L), vancomycin (100 mg/L), and silver nitrate (5 μM) are added after autoclaving (see Notes 14 and 15).

4. Selection I: 4 g/L N6 salts, 1.5 mg/L 2,4-D, 0.7 g/L L-proline, 30 g/L sucrose, 0.5 g/L MES, and 8 g/L purified agar, pH 5.8. Filter sterilized N6 vitamin stock (1 mL/L), cefotaxime (100 mg/L), vancomycin (100 mg/L), silver nitrate (5 μM), and bialaphos (1.5 mg/L) are added after autoclaving.
5. Selection II: The same as Selection I except that bialaphos concentration is increased to 3 mg/L.
6. Pre-regeneration medium (see Note 16): 4.3 g/L MS Salts (39), 1 mL/L (1,000×) MS vitamin stock (modified), 100 mg/L myo-inositol, 0.25 mL/L 2,4-D, 30 g/L sucrose, 3 g/L gelrite, pH 5.8. Filter-sterilized bialaphos (2 mg/L) and cefotaxime (100 mg/L) are added after autoclaving. Use 100 × 15 Petri plates.

2.3.3. Media for Agrobacterium-Mediated Transformation of B104 Inbred Line

All solid media described below use 100 × 25-mm Petri plates and are stored at room temperature. Media is modified from Carvalho et al. (22) and L-cysteine (300 mg/L) is added to the cocultivation medium.

1. Infection (liquid): 4.3 g/L MS salts, 1 mL/L modified MS vitamin stock, 0.5 mL/L dicamba, 0.7 g/L L-proline, 68.4 g/L sucrose, and 36 g/L glucose, pH 5.2. This medium is filter sterilized and stored at 4°C. AS (100 μM) is added prior to use (see Note 17).
2. Cocultivation: 4.3 g/L MS salts, 0.5 mL/L dicamba, 0.7 g/L L-proline, 100 mg/L casein hydrolysate, 100 mg/L myo-inositol, 30 g/L sucrose, and 2.3 g/L Gelrite, pH 5.8. Filter sterilized modified MS vitamin stock (1 mL/L), silver nitrate (88 μM), AS (100 μM), and L-cysteine (300 mg/L) are added after autoclaving (see Note 18).
3. Resting: 4.3 g/L MS salts, 0.5 mL/L dicamba, 0.7 g/L L-proline, 0.5 g/L MES, 100 mg/L casein hydrolysate, 100 mg/L myo-inositol, 30 g/L sucrose, and 2.3 g/L Gelrite, pH 5.8. Filter sterilized modified MS vitamin stock (1 mL/L), silver nitrate (88 μM), and carbenicillin (250 mg/L) are added after autoclaving.
4. B104 Selection I: 4.3 g/L MS salts, 0.5 mL/L dicamba, 0.7 g/L L-proline, 0.5 g/L MES, 100 mg/L casein hydrolysate, 100 mg/L myo-inositol, 30 g/L sucrose, and 2.3 g/L Gelrite, pH 5.8. Filter sterilized modified MS vitamin stock (1 mL/L), silver nitrate (88 μM), bialaphos (2 mg/L), and carbenicillin (250 mg/L) are added after autoclaving (see Note 19).
5. B104 Selection II: The same as B104 Selection I medium except that the bialaphos concentration is increased to 6 mg/L.

2.4. Culture Media for Regeneration of Hi II and B104

Regeneration media, after McCain et al. (40), uses 100×25 mL Petri plates and is stored at room temperature.

1. Regeneration I: 4.3 g/L MS salts, 1 mL/L modified MS vitamin stock, 100 mg/L myo-inositol, 60 g/L sucrose, 3 g/L Gelrite, pH 5.8. Filter-sterilized glufosinate ammonia (6 mg/L) and cefotaxime (100 mg/L) are added after autoclaving.
2. Regeneration II: The same as Regeneration I with the sucrose concentration reduced to 30 g/L and no glufosinate or cefotaxime is added.

2.5. Equipment

1. Horizontal laminar flow benches (The Baker Company, Sanford, ME, USA)
2. Dark biological incubator (I36NL, Percival Scientific, Perry, IA, USA)
3. Light biological incubator (Cu36L5, Percival Scientific)
4. Steriguard 350 bead sterilizers (Inotech Biosystems International, Rockville, MD, USA).
5. Vortex Genie (Fisher Scientific, USA)

3. Methods

3.1. *Agrobacterium*-Mediated Transformation of Immature Zygotic Embryos

3.1.1. *Agrobacterium* Preparation

1. The vector system, pTF101.1 in strain EHA101, is stored as a glycerol stock at −80°C.
2. Every 4 weeks, a "mother" plate is re-initiated from this long-term glycerol stock by streaking the bacteria to YEP (with antibiotics) and growing it for 2 days at 28°C.
3. The "mother" plate is then kept in the refrigerator (4°C) and used as a source plate for plating *Agrobacteria* cells (at 19°C for 3 days) in preparation for twice-weekly experiments (see Note 20).

3.1.2. Embryo Dissection

1. Dehusk the ear, break off the tip of the cob and insert a pair of numbered forceps. This labels the ear while acting as a "handle" for aseptic manipulation during dissection. In a laminar flow bench, place up to 15 prepared ears in a sterile, 4 L beaker. Do not use any ears exhibiting extreme tip rot or discolored kernels (see Note 21).
2. Add ~2 L of sterilizing solution (50% commercial bleach (6% hypochlorite) in ddH_2O + 1 drop of surfactant Tween 20 per liter) to completely submerge the ears while leaving the forceps handles protruding (see Note 22).

3. During the 20-min disinfection, occasionally grasp forceps and swirl the ears in an effort to dislodge air bubbles. Pour off the bleach solution and rinse the ears three times using at least 2 L of sterile ddH_2O at each rinse. The final rinse is drained off and the beaker of ears is left (covered) in the bench until dissections begin.
4. Using aseptic technique, and working in a laminar flow bench, hold onto the end of the forceps, prop the surface-sterilized ear on a large (150 × 15 mm) sterile Petri-plate, and cut off the top 1–2 mm of the kernel crowns with a sharp scalpel blade. Steriguard 350 bead sterilizers are used for sterilization of utensils throughout this protocol.
5. To excise an embryo, insert the end of a sharpened spatula between the endosperm and pericarp at the basipetal side of the kernel and pop the endosperm out of the seed coat. The embryo axis side of the untouched embryo will be visible and the scutellum side will be nested in the endosperm. Gently coax the IZE onto the spatula tip and transfer it directly to liquid infection medium (see Note 23).

3.1.3. Agrobacterium Infection

1. Grow *Agrobacterium* cultures for 3 days at 19°C (or 2 days at 28°C) on solid YEP medium amended with antibiotics.
2. To begin an experiment, scrape one full loop (3 mm) of bacteria culture from the plate and suspend it in 5 mL infection medium supplemented with 100 μM AS in a 50 mL Falcon tube. Affix the tube horizontally to a Vortex Genie (Fisher) platform head using lab tape and shake on lowest setting for 2 h at room temperature. Using liquid infection medium (with AS), adjust to between $OD_{550} = 0.30$ and 0.40 just prior to use.
3. Once this 2 h *Agrobacterium* pre-culture step is complete, dissect up to 100 IZEs directly into a 2-mL Eppendorf tube filled with *Agrobacterium*-free infection medium (with 100 μM AS). These wash tubes are prepared 2 h ahead of time and stored at 4°C until dissection begins.
4. Remove this first wash then wash the embryos a second time with 1 mL of the same medium. After removing the final wash, add 1 mL of *Agrobacterium* suspension ($OD_{550} = 0.30–0.40$).
5. To infect the embryos, gently invert the tube 20 times before resting it on its side (in the dark) for 5 min with embryos submerged in the *Agrobacterium* suspension (see Note 24). These and all subsequent tissue culture steps are carried out in a laminar flow bench using aseptic technique.

3.1.4. Co-Cultivation

1. After the 5 min infection, use a 1-mL Pipetman equipped with a wide-bore pipet tip (see Note 25) to gradually transfer the embryos, along with a minimum amount of *Agrobacterium* suspension, out of the Eppendorf tube and onto the surface of the cocultivation medium. Embryos are collected, a few at

a time, with minimal liquid uptake at each transfer to avoid adhesion of the embryos to the inside of the pipet tip.

2. When embryo transfer is complete, use a 1-mL tip to remove excess *Agrobacterium* suspension from the surface of the co-cultivation medium and the area surrounding each embryo. Collect the used bacterial suspension in a disposable Petri-dish (see Notes 26 and 27).
3. Leave the lid of the cocultivation plate ajar for up to 1 h to let the medium and embryo surfaces dry further before orienting each embryo scutellum side up with the aid of a stereo microscope.
4. Wrap plates with vent tape (air permeable adhesive tape) and incubate at 20°C (dark) for 3 days in a biological incubator.

3.1.5. Resting

1. After 3 days cocultivation, transfer all embryos to resting medium at 28°C (dark) for 7 days.
2. Continue to transfer all embryos throughout the following selection steps. Do not discard any embryos prematurely. Tissue culture plates are incubated in a biological chamber throughout resting and selection steps.

3.2. Selection for Stable Transformation Events

3.2.1. Hi II Events

1. After 7 days on resting medium (see Note 28), use sterile forceps to transfer embryos to Selection I medium (35 IZEs per plate) containing 1.5 mg/L bialaphos, for 2 weeks followed by two more 2-week passages on Selection II medium (3 mg/L bialaphos). Plates are wrapped with Parafilm® and incubated at 28°C in the dark. All Hi II selection steps are done without the aid of a stereo microscope.
2. As early as five and as late as 10 weeks after infection, putative Type II callus events are visible (with the naked eye) emerging from a subset of embryos.
3. Putative events are transferred away from the original experiment plate to a fresh plate of Selection II medium for an additional, 2-week subculture to verify that they are bialaphos resistant. We refer to this latter step as "picking" putative events.
4. If a putative callus event continues to grow rapidly, it is assigned an ID number. Callus events containing stalked somatic embryos (prescreened with the aid of a stereo microscope) are subcultured, one event per plate, to Pre-regeneration medium.
5. If the diameter of the callus clump is greater than 2 cm at this stage, it is divided into smaller pieces (1 cm) at transfer. Petri plates are wrapped with Parafilm® and incubated in the dark (25°C) for 10–14 days.
6. Average TF for Hi II using this protocol is 8%, or 8 independent, bialaphos resistant Type II callus per 100 infected (and selected) IZEs.

3.2.2. B104 Events

1. After 7 days on B104 resting medium (see Note 29), transfer all embryos to B104 Selection I medium (35 IZEs per plate) containing 2 mg/L bialaphos, for 2 weeks followed by two more 2-week passages on B104 Selection II medium containing 6 mg/L bialaphos. Plates are wrapped with Parafilm® and incubated in the dark (28°C).
2. As early as six, and as late as 12 weeks after infection, putative Type I callus events are visible emerging from a subset of selected IZEs (see Note 30).
3. Putative events are transferred away from the original experiment plate to a fresh plate of B104 Selection II medium for 2 additional weeks. Continued vigorous proliferation after this "picking" step verifies that the event is bialaphos resistant.
4. Unlike the corresponding Hi II step in which the clump of Type II callus representing one putative event is kept intact, when a B104 putative event is picked, embryogenic Type I callus is separated from non-embryogenic callus lobes and differentiating leaf or root portions of the callus clump with the aid of a stereo microscope. Only the embryogenic callus is retained and broken into 0.25 cm pieces on the surface of a fresh plate of B104 Selection II medium.
5. After 2 weeks, the Type I embryogenic callus proliferating from some or all of these pieces is regrouped and subcultured in 0.5 cm pieces, again using the stereo microscope, to the surface of B104 Selection II medium in preparation for naming and regeneration.
6. Average TF for B104 using this protocol is 3%, or 3 independent, bialaphos-resistant Type I calluses per 100 infected (and selected) IZEs.

3.3. Regeneration of Transgenic Plants

3.3.1. Hi II

1. With the aid of a stereo microscope, use sterile scalpels or needle nose forceps to transfer 12–15 small pieces (4 mm) of somatic embryo-enriched callus from Pre-regeneration medium to Regeneration I medium. Wrap plates with vent tape and incubate at 25°C (dark, see Note 31).
2. After 2 weeks on Regeneration I medium, somatic embryos appear swollen, opaque and white. In some cases, the coleoptile is already visible emerging from these germinating, somatic embryos.
3. Use a stereo microscope to transfer ~12 individual, mature somatic embryos from Regeneration I medium to the surface of Regeneration II medium for germination in a lighted biological incubator (25°C, 80–100 $\mu E/m^2/s$ light intensity, 16:8 photoperiod).
4. Wrap Petri plates with vent tape. Hi II plantlets sprout leaves and roots on this medium within 1 week and are ready for

transfer directly to soil about 3 days later (10 days after transfer to Regeneration II medium).

3.3.2. B104 (see Note 32)

1. Using a stereo microscope, transfer 15–20, 5 mm embryogenic Type I callus pieces (pried apart, not cut) from the surface of B104 Selection II medium to the surface of Regeneration I medium. Multiple somatic embryos may be fused together in one piece of callus. Wrap Petri plates with vent tape and incubate in the dark (25°C).
2. After 3 weeks, the majority of callus pieces will produce one or more mature somatic embryos. Like the corresponding Hi II regeneration step, the B104 mature somatic embryos will appear opaque and white, but unlike Hi II, they will form at lower frequency and will, in many cases, be fused together.
3. Using a stereo microscope, pry these mature somatic embryos apart from any unhardened callus and from each other where possible without damaging embryo integrity.
4. Transfer these pieces (fused or not), 15 per plate, to Regeneration II medium for germination in the light (25°C, 80–100 μE/m^2/s light intensity, 16:8 photoperiod). Germinated B104 plantlets with roots and shoots are ready for transfer to soil from between 7 and 14 days later (see Notes 33 and 34).

3.4. Growth Chamber and Greenhouse Plant Care

1. A detailed protocol for growing immature embryo donor plants from seed, and for growing regenerated transgenic plantlets to maturity, can be found in the ISU Plant Transformation Facility greenhouse protocol at: http://www.agron.iastate.edu/ptf/protocol/Greenhouse%20Protocol.pdf. Our greenhouse is located in Ames, IA, USA.
2. While this protocol provides helpful guidelines for growing greenhouse maize, it should be noted that conditions for success will vary depending on location and greenhouse conditions.

4. Notes

1. These two parents may differ in vigor. Multiple plantings of both parents ensure constant availability of parent A silks × parent B pollen for the F_1 cross.
2. In general, greenhouse derived Hi II IZEs transform at higher rates (using *Agrobacterium*-mediated methods) than do field embryos using this protocol, although transgenic events have been recovered from both sources.
3. We generally recover at least 120 IZEs from one greenhouse-grown Hi II ear.

4. In general, greenhouse derived IZEs of B104 transform at higher rates using *Agrobacterium*-mediated methods than do field embryos using this protocol, although transgenic events have been recovered from both.
5. Post-infection ECIF for 1.2 mm B104 embryos is lower than for 1.5 mm B104 embryos or 1.2 mm Hi II embryos.
6. We generally recover ~150 IZEs from one greenhouse-grown B104 ear.
7. We have also used strains LBA4404 and GV3101 with this vector with varying degrees of success.
8. Spectinomycin may come out of solution in the freezer and must be resuspended after thawing and before using.
9. Modified MS vitamins (13) contain higher thiamine HCl and lower nicotinic acid concentrations compared to MS vitamins.
10. Use low heat. Do not boil the 2,4-D while dissolving it in KOH.
11. AS will sometimes precipitate after freezer storage and is re-dissolved by vortexing for 15 min.
12. Carbenicillin efficacy may vary by lot number.
13. Cocultivation medium is either 1 or 4 days old when used.
14. Resting medium is made in small batches to ensure that it is as fresh as possible at use (<3 weeks old).
15. Vancomycin and cefotaxime will form a precipitate if mixed together. Add each to media separately and stir well after adding.
16. This medium is used to slow Hi II callus growth and encourage somatic embryo formation.
17. This is our current liquid infection medium for B104 and is modified from Frame et al. (13).
18. This cocultivation media is also used at 1 or 4 days old and when solidified is hazy compared to Hi II cocultivation medium. Stir well before pouring.
19. This is our current selection scheme for B104 and is modified from Frame et al. (13).
20. We compared the effect of using refrigerator-stored (4°C) *Agrobacterium* mother plates, or –80°C stored glycerol stocks to initiate the 19°C/3 day bacteria plate used for infection experiments. The average TF for embryos infected with the vector pTF102 (12) in EHA101 initiated from a 4°C mother plate was 6.4%. For embryos infected with *Agrobacteria* initiated from glycerol stock (–80°C), TF was 5.6%.
21. Pink kernels in particular may be an indication of bacterial contamination.
22. We reuse this bleach once and store it in the dark between uses.

23. A skilled technician can dissect at least 150 IZEs per ½ h. Do not damage embryos at dissection or dig around the ear for embryos that are not easily retrieved, as this increases the probability of introducing contamination into your experiment tube.
24. We do not leave embryos in the wash for extended periods of time. Washing, infection and plating to cocultivation medium steps are all carried out without interruption.
25. One mL filtered pipet tips are trimmed using scissors to make a 3 mm bore hole and re-autoclaved before using.
26. This waste, along with all tissue culture plates, medium (liquid or solid), or plant tissues exposed to *Agrobacterium* and the genetically modified DNA it contains are autoclaved as biohazard waste before disposal.
27. If infecting multiple constructs on the same day, be sure to discard the *Agrobacterium*-liquid disposal dish between constructs so that no back splashing occurs; this may result in cross contamination between constructs.
28. After 1 week on resting medium, ECIF for *Agrobacterium*-infected Hi II embryos cocultivated on medium containing 300 mg/L cysteine is ~85%.
29. After 1 week on resting medium, ECIF for *Agrobacterium*-infected B104 embryos cocultivated on medium containing 300 mg/L cysteine is ~70% (13).
30. B104 Type I putative callus events grow vigorously and often appear to "dig into" the medium surface.
31. Do not overfill the plate and keep pieces small and enriched with stalked embryos – the key to this regeneration method is to induce differentiation of the preformed somatic embryos through desiccation and slowed growth.
32. The regeneration method and media described here were reported in Frame et al. (13) and take 35 days to recover transgenic plants. Regeneration protocol comparisons carried out since 2006 using non transgenic B104 callus indicate that 5 plantlets per plate can be regenerated within 15 days using a regeneration protocol modified from Zhao et al. (14) in which the media includes 6% sucrose and zeatin, and for which all regeneration steps are carried out in the light. To date, we have not compared these regeneration protocols using transgenic B104 callus.
33. Plantlet recovery for B104 may require in vitro pruning. Subculture sprouting plantlets to a fresh plate of Regeneration II medium after 10 days in the light to encourage maximum plantlet recovery.
34. To confirm that the *bar* gene is expressed in B104 regenerated plants, 2–3 weeks after being taken to soil, plantlets are sprayed with 500 mg/L glufosinate prepared from the herbicide Liberty® (Bayer Crop Sciences, USA) and 0.1% Tween 20 (v/v).

Acknowledgments

Our thanks to Jennifer McMurray and Tina Paque for their contributions in the laboratory and greenhouse, and to Dr. Arnel Hallauer for providing the original B104 seed. This work is supported partially by the National Science Foundation (DBI #0110023), the Iowa State University Agricultural Experiment Station, the Office of Biotechnology, the Plant Science Institute, and the Baker Endowment Advisory Council for Excellence in Agronomy at Iowa State University.

References

1. Green CE, Phillips RL (1975) Plant regeneration from tissue cultures of maize. Crop Sci 15:417–421
2. Vasil V, Vasil IK, Lu C (1984) Somatic embryogenesis in long-term callus cultures of *Zea mays* L (Gramineae). Am J Bot 71:158–161
3. Fransz PF, Schel JHN (1990) Cytodifferentiation during the development of friable embryogenic callus in maize (*Zea mays*). Can J Bot 69:26–33
4. Armstrong CL, Green CE (1985) Establishment and maintenance of friable, embryogenic maize callus and the involvement of L-proline. Planta 164:207–214
5. Gordon-Kamm WJ, Spencer TM, Mangano ML, Adams TR, Daines RJ, Start WG, O'Brien JV, Chambers SA, Whitney J, Adams R, Willetts NG, Rice TB, Mackey CJ, Krueger RW, Kausch AP, Lemaux PG (1990) Transformation of maize cells and regeneration of fertile transgenic plants. Plant Cell 2:603–618
6. Fromm ME, Morrish F, Armstrong C, Williams R, Thomas J, Klein TM (1990) Inheritance and expression of chimeric genes in the progeny of transgenic maize plants. Bio/Technology 8:833–839
7. D'Halluin K, Bonne E, Bossut M, De Beuckeleer M, Leemans J (1992) Transgenic maize plants by tissue electroporation. Plant Cell 4:1495–1505
8. Koziel MG, Beland GL, Bowman C, Carozzi NB, Crenshaw R, Crossland L, Dawson J, Desai N, Hill M, Kadwell S, Launis K, Lewis K, Maddox D, McPherson K, Meghji MR, Merlin E, Rhodes R, Warren GW, Wright M, Evola SV (1993) Field performance of elite transgenic maize plants expressing and insecticidal protein derived from *Bacillus thuringiensis*. Bio/Technology 11:194–200
9. Songstad DD, Armstrong CL, Petersen WL, Hairston B, Hinchee MAW (1996) Production of transgenic maize plants and progeny by bombardment of Hi-II immature embryos. In Vitro Cell Dev Biol Plant 32:179–183
10. Ishida Y, Saito H, Ohta S, Hiei Y, Komari T, Kumashiro T (1996) High efficiency transformation of maize (*Zea mays* L.) mediated by *Agrobacterium tumefaciens*. Nat Biotechnol 14:745–750
11. Zhao ZY, Gu W, Cai T, Tagliani LA, Hondred D, Bond D, Krell S, Rudert ML, Bruce WB, Pierce DA (1998) Molecular analysis of T0 plants transformed by *Agrobacterium* and comparison of *Agrobacterium*-mediated transformation with bombardment transformation in maize. Maize Gen Coop Newsl 72:34–37
12. Frame BR, Shou H, Chikwamba RK, Zhang Z, Xiang C, Fonger TM, Pegg SE, Li B, Nettleton DS, Pei D, Wang K (2002) *Agrobacterium tumefaciens*-mediated transformation of maize embryos using a standard binary vector system. Plant Physiol 129:13–22
13. Frame BR, McMurray JM, Fonger TM, Main ML, Taylor KW, Torney FJ, Paz MM, Wang K (2006) Improved *Agrobacterium*-mediated transformation of three maize inbred lines using MS salts. Plant Cell Rep 25:1024–1034
14. Zhao ZY, Gu W, Cai T, Tagliani LA, Hondred D, Bond D, Schroeder S, Rudert M, Pierce DA (2001) High throughput genetic transformation mediated by *Agrobacterium tumefaciens* in maize. Mol Breed 8:323–333
15. Brettschneider R, Becker D, Lorz H (1997) Efficient transformation of scutellar tissue of immature maize embryos. Theor Appl Genet 94:737–748
16. Ishida Y, Saito H, Hiei Y, Komari T (2003) Improved protocol for transformation of maize

(*Zea mays* L.) mediated by *Agrobacterium tumefaciens*. Plant Biotech 20:57–66
17. Lee BK, Kennon AR, Chen X, Jung TW, Ahn BO, Lee JY, Zhang Z (2007) Recovery of transgenic events from two highly recalcitrant maize (*Zea mays* L.) genotypes using *Agrobacterium*-mediated standard-binary-vector transformation. Maydica 52:457–469
18. Tomes DT, Smith OS (1985) The effect of parental genotype on initiation of embryogenic callus from elite maize (*Zea mays* L.) germplasm. Theor Appl Genet 70:505–509
19. Duncan DR, Williams ME, Zehr BE, Widholm JM (1985) The production of callus capable of plant regeneration from immature embryos of numerous *Zea mays* genotypes. Planta 165:322–332
20. Hodges TK, Kamo KK, Imbrie CW, Becwar MR (1986) Genotype specificity of somatic embryogenesis and regeneration in maize. Bio/Technology 4:219–223
21. Vain P, Yean H, Flament P (1989) Enhancement of production and regeneration of embryogenic type II callus in *Zea mays* by $AgNO_3$. Plant Cell Tiss Org Cult 18:143–151
22. Carvalho CHS, Bohorova N, Bordallo PN, Abreu LL, Valicente FH, Bressan W, Paiva E (1997) Type II callus production and plant regeneration in tropical maize genotypes. Plant Cell Rep 17:73–76
23. Lu C, Vasil V, Vasil IK (1983) Improved efficiency of somatic embryogenesis and plant regeneration from tissue cultures of maize (*Zea mays* L.). Theor Appl Genet 66:285–289
24. Vain P, McMullen MD, Finer JJ (1993) Osmotic treatment enhances particle bombardment-mediated transient and stable transformation of maize. Plant Cell Rep 12:84–88
25. Armstrong CL, Green CE, Phillips RL (1991) Development and availability of germplasm with high Type II culture formation response. Maize Genet Coop Newsl 65:92–93
26. Vega JM, Yu W, Kennon AR, Chen X, Zhang Z (2008) Improvement of *Agrobacterium*-mediated transformation in Hi II maize (*Zea mays*) using standard binary vectors. Plant Cell Rep 27:297–305
27. Lupotto E, Conti E, Reali A, Lanzanova C, Baldoni E, Allegri L (2004) Improving in vitro culture and regeneration conditions for *Agrobacterium*-mediated maize transformation. Maydica 49:21–29
28. Huang X, Wei Z (2005) Successful *Agrobacterium*-mediated genetic transformation of maize elite inbred lines. Plant Cell, Tiss Org Cult 83:187–200
29. Ishida Y, Hiei Y, Komari T (2007) *Agrobacterium*-mediated transformation of maize. Nat Protoc 2:1614–1621
30. Hallauer R, Lamkey KR, White PR (1997) Registration of five inbred lines of maize: B102, B103, B104, B105 and B106. Crop Sci 37:1405–1406
31. Paz MM, Shou H, Guo Z, Zhang Z, Banerjee AK, Wang K (2004) Assessment of conditions affecting *Agrobacterium*-mediated soybean transformation using the cotyledonary node explants. Euphytica 136:167–179
32. Hajdukiewicz P, Svab Z, Maliga P (1994) The small, versatile pPZP family of *Agrobacterium* binary vectors for plant transformation. Plant Mol Biol 25:989–994
33. Hood EE, Helmer GL, Fraley RT, Chilton MD (1986) The hypervirulence of *Agrobacterium tumefaciens* A281 is encoded in a region of pTiBo542 outside of T-DNA. J Bacteriol 168:1291–1301
34. White J, Chang S, Bibb MJ, Bibb MJ (1990) A cassette containing the bar gene of *Streptomyces hygroscopicus*: a selectable marker for plant transformation. Nucleic Acids Res 18:1062
35. Carrington JC, Freed DD (1990) Cap-independent enhancement of translation by a plant potyvirus 5′ nontranslated region. J Virol 64:1590–1597
36. Mason HS, DeWald DB, Mullet JE (1993) Identification of a methyl jasmonate-responsive domain in the soybean vspB promoter. Plant Cell 5:241–251
37. An G, Ebert PR, Mitra A, Ha SB (1988) Binary vectors. In: Gelvin SB, Schilperoort RA (eds) Plant molecular biology manual. Kluwer Academic, Dordrecht, pp 1–19
38. Chu CC, Wang CC, Sun CS, Hsu C, Yin KC, Chu CY, Bi FY (1975) Establishment of an efficient medium for anther culture of rice through comparative experiments on the nitrogen source. Sci Sin 18:659–668
39. Murashige T, Skoog F (1962) A revised medium for rapid growth and bioassays with tobacco tissue cultures. Physiol Plant 15:473–497
40. McCain JW, Kamo KK, Hodges TK (1988) Characterization of somatic embryo development and plant regeneration from friable maize callus cultures. Bot Gaz 149: 16–20

Chapter 23

Biolistic-Mediated Transformation Protocols for Maize and Pearl Millet Using Pre-Cultured Immature Zygotic Embryos and Embryogenic Tissue

Martha M. O'Kennedy, Hester C. Stark, and Nosisa Dube

Abstract

Maize (*Zea mays* L.) is the most important cereal food crop in sub-Saharan Africa and Latin America, and a key feed crop in Asia, whereas pearl millet (*Pennisetum glaucum* (L.) R. Br.) is a staple food that supplies a major proportion of calories and protein to large segments of the populations living in the semi-arid tropical regions of Africa and Asia. The limitations of biological gene transfer with *Agrobacterium tumefaciens* specifically related to recalcitrant cereal crops, led to the development of alternative methods of which high-velocity microprojectiles, biolistic genetic transfer is the most successful and also the most widely employed. *Agrobacterium* facilitated transformation is the method of choice especially for deregulation of commercial transgenic food crop products, but biolistic-mediated transformation is still valid for proof of concept and functional genomics applications. Biolistic-mediated transformation and the production of transgenic plantlets via somatic embryogenesis of two maize strains viz. Hi-II (a laboratory strain) and M37W (a South African elite white maize genotype) as well as a pearl millet strain (842B) are described in this chapter. The stages described include: (1) proliferation of immature zygotic embryos for biolistic-mediated transformation, (2) induction and maintenance of transgenic embryogenic tissue on selection medium; (3) maturation (both morphological and physiological) of transgenic somatic embryos; and (4) germination of the somatic embryos to putative transgenic primary events. Maize and pearl millet cultures were regenerated via somatic embryogenesis as they are bipolar structures that shoot and root simultaneously. The culture media described in this chapter rarely induced or regenerated plantlets via organogenesis.

Key words: Immature zygotic embryos, Maize, Pearl millet, Somatic embryogenesis, Transgenic

1. Introduction

Maize is a major world crop and an important model monocot plant for studying genetics, genomics and molecular biology (1). *Agrobacterium* is an indispensable tool for transformation of a large number of model crops (2), which usually generates

Trevor A. Thorpe and Edward C. Yeung (eds.), *Plant Embryo Culture: Methods and Protocols*, Methods in Molecular Biology, vol. 710, DOI 10.1007/978-1-61737-988-8_23,

transgenic plants carrying a single-copy transgene of a "clean" transgene flanked by the left and the right T-DNA border sequences. Recently, a number of reports have shown the presence of multiple transgene copies and vector sequences in up to 75% of *Agrobacterium*-mediated transformation events (2,3). Particle bombardment remains a uniquely advantageous transformation method, and indeed the only one available for many species (4). Recently, Lowe et al. (5) produced more than 1,600 maize (inbred H99) events consistently yielding single copy events at transformation frequencies of 46% using only 2.5 ng cassettes DNA per shot. This chapter, therefore, describes a detailed procedure on the biolistic-mediated introduction of minimal transgene expression cassettes (MTECs) in maize and pearl millet. This approach enhances low copy number integration of transgenes, and minimising transgene rearrangements and gene silencing (5–7). The removal of vector backbones, which have the tendency to promote transgene rearrangements, would minimise the influence of recombinogenic elements on the process of integration (6).

The structural gene (*man*A) from *Escherichia coli* has previously been used to successfully produce transgenic maize (8–10), cassava (11), sugarbeet (12, 13) and pearl millet (14). The *man*A gene is known to be superior to antibiotic or herbicide (*pat* or *bar*) selectable marker genes for plant transformation of maize, wheat and sugar beet (10, 12, 15). The method outlined in this chapter for both maize and pearl millet transformation, describes the use of the mannose selection system which involves the phosphomannose isomerase (PMI)-expressing gene, *man*A, as the selectable marker gene and mannose, which is converted to mannose-6-phosphate by endogenous hexokinase, as the selective agent (Positech marker, Syngenta). The positive mannose selectable marker gene technology is used to (1) limit the number of escapes, (2) improve the transformation efficiency and (3) eliminate the use of antibiotic or herbicide resistant genes as selectable marker genes in maize and pearl millet transformation.

A preliminary risk assessment done by Reed et al. (15) indicated that the PMI protein in transgenic maize was (1) readily digested in simulated mammalian gastric and intestinal fluids, (2) there was no detectable changes in glycoprotein profiles and (3) no statistically significant differences were obtained in grain yield and nutritional composition compared to untransformed maize. Furthermore, the database search revealed no significant homology of the *E. coli man*A gene product to any known toxin or allergen (15).

2. Materials

2.1. Plant Material

A laboratory strain of maize, Hi-II, was obtained from the Maize Genetics Cooperation Stock Center and M37W from the Agricultural Research Council (ARC), South Africa, whereas pearl millet seed, genotype 842B, was kindly provided by ICRISAT, Zimbabwe.

2.2. Tissue Culture Media and Components

1. Tissue culture media composition: MS, N6 and L3 are described in detail (Table 1).
2. $N6_{10}$ callus induction medium contains N6 macro-, micronutrients and organic compounds, 20 g/L sucrose, 6 g/L agarose (Sigma, A0169), 10 mg/L $AgNO_3$, 2 mg/L 2,4-D, 25 mM L-proline, 100 mg/L Casein hydrolysate, pH 5.8 (see Note 1).

Table 1
Composition of MS, N6 and L3 based tissue culture media

Components (mg/L)	MS	N6	L3
Macro nutrients			
KNO_3	1,900	2,830	1,750
NH_4NO_3	1,650		200
$MgSO_4 \cdot 7H_2O$	370	185	350
KH_2PO_4	170	400	200
$CaCl_2 \cdot 2H_2O$	440	166	450
$(NH_4)_2SO_4$		463	
Micronutrients			
H_3BO_3	6.2	1.6	1.25
$MnSO_4 \cdot H_2O$	16.9	3.3	6.25
$ZnSO_4 \cdot 7H_2O$	8.6	1.5	1.875
$NaMoO_4 \cdot 2H_2O$	0.25	0.25	0.0625
$CuSO_4 \cdot 5H_2O$	0.025		0.00625
$CoCl_2 \cdot 6H_2O$	0.025	0.025	0.00625
KI	0.8	0.8	0.1875
$FeSO_4 \cdot 7H_2O$	27.8	55.6	55.6
$Na_2EDTA \cdot 2H_2O$	74.6	74.6	74.6
Organics			
Thiamine-HCl	0.1	1	10
Pyridoxine-HCl	0.5	0.5	1
Nicotinic acid	0.5	0.5	1
myo-Inositol	100		100
Glycine	2	2	
L-Glutamine			420

3. N6E callus induction medium contains N6 macro-, micronutrients and organic compounds, 30 g/L sucrose, 8 g/L agar, 0.85 mg/L $AgNO_3$, 1.5 mg/L 2,4-D, 6 mM L-proline, pH 5.8 (see Note 1).
4. Pearl millet induction medium (medium J, (18)) contains L3 macro-, micronutrients and organic compounds, 30 g/L maltose, 4 g/L Gelrite, 2.5 mg/L 2,4-D, 20 mM L-proline, pH 5.8 (see Note 1).
5. Key organic components such as L-proline, L-sorbitol, L-mannitol, thiamine-HCl, nicotinic acid, *myo*-inositol, glycine, L-glutamine, agarose and Spermidine free base (S4139), preferably cell tissue culture tested, were all purchased from Sigma-Aldrich, Life Sciences.
6. d-(+)-Mannose was purchased from Fluka, Laboratory chemicals and analytical reagents. Agar was purchased from biolab, MERCK.

2.3. Apparatus and Kits

1. The Biolistic helium-driven PDS-1000/He Biorad was used for all experiments and operated according to the manufacturer's recommendations.
2. Gel extraction kits used were: QIAquick (QIAGEN) and Geneclean II kit (Q BIOgene).

3. Methods

3.1. Donor Plants and Explant Source

Seed was germinated between layers of absorbent brown paper and cellulose wadding (Multa seed) before planting. Seed of both maize and pearl millet were planted in a soil mix consisting of red soil, rough sand and compost (1:1:1) and were watered daily with a soluble fertiliser (Hortichem N:P:K at 3:1:5; Ocean Chemicals) until flowering. Cobs were harvested 10–14 days post-pollination for excision of immature zygotic embryos (IZEs).

3.2. Sterilisation of Kernels and Embryo Isolation

Greenhouse-grown cobs or florets of maize or pearl millet, were soaked in 70% (v/v) ethanol for 1 min and sterilised for 15 min in a 2.5% (v/v) sodium hypochlorite solution containing 0.1% (v/v) of the surfactant Tween 20 before being thoroughly rinsed with sterile distilled water. IZEs were aseptically excised from the florets using a dissecting microscope and placed with their axes in contact with the callus induction medium. All tissue culture procedures were performed under aseptic conditions.

Aseptically, holding the kernel with a curved ended watchmaker's forceps, the pericarp of pearl millet was gently sliced at the embryo end with a blade. The kernel was gently squeezed with the forceps and the embryo lifted out with the tip of the blade. Isolation of embryos from maize cobs was done by wearing gloves thoroughly sprayed by 70% (v/v) ethanol, holding the cob in one hand. The tip of the kernels on the silk scar end was shaved off from top to bottom (approximately 1–2 mm depth), the scalpel blade inserted on the outer adaxial side of the kernel and gently pressed down to squeeze out the endosperm. The embryo was then gently removed from the adaxial side of the kernel just below the pericarp (see Note 2).

The embryos (strictly 0.8–1.1 mm in length) were cultured on solid callus initiation medium for 3–7 days prior to bombardment (M37W and 842B) or 1–2 months to produce embryo-derived embryogenic callus (Hi-II maize; see Note 3).

3.3. Preparation of Minimal Transgene Expression Cassettes and Particle Bombardment

Proliferating IZEs (M37W and 842B) or embryo derived embryogenic callus (Hi-II) were used as target tissue for bombardment using a Biolistic helium-driven PDS-1000/He Biorad device.

MTECs (16) were isolated with the appropriate restriction enzymes, whereafter the fragments of interest were gel purified with a QIAquick (QIAGEN) gel extraction kit or Geneclean II kit (Q BIOgene) under sterile conditions. A single purification is sufficient if the fragments are separated efficiently (far apart on the gel) before excision and purification.

On the day of bombardment, the most responsive pre-cultured embryos which proliferates in the scutellum area (M37W and 842B) or embryogenic callus (Hi-II maize) are placed in the centre (0–2 cm diameter) of a 9 cm Petri dish of callus induction medium (see Subheading 3.7.2) supplemented with osmoticum (0.2 M l-sorbitol and 0.2 M l-mannitol) as described (17).

Proliferating IZEs or embryogenic callus were co-bombarded with the selectable marker gene and gene(s) of interest at a ratio (molar) of 1:2 respectively. After 3–4 h on osmoticum pretreatment, particle bombardment proceeds at a helium pressure of 900–1,100 psi (M37W and 842B) or 450–650 psi (Hi-II callus). The bombarded embryos or embryogenic tissue are then spread on the same plate and left for an additional 16 h post bombardment treatment on osmoticum medium in the dark.

Subsequently, bombarded embryos or embryogenic tissue are cultured for 4–6 days on osmoticum-free medium for recovery before they are transferred to mannose-containing selection medium (10 g/L mannose is the osmotic equivalent of 20 g/L sucrose).

Cells containing and expressing the *man*A transgene, PMI, were selected for by using various concentrations of mannose in combination with maltose or sucrose as described in Subheading 3.7.2 below.

3.4. Microcarriers Sterilisation

The bombardment method is described in detail by the manufacturer (Biorad), briefly:

1. Thirty milligram of 0.6 μm gold particles were sterilised by adding 1 mL of 70% ethanol (v/v) and vortexed vigorously for 3–5 min. The particles were allowed to soak in 70% ethanol for 15 min before pelleting by centrifugation for 2 s at 6,000 rpm. The supernatant was discarded.
2. The following wash steps were repeated thrice:
   (a) Add 1 mL of sterile H_2O.
   (b) Vortex vigorously for 1 min.
   (c) Allow the particles to settle for 1 min.
   (d) Pellet by spinning for 5 s at 11,000 rpm.
   (e) Remove and discard the supernatant.
3. Five hundred microlitres of sterile 50% glycerol was added to bring the microparticle concentration to 30 mg 500/μL (assuming no loss of gold particles). The mix was vortexed vigorously and immediately aliquoted as 50 μL (3 mg gold) volumes in 1.5 mL microcentrifuge tubes and stored at 4°C (see Note 4).

3.5. Coating Washed Microcarriers with DNA

1. The microcarriers prepared in 50% glycerol were vortexed vigorously for 5 min (50 μL containing 3 mg gold) to resuspend and disrupt agglomerated particles. In addition, continuous agitation of the microcarriers is needed for uniform DNA precipitation onto microcarriers. The following was added while vortexing:

   5 μL DNA (50–160 ng per shot)

   50 μL 2.5 M $CaCl_2$ (see Note 5)

   20 μL 0.1 M spermidine (see Note 6)
2. The mixture was continuously finger tapped for 2–3 min. The microcarriers were allowed to settle for 1 min and then pelleted by centrifugation for 2 s in a microcentrifuge (6,000 rpm). The supernatant was discarded. A 140 μL of 70% ethanol (HPLC or spectrophotometric grade) was added and then briefly tapping the tube with a finger to resuspend the particles. The coated microcarriers were centrifuged as described above and the supernatant discarded. The coated microcarriers were then washed with 140 μL absolute ethanol, centrifuged as before and the supernatant discarded. Enough absolute ethanol was added to the pellets to provide 12–20 μL per sterile disposable macrocarrier. The pellets were gently resuspended by tapping the side of the tube several times, and then pipetted onto sterile disposable macrocarriers (6 macrocarriers; 50–160 ng/shot; 1:2 ratio of MTECs of selectable marker gene and gene(s) of interest).

3.6. Bombardment Procedure

1. A distance of 7.5 cm (2nd slot) between the macrocarriers and the target tissue was used for all bombardments. The helium supply exceeded 200 psi beyond the indicated pressure of the rapture disk. The helium pressure used was 450–650 psi (Hi-II maize) or 900–1,100 psi for elite maize and pearl millet.
2. The following components were pre-sterilised in a vertical Systec 95 autoclave at 121°C for 20 min: rupture disk retaining cap, microcarrier launch assembly and macrocarrier holders. The rupture disks were sterilised by dipping (no more than 5 s) in 70% isopropanol as extensive exposure to isopropanol will lead to delamination.
3. After spraying the chamber with 70% ethanol, sterile rupture disks were loaded into the sterile retaining cap. The macrocarrier launch assembly and target cells were placed in the chamber, the Petri dish lid removed and the door closed.
4. The chamber was evacuated by holding the vacuum at desired level (24–25 in. of Hg). The fire button was continuously depressed until rupture disk bursts, then released so that the helium pressure gauge drops to zero.
5. The vacuum from the chamber was released immediately after the rupture disk busted, the target cells were removed and covered with the Petri dish lid. Disposable macrocarriers, spent rupture disks and stopping screens were all discarded.

3.7. Post Bombardment and Selection of Putative Transgenic Events

1. The bombarded callus or embryos were spread evenly on the same Petri dish and transferred after approximately 16–18 h (after recovery from bombardment shock) onto callus induction medium omitting the osmoticum.
2. After 2–7 days (preferably 7 days for M37W) culturing on callus induction medium, the proliferating embryos or callus are transferred to callus induction medium supplemented with mannose as selection agent:

   M37W: $N6_{10}$ medium supplemented with 4 g/L sucrose and 8 g/L mannose for a period of 3–4 weeks (see Note 7).

   Hi-II maize: N6E medium supplemented with 6 g/L sucrose and 12 g/L mannose for a period of 2–4 weeks followed by medium supplemented with 4 g/L sucrose and 13 g/L mannose for an additional 2–4 weeks with 2 weekly subcultures.

   Pearl millet: Callus induction medium with 2 g/L maltose and 15 g/L mannose.
3. A maturation step followed on the callus induction medium omitting the osmoprotectant L-proline (for more detail of function see (18, 19)) and 2,4-D (see Note 8), whilst doubling the carbohydrate source, for a period of 9–14 days (see Subheading 3.8).

4. Plants were transferred to regeneration media as described below (see Subheading 3.8). Regenerating putative transgenic plants were subcultured at 2–3 weeks intervals until they reached 8–10 cm in height and subsequently hardening off.
5. The total period in tissue culture, from the excision of IZEs until the time that plantlets were hardened off to the greenhouse, was:

   M37W: Callus selection medium (3–4 weeks), maturation (7–9 days) and regeneration (2 weeks) followed by ½ MS (2 weeks) without mannose at the appearance of a 1 cm shoot in order to develop a strong root system with well developed root hairs (see Note 9).

   Hi-II: Callus selection medium (4–8 weeks), maturation (12–14 days) and regeneration (1–2 months).

   Pearl millet: Callus selection medium (4–6 weeks), maturation (12–14 days) and regeneration (1–3 months).

3.8. Culture Media Details

The composition of the tissue culture media are described in Table 1.

3.8.1. Maize

Callus induction media described below have been previously described (18–20). The media were designated $N6_{10}$ for elite line (M37W) and N6E for the laboratory strain of maize (Hi-II).

3.8.1.1. Elite Maize Line, M37W, Media Regime

Excised IZEs were cultured on N_6 based medium containing 2 mg/L 2,4-D, 25 mM l-proline, 10 mg/L $AgNO_3$, 20 g/L sucrose and 6 g/L agarose; this medium was designated $N6_{10}$. The pH of the medium was adjusted to 5.8 before autoclaving. Cultures were incubated at 25°C in the dark for 4–7 days before transfer to N610 mannose selection medium containing 4 g/L sucrose and 8 g/L mannose for a period of 3–4 weeks. Proliferating IZE cultures were subcultured every 7–14 days to fresh media. After 3 weeks, callus induced was subsequently cultured on maturation medium, containing MS salts and B5 vitamins (10 mg/L Thiamine HCl, 1 mg/L of each Pyridoxine HCl and Nicotinic acid, and 100 mg/L myo-inositol), 8 g/L sucrose, 16 g/L mannose and 6 g/L agarose. Regeneration medium contained MS salts and B5 vitamins, 20 g/L sucrose and 6 g/L, 4 g/L Bacto agar and 2 g/L Gelrite. Regeneration took place in the light (see Note 10).

3.8.1.2. Hi-II, Media Regime

The culture media and procedure without mannose as selection agent was previously described (20–22). IZEs were excised and cultured embryo-axis side down (scutellum side up) on N6E media (N6 salts and vitamins (23), 2 mg/L 2,4-D, 100 mg/L myo-inositol, 2.76 g/L L-Proline, 30 g/L sucrose, 100 mg/L casein hydrolysate, 2.5 g/L Gelrite, pH 5.8 as previously described (24). Silver nitrate (25 μM) was added after autoclaving. The plates were wrapped with vent tape and incubated at 26 ± 2°C in the dark for 2–4 weeks.

Friable Type II callus was bulked up from a few separate embryo explants (20–30) over 8 weeks by subculturing every 2 weeks on the same medium. The callus was subsequently used for biolistic-mediated transformation. After 4 days culturing on N6E medium without osmoticum, callus was transferred to N6E medium supplemented with 6 g/L sucrose and 12 g/L mannose for a period of 2–4 weeks followed by 4 g/L sucrose and 13 g/L mannose for an additional 2–4 weeks with 2 weekly subcultures. Callus was then subjected to maturation conditions by transferring about 12 small pieces (approximately 4 mm) of embryo-enriched embryogenic callus to Regeneration Medium I (MS salts and vitamins; (25)), 100 mg/L myo-inositol, 8 g/L sucrose and 26 g/L mannose, 3 g/L Gelrite, pH 5.8) and incubating for 12–14 days at 25 ± 2°C in the dark. After 2 weeks, matured somatic embryos were identified using a light microscope, and transferred to Regeneration Medium II (as for Regeneration Medium I but with 6 g/L sucrose and 12 g/L mannose), and placed in the light (70–80 μmol/m^2/s) for germination. Plantlets sprouted leaves and roots on this medium (see Note 11).

3.8.2. Pearl Millet

Callus induction medium, was previously described (26) and contains L3 salts and vitamins as described (27), 2.5 mg/L 2,4-D, maltose as a carbon source and mannose as selection agent, 4 g/L Gelrite as solidifier and modified by supplementing the medium with 20 mM L-proline (18). Maturation and regeneration medium were identical to callus induction medium apart from the omission of L-proline and 2,4-D.

Furthermore, cultures initiated were matured on medium containing double the amount of carbohydrates (4 g/L maltose and 30 g/L mannose) for a period of 2 weeks, followed by regeneration medium with 2 g/L maltose and 15 g/L mannose as described above. Cultures on callus induction and maturation media were incubated at 25 ± 2°C under low-light conditions (1.8 μmol/m^2/s), whereas regenerating shoots (≥1 cm) were incubated under full light (80 μmol/m^2/s).

The micro and macro elements were prepared as independent stock solutions; and components $FeSO_4.7H_2O$ and $Na_2EDTA.2H_2O$ were combined in a third stock solution. The organic components were prepared as individual stock solutions and added after autoclaving. Media composition as previously described (18–21, 23, 25, 26). Abbreviation: EDTA, ethylenediamine tetra-acetic acid.

3.9. Regenerated Plantlets Were Hardened-Off as Follows

Maize M37W plantlets 8–10 cm in height were hardened-off by placing the plantlets with well developed roots directly into small pots (15 cm diameter × 12 cm height) containing sterilised moist fertile soil and placed in a growth room with controlled humidity. The plants were covered with perforated margarine plastic tubs (12.5 cm diameter, 7.0 cm high) (see Note 12).

Pearl millet and maize Hi-II plantlets 8–10 cm in height were hardened-off by placing the plantlets in a mix of perlite and vermiculite (1:1) and watering it daily with Hoagland's solution in a growth rooms (70% humidity at 26 ± 2°C). The plantlets were covered with a plastic autoclave bag, sprayed with water in the inside and kept covered for the first 3–5 days to maintain humidity. The plastic bag was gradually opened and eventually removed. Hardened-off plantlets were fertilised with N:P:K 3:2:1 every 6 weeks.

Transgenic plants and their progeny were self-pollinated or cross-pollinated with plants originating from the same transformation event or pollinated with non-transgenic donor genotypes. Pearl millet was flowering in growth rooms (containment level 3) whereas maize was grown in a greenhouse facility (containment level 2).

3.10. Germination of Transgenic Progeny

Progeny of transgenic plants expressing the positive selectable marker transgene phosphomannose isomerase were identified on half-strength MS medium containing half-strength MS salts, 8 g/L agar and the mannose, sucrose or maltose combination as used in regeneration medium of the different cereal crops.

Protocols describing the molecular downstream analysis confirming stable integration and expression of the transgenes, fall outside the scope of this chapter.

4. Notes

1. L-Proline and organic compounds are filter sterilised and added after autoclaving.
2. Take care to excise without damaging the embryos; scalpel blade poked or half cut embryos will not proliferate.
3. Immature embryo derived embryogenic tissue was prepared by excision and culturing of immature embryos on N6E medium for a period of 4–6 weeks. The highly embryogenic tissue was used for bombardment within 2–5 months after embryo excision to ensure regeneration of fertile transgenic plants.
4. Gold microparticles can be stored at room temperature for up to 2 weeks or for 1 month at 4°C.
5. 2.5 M $CaCl_2$ was stored at 350 μL aliquots and used only for one bombardment experiment to avoid contamination.
6. Spermidine free base was stored as 1 M stock solution at −20°C. Dilute to 0.1 M just before use.
7. Regeneration of transgenic callus was more efficient when transfer was done during the softer stages of the callus

(3 weeks); longer culturing on selection may result in formation of hard white callus which fails to regenerate.

8. 2,4-D is the only hormone used in callus induction medium for both maize and pearl millet and is a synthetic and highly active auxin responsible for cell elongation, swelling of tissue and cell division leading to callus formation.

9. Plants that have developed in vitro have a poorly developed cuticle layer due to the relatively high humidity in vitro and poorly developed roots with few or no root hairs. These roots are vulnerable and may lead to difficulties in plant growth in vivo especially in less humid environment. Well developed root hairs help in more water absorption assisting the thin cuticle layered in vitro plant not to dry out and die (28).

10. Cultures on callus induction and maturation media were incubated at 25 ± 2°C in the dark, whereas callus for regeneration was incubated under a 16 h photoperiod with light being supplied by fluorescent tubes at an intensity of 70–80 μmol/m^2/s.

11. M37W plantlets of 1 cm in height were transferred to selection free ½ MS medium to obtain fertile transgenic events (30%), even though 70% proved to be escapees. In order to avoid escapees, the selection pressure can be adjusted to 5 g/L mannose and 25 g/L sucrose (8).

12. Perforated plastic tubs were used to keep the relative high in vitro humidity with the goal of acclimatising the plantlets in vivo. Acclimatised plantlets were later transferred to a greenhouse (N. Dube et al., manuscript in preparation).

References

1. Vega JM, Yu W, Kennon AR, Chen X, Zhang ZJ (2008) Improvement of *Agrobacterium*-mediated transformation in Hi-II (*Zea mays*) using standard binary vectors. Plant Cell Rep 27:297–305
2. Lange M, Vincze E, Møller MG, Holm PB (2006) Molecular analysis of transgene and vector backbone integration into the barley genome following *Agrobacterium*-mediated transformation. Plant Cell Rep 25:815–820
3. Kononov ME, Bassuner B, Gelvin SB (1997) Integration of T-DNA binary vector 'backbone' sequences into the tobacco genome: evidence for multiple complex patterns of integration. Plant J 11:945–957
4. Altpeter F, Baisakh N, Beachy R, Bock R, Capell T, Christou P, Daniell H, Datta K, Dix PJ, Faquet C, Huang N, Kohli A, Mooibroek H, Nicholson L, Nguyen T, Nugent G, Raemakers K, Romano A, Somers D, Stoger E, Taylor N, Visser R (2005) Particle bombardment and the genetic enhancement of crops: myths and realities. Mol Breed 15:305–327
5. Lowe BA, Prakash NS, Way M, Mann MT, Spencer TM, Boddupalli RS (2009) Enhanced single copy integration events in corn via particle bombardment using low quantities of DNA. Transgenic Res 18:831–840
6. Fu X, Duc L-T, Fontana S, Bong BB, Tinjuangjun P, Sudhakar D, Twyman RM, Christou P, Kohli A (2000) Linear transgene constructs lacking vector backbone sequences generate low-copy-number transgenic plants with simple integration patterns. Transgenic Res 9:11–19
7. Kohli A, Leech M, Vain P, Laurie DA, Christou P (1998) Transgene organization in rice engineered through direct DNA transfer supports a two-phase integration mechanism mediated by the

establishment of integration hot spots. Proc Natl Acad Sci U S A 95:7203–7208

8. Negrotto D, Jolley M, Beer S, Wenck AR, Hansen G (2000) The use of phosphomannose-isomerase as a selectable marker to recover transgenic maize plants (*Zea mays* L.) via *Agrobacterium* transformation. Plant Cell Rep 19:798–803
9. Wang AS, Evans RA, Altendorf PR, Hanten JA, Doyle MC, Rosichan JL (2000) A mannose selection system for production of fertile transgenic maize plants from protoplasts. Plant Cell Rep 19:654–660
10. Wright M, Dawson J, Dunder E, Suttie J, Reed J, Kramer C, Chang Y, Novitzky R, Wang H, Artin-Moore L (2001) Efficient biolistic transformation of maize (*Zea mays* L.) and wheat (*Triticum aestivum* L.) using the phosphomannose isomerise gene, pmi, as the selectable marker. Plant Cell Rep 20:429–436
11. Zhang P, Puonti-Kaerlas J (2000) PIG-mediated cassava transformation using positive and negative selection. Plant Cell Rep 19:1041–1048
12. Joersbo M, Donaldson I, Kreiberg J, Petersen SG, Brunstedt J, Okkels FT (1998) Analysis of mannose selection used for transformation of sugar beet. Mol Breed 4:111–117
13. Joersbo M, Petersen SG, Okkels FT (1999) Parameters interacting with mannose selection employed for the production of transgenic sugar beet. Physiol Plant 105:109–115
14. O'Kennedy MM, Burger JT, Botha FC (2004) Pearl millet transformation system using the positive selectable marker gene phosphomannose isomerase. Plant Cell Rep 22:684–690
15. Reed J, Privalle L, Powell ML, Meghji M, Dawson J, Dunder E, Suttie J, Wenck A, Launis K, Kramer C, Chang Y-F, Hansen G, Wright M (2001). Phosphomannose isomerase: an efficient selectable marker for plant transformation. In Vitro Cell Dev Biol Plant 37:127–132
16. Christou P, Kohli A (2005) Transformation method and transgenic plants produced thereby. Patent EP 1407000 B1 (describing a method for producing a population of transgenic plants having a minimal transgene expression cassette integrated into the plant's genomes)
17. Vain P, McMullen MD, Finer J (1993) Osmotic treatment enhances particle bombardment-mediated transient and stable transformation of maize. Plant Cell Rep 12:84–88
18. O'Kennedy MM, Smith G, Botha FC (2004) Improved regeneration efficiency of a pearl millet (*Pennisetum glaucum*) breeding line. S Afr J Bot 70:502–508
19. Armstrong CL, Green CE (1985) Establishment and maintenance of friable, embryogenic maize callus and the involvement of L-Proline. Planta 164:207–214
20. Songstad DD, Armstrong CL, Petersen WL (1991) $AgNO_3$ increases type II callus production from immature embryos of maize inbred B73 and its derivatives. Plant Cell Rep 9:699–702
21. Che P, Love TM, Frame BR, Wang K, Carriquiry AL, Howell SH (2006) Gene expression patterns during somatic embryo development and germination in maize Hi II callus cultures. Plant Mol Biol 62:1–14
22. Armstrong CL, Songstad DD (1994) Method for transforming monocotyledonous plants. Patent EP0586355 (describing a method for delivering foreign genetic material into immature embryos isolated from monocotyledonous plants, particularly those of the Gramineae family)
23. Chu CC, Wang CC, Sun CS, Hsu C, Yin KC, Chu CY (1975) Establishment of an efficient medium for anther culture in rice through comparative experiments on the nitrogen sources. Sci Sin 18:659–668
24. Songstad D, Armstrong CL, Petersen WL, Hairston B, Hinchee MAW (1996) Production of transgenic maize plants and progeny by bombardment of Hi-II immature embryos. In Vitro Cell Dev Biol Plant 32:179–183
25. Murashige T, Skoog F (1962) A revised medium for rapid growth and bioassays with tobacco tissue cultures. Physiol Plant 15:473–497
26. Gless C, Lörz H, Jahne-Gartner A (1998) Establishment of a highly efficient regeneration system from leaf base segments of oat (*Avena sativa* L.). Plant Cell Rep 17:441–445
27. Jähne A, Lazzeri PA, Jäger-Gussen M, Lörz H (1991) Plant regeneration from embryogenic suspensions derived from anther cultures of barley (*Hordeum vulgare* L.). Theor Appl Genet 82:74–80
28. Pierik RLM (1987) In Vitro Culture Of Higher Plants. Kluwer Academic Publishers, Dordrecht, pp 45–82 ISBN 90-247-3531-9

Chapter 24

Agrobacterium tumefaciens-Mediated Genetic Transformation of Cereals Using Immature Embryos

Ashok K. Shrawat and Allen G. Good

Abstract

A critical step in the development of a robust *Agrobacterium tumefaciens*-mediated transformation system for cereal crop plants is the establishment of optimal conditions for efficient T-DNA delivery into target tissue, from which plants can be regenerated. Although, *Agrobacterium*-mediated transformation of cereals is an important method that has been widely used by many laboratories around the world, routine protocols have been established only in specific cultivars within a species and with specific tissues of high regeneration potential. Cocultivation of highly embryogenic callus tissue or healthy immature embryos with *A. tumefaciens* is considered one of the critical factors in successful genetic transformation of crop plants. Immature embryos collected only from vigorously growing healthy and green plants grown in the field or in the well-conditioned greenhouse are the ideal target for genetic transformation of recalcitrant crop species. Here, we describe an *Agrobacterium*-mediated transformation method that uses immature embryos as the starting material for inoculation with *Agrobacterium*. The aim of this chapter is to provide the key steps/components involved in *Agrobacterium*-mediated transformation of cereal crops. However, these steps or components often vary between protocols and from laboratory to laboratory, and can be optimized or modified based on the requirement of a specific cultivar or species.

Key words: *Agrobacterium*, Cereals, Immature embryos, Transgenic plants

1. Introduction

The development of an efficient method for genetic transformation is a prerequisite for the application of transgenic approaches to the improvement of a given crop species. Cereals are an important source of calories and therefore, cereal crops have been primary targets for improvement by genetic transformation. Although microprojectile bombardment has revolutionized the field of genetic transformation of cereals, there is considerable variation

Trevor A. Thorpe and Edward C. Yeung (eds.), *Plant Embryo Culture: Methods and Protocols*,
Methods in Molecular Biology, vol. 710, DOI 10.1007/978-1-61737-988-8_24,

seen in the stability, integration, and expression of the introduced transgene (1). In comparison, *Agrobacterium*-mediated transformation is considered the method of choice for the genetic modification of many plant species because it allows efficient insertion of stable, unrearranged, and single-copy sequences into the plant genome (2, 3). In general, *Agrobacterium tumefaciens* is first transformed with the DNA construct of interest (T-DNA) and then this modified bacterial strain is used to introduce the T-DNA into plants. For the last two decades, dicotyledonous plants have been transformed using the soil phytopathogen *A. tumefaciens.* Initially, it was not clear if this technology could be extended to monocotyledonous plants, as they are not natural hosts of *Agrobacterium*. However, in 1994, a highly efficient method of *Agrobacterium*-mediated transformation of Japonica rice was reported (4). Since 1994, *Agrobacterium*-mediated transformation has emerged as a method of choice for transferring genes of interest not only into rice, but also into the other major cereals including maize, barley, wheat, sorghum, and sugarcane (5, 6). Key factors in these achievements include the optimization of the type of plant material for infection with *Agrobacterium*, the choice of vectors, the choice of *A. tumefaciens* strains, and optimization of tissue culture techniques. For the successful production of transgenic plants in any species, foreign genes must be delivered into cells that are actively dividing and are capable of regenerating plants. The majority of researchers world-wide have used immature embryos as the primary target tissue for the delivery of foreign genes. Thus, the primary determinants of a successful transformation are the response of immature embryos in tissue culture, the types of cells that grow from immature embryos, and subsequent characteristics in growth and regeneration.

The advantages of using immature embryos are especially evident when transforming elite cereal cultivars, which are often quite recalcitrant to tissue culture and transformation. It is important to emphasize that successful transformation using immature embryos depends on the quality of the embryos. In cereal tissue culture, it is well established that somatic embryogenesis and transformation frequency are influenced by the age of the explant and that younger embryos produce comparatively more somatic embryos and consequently more transgenic plants than older explants (7–9). Scutellum size has also been shown to influence culture response and subsequently transformation frequency in cereals. In wheat, in comparison to scutella smaller than 0.5 mm or larger than 1.5 mm, Rasco-Gaunt et al. (10) obtained highest embryogenesis and shoot regeneration from scutella ranging from 0.75 to 1.0 mm in size. Unfortunately, the response from immature embryos are very genotype specific and many agronomically important genotypes of major cereals, especially so-called elite varieties are poor in tissue culture response, and thus only a limited number of genotypes have been efficiently transformed so far (6).

Following the success of rice, Ishida et al. (11) reported stable transformation of maize cultivar A188 and its hybrids after cocultivation of freshly isolated immature embryos with *Agrobacterium* harboring a super-binary vector similar to that developed by Hiei et al. (4). The developed system resulted in transgenic maize plants with transformation frequencies ranging from 5 to 30%. To extend the range of maize genotypes susceptible to *Agrobacterium*, Zhao et al. (12) developed an efficient system for *Agrobacterium*-mediated transformation of maize Hi-II. By optimizing inoculation and cocultivation conditions, they produced transgenic maize plants with a transformation frequency of 7.1%.

A major breakthrough in the *Agrobacterium*-mediated transformation of cereals was reported by Tingay et al. (13), when they used a non-super-virulent strain of *Agrobacterium* carrying a binary vector and produced transgenic barley plants with 4.2% transformation frequency. The major factors which influenced the generation of transgenic barley plants included the wounding of immature embryos and the removal of the axis of immature embryos. Using the same or similar strategy, many other laboratories have successfully produced transgenic barley with *Agrobacterium* (6). In order to increase the transformation frequency in barley, Matthews et al. (14) transformed the barley cultivar Golden Promise following the same strategy as described by Tingay et al. (13), except that immature embryos were infected with *Agrobacterium* on the same day of isolation without prior wounding by biolistic gold particles, and the transformed tissues were selected on hygromycin rather than bialaphos. Their method produced transgenic barley plants with average frequencies of 2–12%. Apart from immature embryos, Kumlehn et al. (15) have recently demonstrated that androgenetic pollen cultures can also be used as an effective target tissue for *Agrobacterium*-mediated transformation of barley. By optimizing a number of factors, such as the pollen preculture time, choice of *Agrobacterium* strain and vector system, *Agrobacterium* population density, medium pH and the concentration of acetosyringone, $CaCl_2$ and glutamine, these authors produced 2.2 fertile transgenic plants per spike. Recently, Shrawat et al. (9) studied a number of factors and found that preculture of immature embryos, cocultivation, presence of acetosyringone and sonication, and vacuum filtration assisted inoculation of 1-day precultured immature embryos produced significant difference in T-DNA delivery. By optimizing these factors for T-DNA delivery, they produced transgenic barley plants with transformation efficiencies ranging from 2.6 to 6.7%.

Following the success of *Agrobacterium*-mediated genetic transformation of rice, maize, and barley, Cheng et al. (16) produced stable transgenic wheat plants within 3 months by cocultivating freshly isolated immature embryos, precultured immature embryos, and embryogenic calli with *Agrobacterium*. In addition to acetosyringone, the presence of a surfactant during inoculation of the tissue with *Agrobacterium* was found to be an important factor for the efficient delivery of T-DNA into wheat. Later, Cheng et al. (17) further exploited the fact that an explant such as an immature embryo with active cell division can enhance T-DNA delivery in order to increase the recovery of stable transgenic plants in wheat (16, 18, 19). Following desiccation of plant tissues after *Agrobacterium* infection and the use of paromomycin and glyphosate selection, they produced stable transgenic wheat plants with frequencies ranging from 4.8 to 19%.

In comparison to rice, maize, wheat, and barley, sorghum is considered the most difficult plant species to manipulate through tissue culture and transformation. For the first time, Zhao et al. (20) attempted to transform sorghum and successfully produced transgenic sorghum plants with an average transformation frequency of 2.1% after cocultivation of immature embryos with *Agrobacterium* carrying a super-binary vector. It was found that the source of the immature embryos had a very significant impact on the transformation efficiency, with field-grown embryos producing a higher transformation frequency than greenhouse grown embryos. Using the *Agrobacterium*-mediated transformation protocol, Gao et al. (21) produced stable transgenic sorghum plants in two inbreds (Tx 430 and C401) and one commercial hybrid (Pioneer 8505) with an average transformation frequency of 2.5% within 4–5 months.

The natural ability of *Agrobacterium* to deliver a discrete segment of DNA into the recipient genome has been exploited in *Agrobacterium*-mediated transformation of cereals (6, 22, 23). Several factors influencing *Agrobacterium*-mediated transformation of cereals have been investigated and discussed (6, 23). These factors include the screening of the most responsive genotype and explant, the *Agrobacterium* strain, the binary vector, the selectable marker gene and promoter, inoculation and coculture conditions, and the tissue culture and regeneration medium. Despite successful reports of *Agrobacterium*-mediated transformation of crop plants, there are still serious handicaps with *Agrobacterium*-mediated transformation of elite cultivars of major cereal crops. *Agrobacterium*-mediated transformation is limited to certain tissues and cultivars (6, 23). A major problem during *Agrobacterium*-mediated transformation of immature embryos is the development of a necrotic response in immature embryos after cocultivation. Immature embryos have been found to be very sensitive to *Agrobacterium* infection and embryo death after cocultivation is considered a limiting step to develop or improve transformation efficiency in cereals (6). Therefore, in order to achieve sufficient number of T-DNA transfer events to occur in the target tissue while maintaining the regenerability of recipient plant cells, a fine balance between the factors affecting the transformation frequency is required. Such a balance may help not only to develop methods to enhance the transformation frequency of economically important plant species, but also to extend the range of *Agrobacterium*-mediated transformation to elite cultivars. In addition, further refinement or optimization of parameters that are considered to be crucial for cereal transformation, such as the screening of highly regenerative tissue, the genotype and the development of an efficient plant tissue culture, and the regeneration system, should broaden the scope for the genetic transformation of economically important crop plants.

2. Materials

2.1. Target Tissue

Immature embryos: quality of immature embryos is one of the key factors for *Agrobacterium*-mediated transformation of cereals (see Note 1).

2.2. Agrobacterium tumefaciens Strain(s)

A number of *Agrobacterium* strains such as LBA4404 (24), EHA101 (25), EHA105 (26), AGL1 (27), or A281 (25) harboring the gene of interest and an appropriate selectable marker gene such as hygromycin resistance gene (*hpt*) for plant selection in a binary vector have successfully been used for cereal transformation. The neomycin phosphotransferase (*nptII*) gene, which confers plants resistant to Geneticin (G418) and paromomycin, has also been used in genetic transformation of cereals. Recently, positive selection marker, such as the phosphomannose-isomerase (*pmi*) gene, has been reported to be an efficient selective marker for rice and maize transformation. In this system, selection is carried out on media that contain mannose as the main carbon source (28). If paromomycin is used as a selective agent, Gelrite should be replaced with agar or agarose because paromomycin is insoluble in media containing Gelrite. If the phosphoinothricin acetyl transferase (*bar*) gene is employed as a selectable marker, glutamine must be removed from the selection media because it may neutralize the mode of action of the selective agent phosphoinothricin, which is a potent inhibitor of glutamine synthetase. Choice of promoter to drive gene of interest and marker gene vary from laboratory to laboratory and according to the need of the project. In cereal transformation, *35S* cauliflower mosaic virus (*CaMV*) and *Ubiquitin-1* promoter from maize have commonly been used to drive the expression of a selectable marker gene.

2.3. Reagents

1. Sterilized distilled water.
2. 70% (vol/vol) ethanol in distilled water – (see Subheading 2.5).
3. 50% Bleach in distilled water – (see Subheading 2.5).
4. 2,4-Dichlorophenoxy acetic acid (2,4-D) – (see Subheading 2.5).
5. 6-Benzylaminopurine acid (6BA) – (see Subheading 2.5).
6. Indole-3-butyric acid (IBA) – (see Subheading 2.5).
7. Acetosyringone – (see Subheading 2.5).
8. Hygromycin B solution – (see Subheading 2.5).
9. 100 mM 5-bromo-4-chloro-3-indoxyl-β-d-glucuronic acid cyclohexylammonium salt (X-gluc) – (see Subheading 2.5).
10. Cefotaxime – (see Subheading 2.5).
11. Timintin – (see Subheading 2.5).
12. YEP plates – (see Subheading 2.5).
13. AB medium – (see Subheading 2.5).
14. Gelrite or phytagel.
15. Tissue culture and transformation media – (see Subheading 2.5).
16. Infection medium – (see Subheading 2.5).
17. Coculture medium – (see Subheading 2.5).
18. Selection medium – (see Subheading 2.5).
19. Preregeneration medium – (see Subheading 2.5).
20. Regeneration medium – (see Subheading 2.5).
21. Rooting medium – (see Subheading 2.5).

2.4. Equipment

1. Stereomicroscope for isolation of immature embryos.
2. Laminar flow hood with Bunsen burner to carry out transformation and tissue culture steps.
3. Autoclave for sterilization of tissue culture and transformation media.
4. Incubator for *Agrobacterium* preculture and coculture steps.
5. Controlled tissue culture room for regenerating transgenic plants under light/dark conditions.
6. Greenhouse for growing transgenic plants in soil.
7. Tabletop shaker.
8. Balance.
9. pH meter.
10. Fridge (4°C) and freezer (–20 and –80°C).
11. Scalpel blade: sterilization is required.

12. Forceps/fine forceps: sterilization is required.
13. Micropore surgical tape.
14. Parafilm.
15. Pipetting aid (1–50 mL).
16. Micropipettes and micropipette tips.
17. Microfuge tubes.
18. Glassware: sterilization is required.
19. Sterile plastic deep Petri plates (100 × 20 mm).
20. Syringe filter.
21. 0.22 μm filter membrane for sterilizing stock solution of hormones.
22. Magenta jars: sterilization is required.
23. Majenta stirring bar.

2.5. Reagent Setup

1. 2,4-D (1 mg/mL stock solution of 2,4-D): add 1 N NaOH or 70% ethanol dropwise to 2,4-D powder until completely dissolved. Make up final volume by adding distilled water, filter sterilize, and store the stock solution at 4°C. Caution – 2,4-D is toxic and therefore, extreme caution is required whether preparing stock solution or using in tissue culture medium. The use of a fume hood is recommended while weighing and making stock solution of 2,4-D.
2. 6-Benzylaminopurine (BAP; 1 mg/mL stock solution of BAP): add 1 N NaOH dropwise to the powder of BAP until completely dissolved. Make up final volume by adding distilled water, filter sterilize, and store the stock solution at 4°C.
3. IBA (1 mg/mL stock solution of IBA): add 1 N NaOH or 70% ethanol dropwise to the powder of IBA until completely dissolved. Make up final volume by adding distilled water, filter sterilize, and store the stock solution at 4°C.
4. Acetosyringone (3′,5′-dimethoxy-4′-hydroxyacetophenone): dissolve acetosyringone (SIGMA Product number D134406) powder in 100% dimethyl sulfoxide (DMSO) and sterilize through filtration. Powder should be stored in a tightly closed container and wrapped with aluminum foil.
5. Hygromycin B solution: dissolve 50 mg hygromycin B powder in 50 mL distilled water (stock solution), filter sterilize, and store at 4°C. For selection of transformed cells, 50 mg/L hygromycin is used in the selection medium. CAUTION – hygromycin is highly toxic and therefore, extreme precaution is needed while preparing hygromycin B solution. Wear protective clothing and gloves to prevent contact with skin and eyes. Alternatively, purchase hygromycin B solution (50 mg/mL) from Sigma (Product number-H0654).

6. 100 mM X-gluc: dissolve 52 mg X-gluc in 1 mL of ethylene glycol monomethyl ether. Store the solution in the dark at −20°C. X-gluc powder is also stored in tightly closed container and desiccated at −20°C. CAUTION – X-gluc is harmful if swallowed, inhaled, or absorbed through skin. CAUTION – ethylene glycol monomethyl ether is very toxic. Wear gloves and eye/face protection.
7. Cefotaxime (200 mg/mL stock solution): dissolve cefotaxime powder in distilled water, filter sterilize, and store the stock solution at −20°C in 1 mL aliquots. CAUTION – powder is irritating to eyes, respiratory system, and skin. Do not breathe dust and wear suitable protective clothing and gloves.
8. Timintin (100 mg/mL stock solution): dissolve timintin powder (Ticarcillin/Clavulanic, 15:1) in distilled water, filter sterilize, and store at −20°C in 1 mL aliquots.
9. YEP medium (for *A. tumefaciens* culture): dissolve 5 g yeast extract, 10 g peptone, and 5 g sodium chloride in 900 mL of distilled water and adjust pH to 7 with NaOH. After adjusting the final volume to 1,000 mL, add 15 g bacto agar and autoclave at 121°C for 20 min for solidifying the YEP medium. Cool the medium to ~50°C, add appropriate antibiotics, and pour 20 mL medium into Petri plates (100 × 20 mm). Use of antibiotics in the medium depends on the type *Agrobacterium* strain and type of plasmid.
10. AB medium (for *A. tumefaciens* culture): dissolve 5 g glucose and 15 g agar in 800 mL distilled water and adjust the volume to 900 mL with distilled water before autoclaving at 120°C for 20 min. After autoclaving, allow medium to cool down ~50°C and then add 50 mL of 20× AB buffer, 50 mL of 20× AB salts, and appropriate antibiotics. Mix buffers and antibiotics well before pouring into sterile Petri plates. Use of antibiotics in the medium depends on the type and *Agrobacterium* strain and type of plasmid.
11. 20× AB buffer: dissolve 60 g dipotassium hydrogenphosphate and 20 g sodium dihydrogenphosphate dehydrate in 800 mL distilled water. Adjust the final volume to 1,000 mL with distilled water after adjusting the pH to 7.0. Store the buffer at room temperature after autoclaving at 121°C for 20 min.
12. 20× AB salts: dissolve 20 g ammonium chloride, 6 g magnesium sulfate heptahydrate, 3 g potassium chloride, 240 mg calcium chloride dehydrate, and 50 mg iron (II) sulfate heptahydrate in 800 mL distilled water. Adjust the final volume to 1,000 mL, autoclave, and store the solution at room temperature.
13. Tissue culture and transformation media: medium for tissue culture and *Agrobacterium*-mediated transformation vary from genotype to genotype, species to species, and laboratory

to laboratory. An example of media used for *Agrobacterium*-mediated transformation of immature embryos of barley is given in Table 1. Optimization of plant tissue culture and transformation media is required if trying to establish *Agrobacterium*-mediated transformation protocol for an elite cultivar.

14. Infection medium (liquid coculture medium): infection medium varies from genotype to genotype, species to species, and laboratory to laboratory. Typically, infection medium

Table 1
An example of the media we have used for *Agrobacterium*-mediated transformation of immature embryos of barley

Medium	Composition of the medium
Callus induction medium	PL medium (34) + 187.5 mg/L L-glutamine + 37.5 mg/L L-proline + 25 mg/L L-asparagine + 2.0 mg/L 2,4-D + 30 g/L maltose, pH 5.8, 0.3% Gelrite. *Note*: Filter sterilize the medium before adding the Gelrite
Liquid coculture medium (infection medium)	Callus induction medium without maltose + 0.25 M glucose + 100 μM acetosyringone + 1.0 mg/L 2,4-D, pH 5.2. *Note*: Filter sterilize
Solid coculture medium (cocultivation medium)	Liquid coculture medium + 0.25 M glucose + 100 μM acetosyringone + 400 mg/L L-cysteine + 2.0 mg/L 2,4-D pH 5.2, 0.3% Gelrite. *Note*: Filter sterilize the medium
Selection medium I	Callus induction medium + 3.0 mg/L 2,4-D + 5 mg/L bialaphos or 50 mg/L hygromycin + 250 mg/L cefotaxime + 150 mg/L timintin, pH 5.8, 0.3% Gelrite. *Note*: Filter sterilize the medium
Selection medium II	Callus induction medium + 3.0 mg/L 2,4-D + 0.1 mg/L BAP + 5 mg/L bialaphos or 50 mg/L hygromycin + 250 mg/L cefotaxime + 150 mg/L timintin, pH 5.8, 0.3% Gelrite. *Note*: Filter sterilize the medium
Selection medium III	Callus induction medium + 2.0 mg/L 2,4-D + 0.1 mg/L BAP + 5 mg/L bialaphos or 50 mg/L hygromycin + 250 mg/L cefotaxime + 150 mg/L timintin, pH 5.8, 0.3% Gelrite. *Note*: Filter sterilize the medium
Preregeneration medium	FHG or MS medium + 5 mg/L ABA + 1 mg/L BAP + 0.5 mg/L NAA + 5 mg/L bialaphos or 50 mg/L hygromycin + 100 mg/L cefotaxime + 100 mg/L timintin, pH 5.8, 0.3% Gelrite. *Note*: Filter sterilize the medium
Regeneration medium	Preregeneration medium + 3 mg/L BAP + 0.5 mg/L NAA + 2.5 mg/L bialaphos or 25 mg/L hygromycin + 100 mg/L cefotaxime + 50 mg/L timintin, pH 5.8, 0.4% Gelrite. *Note*: Filter sterilize the medium before adding the Gelrite
Rooting medium	1/2 MS medium, pH 5.8, 0.3% Gelrite. *Note*: Autoclave

contains a high concentration of glucose. In some protocols, high concentration of sucrose and glucose is added into the infection medium. For example, in *Agrobacterium*-mediated transformation of maize and rice, infection medium enriched with 68.46 g of sucrose and 36.04 g of glucose support high transformation frequency (29). In addition, infection medium also contains a number of amino acids such as l-glutamine, aspartic acid, and casamino acids. The type and concentration of amino acids in the infection medium varies from protocol to protocol. Sterilization of infection medium with a 0.22 μm cellulose-acetate filter is critical for high frequency of transformation. Infection medium containing 100–500 μM acetosyringone has been found to support efficient T-DNA transfer to the target tissue. Acidic pH of the infection medium is very crucial for successful transformation. In the majority of the *Agrobacterium*-mediated transformation protocols, pH 5.2 has been found to support high frequency of transformation.

15. Coculture medium (solid coculture medium): coculture medium is similar to the infection medium except it is solidified with gelling agent such as Gelrite before pouring into the sterile Petri plates.
16. Selection medium (for selection of transformed cells): type of selection medium varies from genotype to genotype and species to species. For example, in maize, better growth of transformed tissues has been reported on LS medium (29). While, for rice, N6 medium has been reported to support better growth of transformed tissue (30). Typically, selection medium is enriched with myo-inositol, casein hydrolysate, L-proline, and L-glutamine (see Table 1). Type and concentration of antibiotics to check the growth of *Agrobacterium* vary from protocols to protocols. Commonly, 250 mg/L carbenicillin or 250 mg/L cefotaxime with or without another antibiotic such as timintin has been effectively used in selection and regeneration medium. In our laboratory, for selection and regeneration of barley and rice transgenic plants, we used a combination of cefotaxime (250 mg/L) and timintin (150 mg/L) to check *Agrobacterium* growth. Depending on the type of selectable marker gene(s) in the plasmid, 5 mg/L phosphinothricin for *bar* selection or 50 mg/L hygromycin for *hpt* selection is used for regenerating basta or hygromycin-resistant transgenic plants. It is important to use high concentration of selective agent from the beginning to obtain a strong selection. However, the appropriate concentration of selective agents in the selection medium should be established. It is important to keep low numbers (5–7) of transformed calli on selection medium because overcrowding at this stage negatively influences the growth of resistant tissue. To enhance

good selection, it is important to detach the proliferation or sub calli from the mother calli and spread them around to ensure that they are in good contact with the medium containing selective agent.

17. Regeneration medium: the type of regeneration medium varies from species to species and laboratory to laboratory. For example, FHG medium (31) has been found to support high-frequency plant regeneration in barley. While, for maize, LS medium (29) and, for rice, N6 medium (30) have been reported to support efficient plant regeneration from transgenic callus tissues. Depending on the type of selectable marker gene(s) in the plasmid, 5 mg/L phosphinothricin for *bar* selection or 50 mg/L hygromycin for *hpt* selection is used for regenerating basta or hygromycin-resistant transgenic plants. It is important to use high concentration of selective agent (phosphinothricin or hygromycin) to avoid or minimizing the regeneration of escapes. The appropriate concentration of selective agents in the regeneration medium should be established. It is important to keep only 5–6 resistant calli on regeneration medium because overcrowding will negatively influence neoformation, especially in recalcitrant varieties.
18. Rooting medium: rooting medium varies from species to species and laboratory to laboratory. Typically, MS medium (32) containing 1 or 2 mg/L IBA is used for inducing/promoting strong root system in transgenic plants.

3. Methods

3.1. Preparation of Immature Embryos for Transformation

1. Grow plants in a growth chamber as per the specific temperature and photoperiod conditions for individual cereal plants. For example, for growing rice plants (cv. Nipponbare), maintain the greenhouse at 28°C day and night under a 14/10-h day and night photoperiod. For maize (cv. A188), maintain daytime temperature between 30 and 35°C and night time temperature between 20 and 25°C. For barley (cv. Golden Promise), maintain day time temperature between 18 and 20°C and night time temperature between 13 and 15°C and for wheat (cv. Bobwhite), greenhouse is maintained at 18–20°C day and 14–15°C night temperatures with a relative humidity of 50–70% under a 16-h photoperiod. Light intensity also varies from species to species.
2. Harvest the spikes or ear containing the correct developmental stage of immature embryos (usually 10–15 days after anthesis) (see Note 1).

3. Spikes are surface sterilized in 70% (v/v) ethanol for 30 s and then 5–10 min in 20% (v/v) bleach (5.25% sodium hypochlorite) with gentle shaking. Rinse with sterile distilled water at least five times. Immature embryos are then dissected from young caryopses under a stereomicroscope in a sterile environment using a sharp scalpel and fine forceps. In case of maize, kernels are detached from cob by cutting the base of the kernel with a scalpel. Immature embryos are then removed by inserting a scalpel into the detached kernel. Embryonic axis can be removed from immature embryos using fine forceps under stereomicroscope. Immature embryo without embryo axis is referred to as the scutellum. Immature embryos or scutella with axis side (now removed) down can be cultured onto solid inoculum medium plate. Due to asynchronous development, all the seeds on any spike or ear will not be suitable for the isolation of immature embryos. Generally, the seeds nearest to the peduncle are younger and smaller. In barley, removal of embryonic axis from immature embryos has supported higher frequency of transformation (9, 13).

3.2. Preparation of Agrobacterium for Transformation

1. Streak a single colony of *Agrobacterium* carrying the gene of interest in a binary vector on AB medium containing appropriate antibiotics for the selection of *Agrobacterium* strain and binary vector. Incubate the cultures at 28°C for 3 days.
2. After 3 days, collect *Agrobacterium* cells using a sterilized microspatula and suspend in liquid coculture medium (infection medium) at a density of 0.5×10^9 colony forming units (OD = 1.0 at 600 nm) (see Note 2). The OD of the medium is adjusted using liquid coculture medium. Inoculum should be prepared fresh (see Note 3). The suspension should be homogenized gently and ensure that no cells aggregate remain in the suspension medium. Complete homogenization of suspension medium improves transformation frequency by preventing excess growth of *Agrobacterium* during cocultivation.

3.3. Transformation of Immature Embryos or Scutella

Common steps in *Agrobacterium*-mediated transformation of immature embryos are illustrated in Fig. 1.

1. Immerse the immature embryos or scutella in 2 mL of liquid coculture medium (infection medium) containing freshly made acetosyringone (see Note 4) at room temperature until the remaining embryos have been isolated. A skilled person is required to isolate the immature embryos within a very short period of time without damaging them.
2. Pour the *Agrobacterium* cell suspension prepared in step 3.2 into the sterile Petri plates. 40–50 immature embryos are immersed into 10–15 mL liquid coculture medium (infection medium) containing 100 μM acetosyringone, briefly shaking

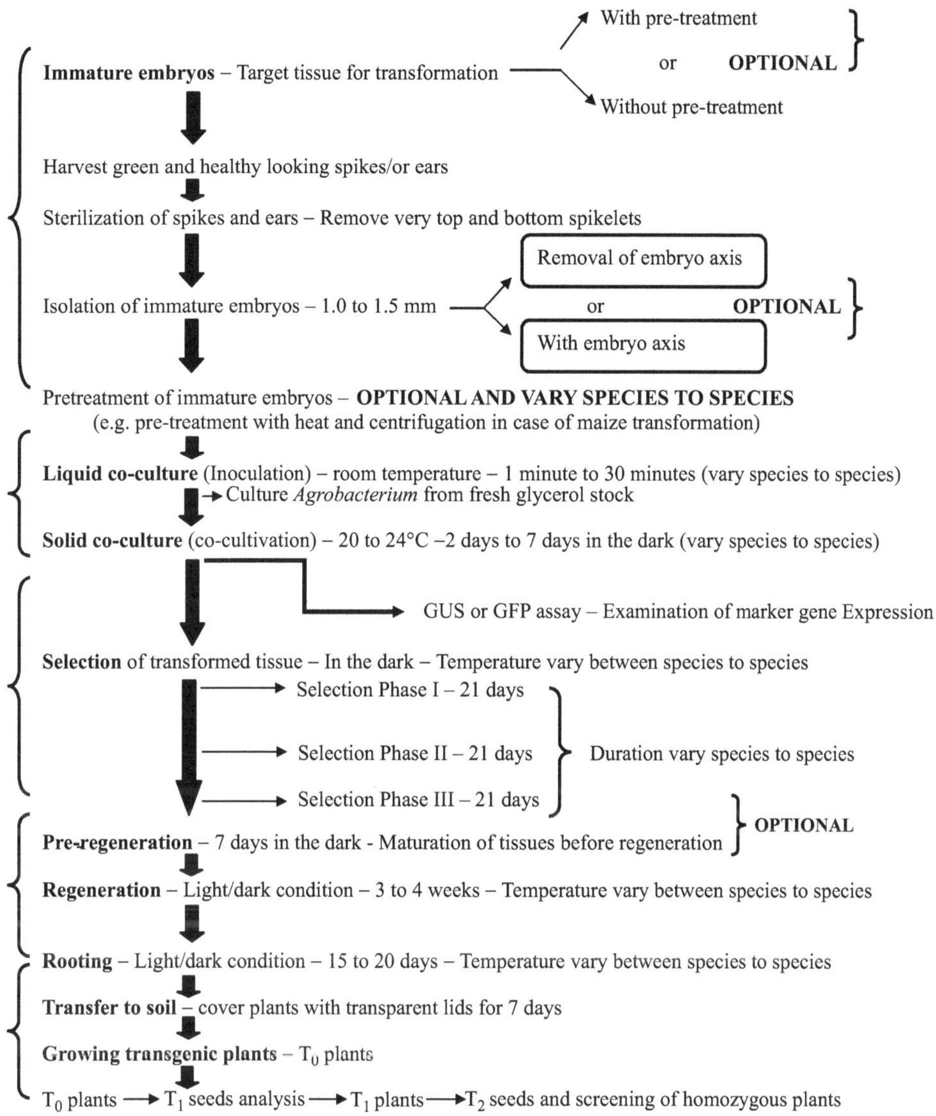

Fig. 1. General steps in *Agrobacterium*-mediated transformation of cereals.

the Petri plates. Infection periods vary from explant to explant, genotype to genotype, and species to species and therefore, it is important to optimize inoculation time. Generally inoculation time ranges from 1 to 30 min. Removal of excess *Agrobacterium* from immature embryos is critical for minimizing the *Agrobacterium* contamination during cocultivation (see Note 5).

3. Remove the infection medium and transfer the embryos (without washing, see Note 5) onto fresh solid coculture medium with the scutellum face up and seal the Petri plates with parafilm. For cocultivating immature embryos with

Agrobacterium, incubate the plates in the dark at 25°C for 2–3 days. Cocultivation time and duration also vary between genotype to genotype and species to species. In general, for cereal crops, cocultivation for 2–3 days at 20–25°C has been found optimal for achieving successful transformation. However, in maize, 7-day cocultivation has been reported to produce high transformation frequency (29). Gene delivery efficiencies can be determined by assaying the transient activity of *uidA* gene by histochemical assays and by visualizing the expression of green fluorescent protein gene (*gfp*) under the fluorescence stereomicroscope following cocultivation (see Note 8). Jefferson (33) recommended the establishment of optimal conditions for gene transfer through preliminary experiments of transient gene expression using reporter genes. Therefore, transient *gus* or *gfp* expression-based studies will be helpful for optimizing conditions affecting the transgene(s) expression and transformation process in crop plants.

3.4. Selection of Transformed Calli

1. First selection – after 3 days of cocultivation period, transfer the uncontaminated embryos to selection medium containing a selection component such as hygromycin for selection of transformed cells and a combination of antibiotics such as cefotaxime and timintin to kill the *Agrobacterium* (see Table 1). Incubate the cultures in the dark at 25°C for 10 days. It is important to transfer only 7–10 embryos per plate for better selection. Do not rinse the embryos or developing calli as rinsing with an antibiotic solution tends to result in poor growth of cells on selection medium (see Note 6).
2. Second selection – transfer the scutellum-derived resistant calli onto fresh selection medium plate and incubate the cultures in the dark for 3 weeks. Calli showing *Agrobacterium* overgrowth should be discarded at this stage (see Note 6).
3. Third selection – transfer the actively proliferating resistant calli onto fresh selection medium and incubate the cultures in the dark for another 3 weeks. During final round of selection, cutting scutellum-derived embryogenic calli into small species and spreading around the mother calli on selection medium is critical to ensure good selection of transformed cells. It is important to transfer only 5–10 calli onto selection medium during third and final round of selection (see Note 7). It is very important to examine the level of expression of a transgene in the immature embryos after cocultivation duration of 2–3 days and in the stable calli after first round of selection for both developing and optimizing the protocol (see Note 8).

3.5. Regeneration of Transformed Plants

1. For regeneration of transgenic plants, transfer the resistant calli to preregeneration medium (see Table 1) containing a cytokinin such as BAP, a selection component such as

hygromycin, and a combination of antibiotics such as cefotaxime and timintin (see Table 1 and Note 9). Incubate the tissues in the dark for 10 days. Resistant calli are transferred on preregeneration medium for maturation.

2. After 10 days, resistant calli are transferred to regeneration medium under appropriate temperature and photoperiod conditions for 3 weeks (see Table 1). Temperature and photoperiod varies between genotype to genotype and species to species. For example, maize regeneration can be carried out at 25°C under continuous illumination (5,000 lx) (29), barley regeneration can be carried out at 24°C under a 16/8-h (light/dark) photoperiod (9), and rice regeneration can be achieved at 28°C under 12/12-h (light/dark) photoperiod (30). Generally, fluorescent light provided at an intensity of 55 μmol/m^2/s is sufficient for regenerating plants from transformed calli.

3.6. Rooting of Regenerated Plants and Transfer to Soil

1. Plantlets, reaching a length of 2 cm, are transferred on rooting medium for 2 weeks to permit vigorous root development before being transferred to soil (see Note 10).
2. After 2 weeks on rooting medium, plantlets with strong rooting system are transferred to soil (see Note 10).

4. Notes

1. The use of immature embryos at the right developmental stage (usually 10–15 days after anthesis) is one of the critical factors in genetic transformation of cereals. Day after anthesis time may differ between species to species and genotype to genotype and therefore, it is important to examine the size of the immature embryos carefully based on the time of their collection after anthesis. Healthy embryos are obtained from healthy and vigorously grown plants. The size of the embryos is a very good indicator of the right stage. Immature embryos that are between 1.0 and 1.5 mm and milky translucent in color in length are optimal for cereal transformation. The color of the embryos is another good indicator of the right developmental stage of immature embryos. If transformation frequency is low, efforts should make to optimize the conditions of greenhouse before troubleshooting other parameters of the protocol. For continuous harvesting of healthy immature embryos, it is very important to seek an advice of a breeder to ensure the proper condition of growth chamber, soil condition, fertilizer and watering, etc.

2. It is also crucial that the bacteria must be rapidly growing and that the bacteria should not grow greater than 10^9 cells per milliliter. Bacteria near the stationary phase of growth change their physiology and do not induce well after acetosyringone treatment.
3. For cereal transformation, we always prepare fresh bacterium inoculum by streaking from a glycerol stock frozen at −80°C.
4. It is very important to keep the acetosyringone stock solution (Sigma, catalog number D13440-6) in the dark at −20°C. Instead of making stock of acetosyringone, we use freshly prepared stock of acetosyringone because it oxidizes during freeze and thaw and lose its potency.
5. Removal of the excess *Agrobacterium* suspension is critical to avoid the excess growth of *Agrobacterium* during 3 days of cocultivation. Flame forceps thoroughly before using between different constructs to avoid cross-contamination.
6. Do not rinse the embryos with antibiotic solution after cocultivation with *Agrobacterium* because this tends to results in poor growth of cells and subsequently poor growth of callus formation. If there is excess growth of *Agrobacterium* after cocultivation duration, decrease the concentration of *Agrobacterium* in infection medium.
7. It is important not to wash the infected immature embryos or developing calli after first selection, but rather to discard *Agrobacterium*-infected immature embryos or calli from the selection plates before collecting the calli for subculture. The callus selection is a key point for efficient transformation. Subculturing healthy and well grown callus pieces tends to produce healthy and green transgenic plants.
8. Examining the level of expression of a transgene in the immature embryos after cocultivation duration of 2–3 days and in the stable calli after first round of selection is considered very useful for both developing and optimizing the protocol, especially for developing a protocol in recalcitrant varieties. Histochemical assay to assess the expression of the reporter genes such as β-gucurondise (*uid*A) gene in cereal transformation can be carried out by staining of transformed tissues with 5-bromo-4-chloro-3-indolyl-b-d-glucuronide acid (X-Gluc) as substrate (33). After chlorophyll extraction with 70% ethanol for at least 3 h, histochemical GUS expression can be determined in leaf samples of the primary regenerants and control plants. Inheritance of another popular reporter gene, green fluorescence protein (*gfp*) gene, can be measured by visual scoring of GFP expression under a fluorescence stereomicroscope. T-DNA delivery can be assessed by counting

immature or calli that had at least one *gus* or *gfp* focus. Genomic DNA can be isolated from the leaf of transgenic plants 10 days after transfer to soil. By designing specific primers for gene(s) of interest, PCR can be performed to find out whether regenerated plants are transgenic or not.

9. It is important to culture only 5–6 resistant calli on regeneration medium because overcrowding at this stage tends to influence neoformation, especially with recalcitrant varieties.
10. Do not keep the regenerated plantlets in the magenta jars for long period of time because this will results in reduced growth of plants in the greenhouse and precocious flowering of small size panicles. After transferring into soil, transgenic plants need to be covered with transplant plastic container to maximize the humidity during acclimatization of transgenic plants into soil.

References

1. Kohli A, Gahakwa D, Vain P, Laurie DA, Christou P (1999) Transgene expression in rice engineered through particle bombardment: molecular factors controlling stable expression and transgene silencing. Planta 208:88–97
2. Shou H, Frame BR, Whitham SA, Wang K (2004) Assessment of transgenic maize events produced by particle bombardment or *Agrobacterium* mediated transformation. Mol Breed 13:201–208
3. Barkat A, Carels N, Bernardi G (1997) The distribution of genes in the genome of Gramineae. Proc Natl Acad Sci USA 94:6857–6861
4. Hiei Y, Ohta S, Komari T, Kumashiro T (1994) Efficient transformation of rice (*Oryza sativa* L.) mediated by *Agrobacterium* and sequence analysis of the boundaries of the T-DNA. Plant J 6.271–282
5. Bajaj S, Mohanty A (2005) Recent advances in rice biotechnology – towards genetically superior transgenic rice. Plant Biotechnol J 3:275–307
6. Shrawat AK, Lörz H (2006) *Agrobacterium*-mediated transformation of cereals: a promising approach crossing barriers. Plant Biotechnol J 4:575–603
7. He DG, Yang YM, Scott KJ (1990) A comparison of scutellum callus and epiblast callus induction in wheat: the effect of genotype, embryo age and medium. Plant Sci 68:103–111
8. Pastori GM, Wilkinson MD, Steels SH, Sparks CA, Jones HD, Parry MAJ (2001) Age-dependent transformation frequency in elite wheat varieties. J Exp Bot 52:857–863
9. Shrawat AK, Becker D, Lörz H (2007) *Agrobacterium tumefaciens*-mediated genetic transformation of barley (*Hordeum vulgare* L.). Plant Sci 172:281–290
10. Rasco-Gaunt S, Riley A, Cannell M, Barcelo P, Lazzeri PA (2001) Procedures allowing the transformation of a range of European elite wheat (*Triticum aestivum* L.) varieties via particle bombardment. J Exp Bot 52:865–874
11. Ishida Y, Saito H, Ohta S, Hiei Y, Komari T, Kumashiro T (1996) High efficiency transformation of maize (*Zea mays* L.) mediated by *Agrobacterium tumefaciens*. Nat Biotechnol 14:745–750
12. Zhao ZY, Gu W, Cai T, Tagliani L, Hondred D, Bond D, Schroeder S, Rudert M, Pierce D (2001) High throughput genetic transformation mediated by *Agrobacterium tumefaciens* in maize. Mol Breed 8:323–333
13. Tingay S, McElroy D, Kalla R, Fieg S, Wang M, Thornton S, Brettell R (1997) *Agrobacterium tumefaciens*-mediated barley transformation. Plant J 11:1369–1376
14. Matthews PR, Wang MB, Waterhouse PM, Thornton S, Fieg SJ, Gubler F, Jacobsen JV (2001) Marker gene elimination from transgenic barley using co-transformation with adjacent "twin T-DNAs" on standard *Agrobacterium* transformation vector. Mol Breed 7:195–202
15. Kumlehn J, Serazetdinova L, Hensel G, Becker D, Lörz H (2006) Genetic transformation of barley (*Hordeum vulgare* L.) via infection of androgenetic pollen cultures with *Agrobacterium tumefaciens*. Plant Biotechnol J 4:251–261

16. Cheng M, Fry JE, Pang S, Zhou H, Hironaka C, Duncan DR, Conner TW, Wan Y (1997) Genetic transformation of wheat mediated by *Agrobacterium tumefaciens*. Plant Physiol 115:971–980
17. Cheng M, Hu T, Layton J, Liu C-N, Fry J (2003) Desiccation of plant tissues post-*Agrobacterium* infection enhances T-DNA delivery and increases stable transformation efficiency in wheat. In Vitro Cell Dev Biol Plant 39:595–604
18. Hu T, Metz S, Chay C, Zhou HP, Biest N, Chen G, Cheng M, Feng X, Radionenko M, Lu F, Fry J (2003) *Agrobacterium*-mediated large-scale transformation of wheat (*Triticum aestivum* L.) using glyphosate selection. Plant Cell Rep 21:1010–1019
19. Wu H, Sparks C, Amoah B, Jones HD (2003) Factors influencing successful *Agrobacterium*-mediated genetic transformation of wheat. Plant Cell Rep 21:659–668
20. Zhao ZY, Cai T, Tagliani L, Miller M, Wang N, Pang H, Rudert M, Schroeder S, Hondred D, Seltzer J, Pierce D (2000) *Agrobacterium*-mediated sorghum transformation. Plant Mol Biol 44:789–798
21. Gao Z, Jayaraj J, Muthukrishnan S, Claflin L, Liang GH (2005) Efficient genetic transformation of *Sorghum* using a visual screening marker. Genome 48:321–333
22. Repellin A, Båga M, Jauhar PP, Chibbar RN (2001) Genetic enrichment of cereal crops via alien gene transfer: new challenges. Plant Cell Tissue Organ Cult 64:159–183
23. Cheng M, Lowe BA, Spencer TM, Ye X, Armstrong CL (2004) Factors influencing *Agrobacterium*-mediated transformation of monocotyledonous species. In Vitro Cell Dev Biol Plant 40:31–45
24. Hoekema A, Hirsch PR, Hooykaas PJJ, Schilperoort RA (1983) A binary plant vector strategy based on separation of vir- and T-region of the *Agrobacterium tumefaciens* Ti-plasmid. Nature 303:179–180
25. Hood EE, Helmer GL, Fraley RT, Chilton MD (1986) The hypervirulence of *Agrobacterium tumefaciens* A281 is encoded in a region of pTiBo542 outside of T-DNA. J Bacteriol 168:1291–1301
26. Hood EE, Gelvin SB, Melchers S, Hoekema A (1993) New *Agrobacterium* helper plasmids for gene transfer to plants (EHA105). Trans Res 2:208–218
27. Lazo GR, Stein PA, Ludwig RA (1991) A DNA transformation-competent Arab idopsis genomic library in *Agrobacterium*. Biotechnology 9:963–967
28. Sunar IK, Sakthivel N (2008) Advances in selectable marker genes for plant transformation. J Plant Physiol 165:1698–1716
29. Ishida Y, Hiei Y, Komari T (2007) *Agrobacterium*-mediated transformation of maize. Nat Protoc 2:1614–1620
30. Nishimura A, Aichi I, Matsuoka M (2006) A protocol for *Agrobacterium*-mediated transformation in rice. Nat Protoc 1: 2796–2802
31. Hunter CP (1988) Plant regeneration from microspores of barley, Hordeum vulgare. Ph.D. Thesis, Wye College, University of London, Ashford
32. Murashige T, Skoog F (1962) A revised medium for rapid growth and bioassays with tobacco tissue culture. Physiol Plant 15:473–497
33. Jefferson RA (1987) Assaying chimeric genes in plants: the genes fusion system. Plant Mol Biol Rep 5:387–405
34. Wan Y, Lemaux PG (1994) Generation of large numbers of independently transformed fertile barley plants. Plant Physiol 104: 37–48

INDEX

Trevor A. Thorpe and Edward C. Yeung (eds.), *Plant Embryo Culture: Methods and Protocols*, Methods in Molecular Biology, vol. 710, DOI 10.1007/978-1-61737-988-8,

D

E

MIX
Papier aus verantwortungsvollen Quellen
Paper from responsible sources
FSC® C105338

If you have any concerns about our products,
you can contact us on
ProductSafety@springernature.com

In case Publisher is established outside the EU,
the EU authorized representative is:
Springer Nature Customer Service Center GmbH
Europaplatz 3, 69115 Heidelberg, Germany

Printed by Libri Plureos GmbH
in Hamburg, Germany